Overpotential

Studies in Modern Science, Technology, and the Environment
Edited by Mark A. Largent

The increasing importance of science and technology over the past 150 years—and with it the increasing social, political, and economic authority vested in scientists and engineers—established both scientific research and technological innovations as vital components of modern culture. Studies in Modern Science, Technology, and the Environment is a collection of books that focuses on humanistic and social science inquiries into the social and political implications of science and technology and their impacts on communities, environments, and cultural movements worldwide.

Mark R. Finlay, *Growing American Rubber: Strategic Plants and the Politics of National Security*

Jill A. Fisher, *Gender and the Science of Difference: Cultural Politics of Contemporary Science and Medicine*

Finn Arne Jørgensen, *More than a Hole in the Wall: The Story of What We Do With Our Bottles and Cans*

Gordon Patterson, *The Mosquito Crusades: A History of the American Anti-Mosquito Movement from the Reed Commission to the First Earth Day*

Jeremy Vetter, *Knowing Global Environments: New Historical Perspectives on the Field Sciences*

Overpotential

Fuel Cells, Futurism, and the
Making of a Power Panacea

MATTHEW N. EISLER

RUTGERS UNIVERSITY PRESS

NEW BRUNSWICK, NEW JERSEY, AND LONDON

LIBRARY OF CONGRESS CATALOGING-IN-PUBLICATION DATA

Eisler, Matthew N., 1970–
  Overpotential : fuel cells, futurism, and the making of a power panacea /
Matthew N. Eisler.
     p. cm. — (Studies in modern science, technology, and the environment)
  Includes bibliographical references and index.
  ISBN 978-0-8135-5177-7 (hardcover : alk. paper)
  1. Fuel cells—Research—History.   2. Fuel cells—Public opinion.   3. Fuel cells—
Political aspects.   I. Title.
  TK2931.E38 2011
  621.31′2429—dc23   2011012792

A British Cataloging-in-Publication record for this book is available
from the British Library.

Visit our Web site: http://rutgerspress.rutgers.edu

Manufactured in the United States of America

For Mom and Dad

CONTENTS

ACKNOWLEDGMENTS

I have much gratitude for the many persons and organizations that helped make this book possible. I gratefully acknowledge the support of the Social Sciences and Humanities Research Council of Canada, the Chemical Heritage Foundation, the IEEE History Center, and the Institute of Historical Research/ Andrew W. Mellon Foundation. I am also indebted to the staffs of the Churchill College Archives Centre at Cambridge University, the Smithsonian Institution Archives, the NASA Headquarters History Office in Washington, DC, and the National Archives at College Park, Maryland. And I owe special thanks to the Department of History at the University of Western Ontario and the Center for Nanotechnology in Society at the University of California at Santa Barbara for postdoctoral fellowships that allowed me to revise and complete a manuscript that originated as an idea for a doctoral dissertation at the Department of History and Classics at the University of Alberta in the summer of 2003.

Many people provided insight, inspiration, and support, without which this book could not have been realized. I thank Greg Anderson, Kathryn Rice Bullock, Charles Chamberlin, Martin J. Collins, Gwen D'Arcangelis, John M. DeCicco, Erika Dyck, David Edgerton, Michael Egan, Colin Fries, Michael N. Geselowitz, Will Hively, Karl V. Kordesch, John Langdon, Roger D. Launius, Christophe Lécuyer, Peter Lehman, Stuart W. Leslie, Robert MacDougall, David Marples, Cyrus C. M. Mody, Suzanne Moon, Yasuyuki Motoyama, Fredrik Nebeker, Allan A. Needell, Michael J. Neufeld, Jane Odom, Gabriella M. Petrick, Daniel Sperling, Charles Stone, Rick Szostak, John Vardalas, Hal Wallace, and especially Sara Eisler, Barbara Herr Harthorn, Suzanne Kellam, W. Patrick McCray, Peter Mickulas, Susan L. Smith, Robert W. Smith, Doreen Valentine, and my mother and father.

Overpotential

Introduction

Fuel Cell Futurism

For a time in the late 1990s and early 2000s, it was nearly impossible to read popular literature on science and technology without encountering praise for the fuel cell. Lauded by engineers, scientists, and policymakers, the device, which converts chemical energy into electrical energy, was then a virtual byword for sustainable power. The technology had an unusually broad appeal. At the core of its popularity was the belief that it was a kind of electrochemical engine, a universal chemical energy converter capable of running on any hydrogenous fuel, combining the best features of the internal combustion engine and the galvanic battery without their handicaps.[1] Compelled by the California Air Resources Board and its Zero-Emission Vehicle mandate of 1990 to market electric passenger vehicles in ever-increasing quantities, the automobile industry saw the fuel cell as a power source superior to the conventional galvanic battery, both because it seemed to afford a much longer range and because it seemed to promise less disruption of the established liquid-fuel automotive culture. Instead of having to reorder their lives around the lengthy recharge period of a conventional battery, consumers could top up their fuel cell electric vehicles with a liquid fuel in minutes. For this reason, fuel cells were also investigated by the oil industry, which hoped to supply the specialized fuels required by the energy conversion technology if it became commercialized and ubiquitous.

Dramatic advances in the state of the art were often followed by predictions of an impending revolution in power source technology, one that would allow consumers to continue to enjoy the comfort and convenience of the modern automobile while accommodating their green sensibilities.[2] Even the White House promoted the technology, framing the hydrogen fuel cell applied in electric vehicles as a power panacea.[3] Billions of dollars flowed into research, mainly from the automobile industry and also from the United States Department of

Energy. But although laboratory experiments and precommercial prototypes showed promise, the fuel cell proved extraordinarily difficult to commercialize. Early in the new millennium, government and industry indefinitely postponed plans to market it.

This was not the first fuel cell boom to go bust. Scientists and engineers have perceived the device as the holy grail of power sources following its invention in the mid-nineteenth century. To many, the fuel cell seemed exempt from the Carnot-cycle limitation on the efficiency of heat engines. Such devices convert chemical energy into heat, using the resulting kinetic motion of molecules in a hot gas to drive a mechanical device such as a piston or turbine. But much of that heat dissipates into the environment without doing useful work. In contrast, fuel cells directly convert chemical energy into electricity without randomizing that energy as heat in the manner of a heat engine.

The fuel cell concept was first explored by European researchers experimenting with reverse electrolysis. Instead of using electricity to dissociate water into hydrogen and oxygen, they attempted to combine oxygen and hydrogen to produce electricity. Working independently, the British lawyer and amateur scientist William Robert Grove and the Swiss physicist Christian Friedrich Schönbein used platinum foil to catalyze a reaction between hydrogen and oxygen in 1839. Grove further elucidated the chemical basis of this reaction in subsequent experiments published in 1843 and 1845. Little more was done until the 1890s, when European scientists revived the concept in a new series of experiments using coal and coal-derived gases as a source of hydrogen. One of the more enthusiastic proponents was the German physical chemist Wilhelm Ostwald. In a passionate address before a group of engineers in 1894, he held that the direct electrochemical conversion of solid coal to electricity could replace the steam engine, a technology Ostwald saw as "incomplete" because it converted only a fraction of the energy in coal to useful work and turned the rest to waste heat, soot, and smog.[4] A number of such devices were built in the late nineteenth and early twentieth centuries, but none were practical. In 1939, the German researcher Emil Baur remarked how strange it was that advances in fuel cell theory and in the state of the technological art in the 1890s had borne no fruit almost half a century later.[5]

But the idea of the fuel cell as a universal chemical energy converter did not die. Indeed, it proved remarkably resilient. After the Second World War, it was revived by government agencies, power source companies, and, later, electric utilities. Thereafter the idea tended to increase in popularity as researchers made rapid progress and wane as new technical problems arose and political and economic conditions changed. But it never disappeared entirely. Some types of fuel cell were successfully developed for special purposes, most notably in aerospace and distributed-power applications, roles in which they performed reasonably well. Still, what galvanized most sponsors of fuel cell research and development, if not necessarily all researchers involved in such projects, was the possibility that the technology could be made into a popular consumer product that would significantly benefit the economy, the environment, and the general quality of life. This vision of the fuel cell as a power panacea proved frustratingly elusive, yet endured for decades.

This book explains the persistence of this vision. As a study of a family of artifacts that long hovered on the margins of society, yet came to represent the apotheosis of power source technologies, it revisits and reconceptualizes the idea of innovation. At the same time, it relates the social relations and political economy of frontier science and engineering with the social phenomenon of futurism.

The practice of anticipating the future has existed as long as civilization, always powerfully linked to the contemporary material concerns and interests of those engaged in prophecy and prediction and those who sought their services. The social relations of what has been termed futurism, however, are rather ill-defined. Some scholars have referred to futurism as a "field," distinguishing nonrational and rational varieties, the latter informed by Enlightenment thought and the development of modern scientific method from the seventeenth century.[6] Successful prediction, of course, is the most convincing test of scientific knowledge. The predictability of phenomena on the physical, biological, and social levels has long been and seems likely to continue to be disputed. When they consider future events, however, both physical and social scientists tend to consider matters in terms of probabilities, rather than predictions.[7] And a tremendously wide range of actors—politicians, planners, activists, advocates, religious figures, and revolutionaries of various sorts—actively work to shape social landscapes in the near future in projects of varying ambition.

As a social phenomenon, modern futurism has affinities with all of these practices. Manifesting across a broad spectrum of professions and social milieus, futurism has never possessed the epistemic coherence suggested by the term "field." It is perhaps best understood as a historically contingent social phenomenon, a preoccupation with inferring future scenarios from recent social events as they relate to dramatic developments in the physical sciences and engineering and the complications of industrial civilization. It is possible to discern types of more or less professional futurists emerging in certain periods such as writers of science fiction from the late nineteenth century, the artists and intellectuals of the "futurist" movement that flourished in Europe in the early twentieth century, a species of strategic planner that appeared in the United States and other advanced industrial countries after the Second World War, and assorted analysts and experts in corporate, governmental, and nongovernmental circles that from the late 1960s began practicing what became variously known as forecasting, future studies, and futurology. Eventually giving rise to an interdisciplinary academic "field," these occupations achieved considerable prestige in the United States in the 1970s and early 1980s.[8]

It is also possible to identify narrower groups of science and engineering experts and their supporters who for various reasons incidentally engaged in futurism during the course of their professional work. The players in the story I am about to tell belong in this category. Scholars sometimes identify techno-futurism as a key characteristic of the social psychology of scientists, engineers, and their patrons in late capitalist society.[9] Communications specialists Marita Sturken and Douglas Thomas refer to technological prediction as an "an impulse, a dilemma, and an economic strategy" revelatory of the ethos of the age far more than the actual potential of the technologies in question.[10] The political scientist Langdon Winner observed that such thinking occludes the nature and effects of technology in society and stultifies the critical reflection necessary for democratic politics.[11]

Like other communities engaged in the production and commodification of technoscience, fuel cell researchers and their patrons made problematic physical and social claims in conjuring the future to justify ongoing projects.[12] And as in other forms of techno-futurism, fuel cell futurism developed and changed over time. The typology of technological prediction developed by the historian David E. Nye is helpful in sketching this historical arc. Predictions tend to be the purview of inventors of wholly new ideas or concepts, holds Nye, and apply to the deep future; forecasts are made by engineers and entrepreneurs in support of innovations, which are improvements of or accessories to original ideas, with a view toward application in the medium term; and projections are made for new models of established technologies by designers and marketers. Nye refers to these kinds of prediction as "little narratives about the future," stories that illustrate a better world to come, not full-blown utopias. When the public believes such stories as told by marketers and public relations experts, they often become self-fulfilling.[13]

The history of the fuel cell features rather different narrative threads and outcomes. Modern fuel cell researchers and their patrons were primarily involved in constructing the technological *forecast*, as Nye defines it. In first decades after the Second World War, these forecasts and the narratives of the future they supported tended to be relatively modest. But as researchers advanced the state of the art over the years, their forecasts increasingly depicted visions of social landscapes with utopian qualities.

Here we should distinguish futurism from utopianism, the practice and technique of reimagining present imperfect societies transformed into planned perfect societies. Not all forms of futurism are utopian, but "utopia" has come to be understood as an unambiguously good society set in the future.[14] The original literary utopia, to be sure, had a rather different premise. As envisaged by Thomas More, lawyer and advisor to Henry VIII, utopia was a problematic communist society with attractive and unattractive features set in contemporary time. In this way, More critically engaged the pressing sociopolitical and cultural issues of 1515.[15] The literary utopia or dystopia became a well-established form of social critique. But over time, such thought experiments, and the concept of the utopia itself, became associated with quite dissimilar objectives. The egalitarian societies sought by non-Marxian and Marxian socialist movements in the nineteenth and twentieth centuries were often described as utopias, usually by detractors of these projects. In a famous injunction, Marx counterposed the hitherto abstract ends of philosophers in analyzing the world with the political application of analysis in service of revolutionary social change.[16]

On the other hand, writers in the genre of the technological utopia that flourished in the United States in the late nineteenth and early twentieth centuries were not nearly as equivocal and reflective as More or as sweeping and programmatic as Marx. Most were less interested in critical discussion of contemporary American society than in reimagining it fully equipped with technological systems that were already developed or on the verge of entering large-scale use, enabling human happiness and comfort. And these authors did not specify precisely how these worlds had come into existence.[17] The thought experiments of the literary technological utopias were, hence, congenial to American elites. During the Depression, these works inspired the design of major international exhibitions in Chicago (1933) and New York (1939), conceived as capitalist consumer utopias by corporate and governmental leaders eager to riposte the radical political feeling of the day. Such visions glossed social relations and, like the prose fiction that inspired them, focused less on innovations than on existing, if not always ubiquitous, pieces of technology and architecture whose exteriors were frequently treated in modernist style. Designed to highlight science as the driving force of society, the New York World's Fair of 1939, for example, featured television as perhaps its most prominent innovation. But the star attractions were dioramas of automobile-centered suburban and skyscraper cities (the Democracity and Futurama installations in the Perisphere and General Motors pavilions respectively), vistas already familiar to locals and visitors familiar with film and print media. The consumer utopia of the near future required only "enlightened administration" to be realized.[18]

This, of course, meant planning. The imagined world on display in New York in 1939 was largely realized after the Second World War thanks to government policies that subsidized home loans, interstate highway infrastructure, and energy, complementing the pent-up purchasing power accrued by workers during the war. But the postwar period also witnessed the rise of more speculative technological projects, triggered in part by the complications and contradictions of postwar military-industrial economic and energy planning. Here, then, we begin to see the historical relationship between the ideology of American-style utopian futurism, on the one hand, and the socioeconomic and political processes responsible for transforming innovations into technologies-in-use on the other.[19] Over time, the locus of utopian (and dystopian) futurist discourse emerged in the dialogue between developers of powerful industrial and consumer technologies and the developers of innovations (often but not always the same groups), the latter projects often being promoted as a solution to the problems created by technologies-in-use. These social relations fall under the rubric of what Winner called "technological politics," in which "the rule of technological circumstances . . . supplant[s] other ways of building, maintaining, choosing, acting, and enforcing, which are more commonly considered political."[20]

Fuel cell futurism sprang from such a polity. It originated at a time when there was no real demand for a power panacea. In the 1950s, primary energy was plentiful and the science of electrochemistry essential to fuel cell engineering was moribund, at least in the United States, a result, in part, of the triumph of the fossil fuel system of automobility in the 1920s. The fuel cell found a home in corporate and government laboratories, partly because Cold War planners, drawing from their experiences during the Second World War, developed an appetite not simply for improved technologies but for "breakthrough" technologies. Like other groups of scientists and engineers, fuel cell researchers extrapolated from existing phenomena, appealing to future technological capabilities as a means of mobilizing expectations in service of existing enterprises.

Over time, fuel cell futurism aligned with and reinforced other forms of U.S.-style utopian techno-futurism. One such enterprise was the space program, an engineering project that used aerospace technology not only to demonstrate the superiority of the contemporary American way of life, broadly construed, but also to invoke a vision of a future advanced consumer society. Fuel cell futurism further developed in response to the crisis of America's high-energy regime, the expression Nye coined to describe the techno-economic system built between the late 1930s and the early 1970s that enabled an increase in energy consumption of some 350 percent.[21] Energy and economic policies, notes political scientist Timothy Mitchell, configured this system primarily as a hydrocarbon utopia. Aiming to stimulate the consumption of cheap petroleum after the Second World War, American energy planners decoupled price from demand and the relative state of resource depletion, creating the illusion of limitless energy. Keynesian economists contributed to this illusion by conceiving national wealth in terms of aggregate money transactions (the "gross national product"), making no distinction between beneficial and harmful costs. The chimera that national income faced no physical or territorial limits to growth was shattered by the recession of the 1970s.[22] No longer able to ignore the escalating social and environmental costs of the high-energy regime, politicians promulgated initiatives including amendments to the Clean Air Act and the Corporate Average Fuel Economy (CAFE) standard in an attempt to regulate the efficiency of the energy conversion chain. Designed to force technological fixes, such measures did not have inconsequential effects. But they did not fundamentally change the structure of the high-energy regime. They did, however, help create new hearings for advocates of fuel cell power in academe, government, and corporate boardrooms.

The history of fuel cell futurism thus coalesces with the history of innovation and the history of technology-in-use. The twentieth-century variation of the classic Enlightenment notion of progress, the idea of innovation (encapsulating both invention and the reduction of invention to use, as Nye understands these terms) held almost totemic potency for politicians, economists, scientists, and engineers after the Second World War in a variety of countries.[23] So deeply institutionalized did this notion become in the United States that policies supporting innovation (a term used frequently in science and technology policy quarters as a euphemism for "research and development") became synonymous with industrial and economic policies and economic growth. The astonishing rise of the information technology sector in the late 1960s and early 1970s at a time when American fossil fuel–based heavy industry began its slow decline helped validate this idea. As all sectors of American manufacturing shifted an increasingly large proportion of production abroad to take advantage of lower wages and weaker labor rights in the 1980s and 1990s, influential observers came to believe that the nation's comparative economic advantage increasingly resided in its educational and research and development complex. Moreover, the policies presumed responsible for success in the information technology sector (government-backed education in the physical sciences, incentives for innovation, and collaborative research and development between academe, industry, and federal agencies) began to be interpreted in science and engineering circles as the basis of a general-purpose model of innovation equally suited for greening the systems of the high-energy regime.[24]

In fact, the process of transforming inventions and innovations into commercial products differed greatly in these respective fields. Progress occurs much more slowly in energy conversion technology than in electronics, noted Department of Energy science undersecretary Steven E. Koonin in 2010, partly because it is physically more difficult to move molecules than bytes. Historically, the pace of miniaturization in power sources lagged far behind electronics. Nuclear power accumulated a mixed record everywhere it was developed. But in the United States, added Koonin, there has also been a disjunction between processes of inventing advanced power sources and processes of commercially producing them. He noted the rapid decline in the market share of American manufacturers in this sector since the early 1990s, even in areas of technology pioneered by domestic industry such as the photovoltaic cell, a product of the iconic Bell Laboratories.[25]

Indeed, American scientists, engineers, politicians, and entrepreneurs have never really lacked ideas for energy and power inventions and innovations. In the early twenty-first century, the broader problem was how to adapt an energy regime based largely on the technology and social relations of the previous century to rapidly changing economic and environmental circumstances, a question that, in essence, was political. In his history of the battery electric car in the United States, David A. Kirsch touched on the relationship between innovation, technologies-in-use, and futurism in an observation that is germane to the history of the fuel cell. He found that as the mounting costs of the fossil fuel–based system of automobility forced a public debate on the viability of alternative technologies including battery electric power in the late 1960s and early 1970s, the auto industry responded by forcing battery engineers to accept standards of cost-effective performance derived from the internal combustion engine, a device manufacturers continually refined on a vast scale. With far fewer resources, battery researchers stood little chance of drawing even with continually shifting performance benchmarks. For them, consequently, the ideal battery lay perpetually in the future.[26] The history of fuel cell research and development featured a similar pattern. Here, too, futurism was a by-product of the social relations of innovation in the field of electrochemical energy conversion, emerging at the interstices of relatively unstable laboratory communities embedded within a relatively stable industrial energy conversion order. In a series of vignettes chronicling the efforts of a variety of interest groups to adapt fuel cell power to practically every known application using a variety of chemical fuels across seven decades following the Second World War, I elaborate the point that different fields of industrial research and development lacked epistemic unity (no less than with different fields of academic basic science and engineering, as historians, sociologists, and philosophers have long recognized), underscoring field-specific patterns of innovation and commercialization.[27] Through a case study of a "failed" innovation, then, this book comprehensively explores the political economy of one species of technological futurism, casting into relief the combined socio-physical factors that perpetuated imaginary futures over a sustained period of time. In so doing, it serves another, perhaps more ambitious goal. By tracing the genealogy of the idea of the fuel cell as a power panacea, I explore the history of certain key technologies-in-use in the primary energy conversion and transportation systems, and the social relations embedded within them, from a new perspective. Assessing continuity and change in the energy conversion chain, I consider the material and ideological consequences of fuel cell futurism for U.S. energy, innovation (research and development), and environmental politics. In this way, this book contributes fresh insights to critiques of the project to renovate high-energy civilization.

Batteries and Fuel Cells

Of all power sources, conventional batteries have the most in common with fuel cells. Both are electrochemical energy conversion devices, electrochemistry being the science and technology of the two-way passage of ions and electrons across a conducting medium between two electrodes, producing chemicals by means of electricity and producing electricity by means of chemicals. Both fuel cells and batteries can function as electricity producers or galvanic (primary) cells, converting chemical energy into electricity, or as electrolytic or rechargeable (secondary) cells, converting electricity into stored chemical energy. As galvanic devices, fuel cells and batteries cause atoms to shed their electrons, which pass into an external circuit, producing electricity, heat, and waste. As rechargeable devices, fuel cells and batteries can in principle function both in the galvanic mode, combining chemical reactants to release stored electricity, and in the electrolytic mode, using externally produced electricity to decompose reactant waste into the original chemical constituents, introducing electrons that turn ions back into regular atoms. When equal rates of electron gain or loss occur between the interface of electrodes and an electrolyte, electrochemical devices are said to be in a state of equilibrium. Electricity and chemical products are produced only during net electron loss or gain respectively, which occurs when an electrode departs from its equilibrium value. This nonequilibrium state is known as overpotential. Electrolytic cells are known as driven electrochemical systems because they require power in order to work. In contrast, galvanic devices are considered self-driving, yielding electrons through the spontaneous reaction of materials.

The amount of useful power that batteries and fuel cells produce is a function of Ohm's law. This holds that an electric current between two points is directly proportional to voltage (the potential difference between two points in an electric field that causes current to flow) and inversely proportional to resistance. A common analogy used to illustrate voltage is the pressure differential in a pipe that causes a fluid to move. The greater the voltage, the greater the flow of electric current, or the quantity of charge carriers that move past a given point. This is distinct from current density, or the rate of electrochemical reaction. Boosting the current density of a fuel cell requires an increasingly larger share of the energy converted by the reaction. As current density is increased, there is a linear fall in voltage until, at very high current densities, there is a rapid loss of power. As a result, practical fuel cells are unsuitable for high-voltage applications.

Both batteries and fuel cells can in principle function reversibly, that is, in both galvanic and electrolytic modes. But fuel cells have been pursued largely in galvanic form, a consequence, in part, of the unique way fuel cells handle chemical fuel. Unlike batteries, fuel cells store their reactants externally and, hence, have no energy density (the capacity to store electricity for a given period of time relative to volume). As indicators of performance, researchers instead measure current density in a given surface area of a fuel cell (typically in terms of milliamperes per square centimeter or amperes per square meter) and power density (the ratio of power available for useful work to the weight or volume of a fuel cell, expressed as watts per kilogram or liter). Importantly, these units of measurement are not expressed as a time relationship. This inspired many engineers to assume that as long as chemical reactants were supplied, fuel cells would continue to operate invariantly, that is, with no internal chemical deterioration over time, unlike primary and secondary batteries.

The most efficient way for fuel cells to function reversibly, that is, as a "storage" device, is by using pure hydrogen as the energy carrier and storage medium because water, the source of pure hydrogen, is the least problematic chemical reaction product, being relatively easily dissociated. But pure hydrogen is a notoriously tricky and costly fuel that is difficult to store and transport. These factors helped inspire engineers to think of the fuel cell as an electrochemical engine consuming cheap, common hydrogenous carbonaceous fuels in an irreversible electrochemical reaction.

In semiconductors, "the material [is] the device,"[28] and in large measure this is also true of electrochemical power sources. Containing no moving parts in and of themselves, batteries and fuel cells are volatile mixtures of chemicals that conduct and catalyze matter and degrade over time, even when not under load, often undergoing unexpected side reactions that are difficult to control or negate. In the 1940s and 1950s, electrochemical theory was underdeveloped, and fuel cell researchers worked empirically.[29] For years, they experimented with various combinations of electrolytes and catalysts, each of which offered offsetting advantages and disadvantages in cost and performance. The device generally acknowledged by electrochemical authorities as the first practical fuel cell, developed in the late 1950s in Britain, used a liquid alkaline electrolyte and consumed pure hydrogen. Because carbon dioxide poisons such an electrolyte, reacting with it to form a solid carbonate, engineers hoped to develop fuel cells with more robust electrolytes composed of acids, molten carbonates, or solid oxides. Operating at around 700°C and 1,000°C respectively, the last two types suffer from thermal expansion and corrosion at such high temperatures. As a result, most efforts to develop the fuel cell as a universal chemical energy converter after the Second World War were devoted to designs using liquid and solid acidic electrolytes at roughly 100°C to 200°C.

Success, Failure, and the Future Machine

In some important respects, the histories of batteries and fuel cells have followed similar paths. Both were indelibly shaped by developments in transportation and heat-engine technologies that delimited commercial success in crucial ways. Richard Schallenberg provides important context in his magisterial study of the golden age of batteries in the United States in the late nineteenth and early twentieth centuries. Early electrical plant operators often employed large storage batteries to smooth the often-uneven power output of early generators until the advent of reliable electromagnetic generating systems in large central stations. For a time, the electric motor powered by the lead-acid battery was a popular form of automobile propulsion in urban areas in Western Europe and the eastern United States. The oldest type of rechargeable battery, invented in 1859 by Gaston Planté, the lead-acid combination was widely used in fleet vehicles such as taxis, delivery trucks, and short-range city cars that could be easily serviced from central depots. Such vehicles became increasingly rare with the era of cheap gasoline and the mass-produced internal combustion engine. By the early 1920s, the battery-powered electric motorcar had virtually disappeared from U.S. public roads, although it continued to be used in taxi and delivery truck fleets in some European cities for years afterward. As these trends occurred, notes Schallenberg, American electrical engineers were no longer stimulated to think in terms of electrochemical solutions to problems.[30] In the United States, the academic discipline of electrochemistry languished for decades, trailing physics and chemistry in prestige and resources.

That batteries became ubiquitous in the twentieth century and fuel cells did not was, in part, a consequence of the ability of their respective developers and promoters to reconcile the physical qualities and requirements of the technology with commercial expectations in light of economic, cultural, and political trends that emerged in the wake of the triumph of internal-combustion-engine automobility. Battery makers had lost the largest market for the lead-acid battery, but almost immediately they secured a new one when the power source began to be used to supply power for lighting and, even more importantly, starting the gasoline automobile. The battery-powered electric starter allowed car designers to eliminate the difficult and dangerous hand crank and the monopoly in ease of starting previously enjoyed by battery-powered electric vehicles.[31] Standardized in the 1920s, the starting, lighting, and ignition battery helped secure the dominance of both the gasoline automobile and the lead-acid battery in this auxiliary role owing to its unparalleled cost-effectiveness.

For the next half-century, scientists and engineers made only modest progress in advancing battery technology. The deluge of portable electric and electronic devices flooding the market after the Second World War did spur development of alkaline rechargeable and disposable batteries, which became commercially available in the 1960s. Disposable lithium-based batteries became available in the early 1970s. Beginning in the late 1960s and early 1970s, serious research in more powerful and sophisticated chemistries for rechargeable batteries unfolded in diverse settings around the world. Researchers at the Ford Motor Company invented the sodium-sulfur battery in 1967. In the 1970s and the 1980s, workers at Argonne National Laboratory in Illinois, the National Physical Research Laboratory in Pretoria, South Africa, and Oxford University investigated lithium-based battery chemistries. In the late 1980s and early 1990s,

Japanese industry dominated the research, development, and manufacturing of advanced batteries, above all, lithium ion secondary batteries, mainly because Japanese firms had created a demand in building a vast market for portable consumer electronics. No similar synergy manifested in the U.S. industrial landscape. Conservative American battery manufacturers dabbled in but then abandoned the field of lithium ion rechargeables in the early 1990s, when this market amounted to only a few hundred million dollars.[32] By 2011, it was worth anywhere from $10 billion to $14 billion. Consequently, American battery makers were not well placed to take advantage of the resurgence in interest in battery and hybrid electric automobility at the turn of century. From the late 1990s, automakers used nickel–metal hydride batteries in experimental pure electric and commercial hybrid electric vehicles and, from the mid-2000s, began to test lithium ion batteries in these applications, putting them in commercial service in hybrid and pure battery electric autos in 2010.[33]

Applications were not especially obvious to researchers when they tentatively revisited the fuel cell in the 1940s and early 1950s. Like the battery in its early days, it was a technology (i.e., a solution) "in search of a problem."[34] The escalation of the Cold War after the launch of the first Sputnik satellite in 1957 supplied several when the rapid expansion of the electronics and aerospace industries in the United States created a demand for advanced power sources. However, the modern story of the fuel cell begins with an important European chapter, one that illustrates that assumptions concerning the relationship between innovation and economic growth were not unique to the United States. The first link in the chain of expectations for a power panacea after the Second World War was forged in Britain by the mechanical engineer Francis Thomas Bacon and a series of public-sector patrons of research and development. With their support, he devised and developed what is widely acknowledged as the first practical fuel cell. The "Bacon cell" became an orphan in the land of its invention because it had shortcomings British industry was unwilling to overlook. On the other hand, American entrepreneurs were interested in the technology, mainly because the federal government was prepared not only to subsidize research to improve it but also to procure it for quasi-military applications.

The permanent national security emergency enabled American industry and government to justify investigating and investing in a number of other types of fuel cell in addition to the Bacon cell. And it also had important effects on the ways practitioners understood the science and engineering of fuel cell electrochemistry. Like their counterparts in other science-based technology fields, notably semiconductors, fuel cell workers struggled with the epistemological consequences of the organizational division of labor into discrete units of "research" and "development."[35] This complex historical phenomenon has been the subject of much social science research. Prior to the Second World

War, most industrial firms considered research and development as a type of investigation conducted in one department by one kind of expert, the engineer.[36] Applying scientific methodology to the study of the basic physical principles of materials and machines, engineers created a semiautonomous interdisciplinary branch of knowledge known as engineering research or science. But in the years before the Second World War, observed science and technologies studies analyst Ronald Kline, eminent American engineers seeking to improve the social standing of their discipline began to refer to their work as "applied science" in order to accrue some of the prestige associated with what had long been termed "pure science." In turn, they relabeled the latter "fundamental" or "basic" science to avoid tarnishing their activities with the "implication of impurity." By erecting social boundaries between academic science, industrial research, and engineering, these researchers helped lay the basis for linear social relations of innovation.[37]

This idea resonated deeply in government, industry, and academic quarters after the war thanks to the success of the wartime state-backed, science-based weapons programs, the growing complexity of advanced technology, the influence of applied-science ideologues such as Vannevar Bush, and the crystallization of the military-industrial complex. Corporations began to separate research from production activities partly because firms in the burgeoning semiconductor industry like RCA increasingly saw basic research as the foundation of future generations of technology, as the historian Hyungsub Choi has noted.[38] And federal promotion of advanced military research further encouraged the segregation of innovation, either because the scope of such programs prompted industrial contractors to place them in their research divisions or because federal rules compelled them to classify, account for, and, hence, separate research from development and manufacturing.[39]

As in the semiconductor field, the linear management imperative had important implications for innovation in the fuel cell sector. Here, sequential development was driven both by government protocols and by the ideology of applied science. But it was also propelled by the unique political economy of fuel cell research communities. Social scientists began studying laboratory cultures in the 1970s, learning a good deal over the years of how external social forces shaped the production of knowledge and discourse in academic, industrial, and state laboratories.[40] In important ways, however, the fuel cell laboratory was distinct from these environments. It occupied an oddly ambiguous institutional position that reflected the relatively weak institutional standing of electrochemistry. Properly speaking, a true fuel cell industry did not exist for most of the twentieth century. Most of the first major fuel cell projects were instead sponsored by government agencies that set up research units within existing state or state-supported laboratories or contracted work to power source companies. Accordingly, most fuel cell laboratory communities were distributed and subsumed within larger governmental and corporate research structures until the 1990s, when dedicated fuel cell research companies with their own labs began to emerge. Over the years, a few fuel cell lab communities became established and developed distinct identities. However, the majority were marginal and many ephemeral, tending to have limited resources and tight timelines for results.

These pressures, in turn, shaped a particular discourse and conduct of research. Like all forms of technology testing, fuel cell trials were an interpretive process where generalized results reflected the worldviews of laboratory workers as much as the objective properties of materials and devices. These worldviews were mediated by a wide variety of nontechnical factors including tradition, experience, and political and economic interests.[41] Because test designers desired unambiguous results and total control over "potential disturbing factors," they tended to simulate experimental conditions significantly unlike real-world situations, often omitting all reference to the circumstances in which the test was executed. The resulting epistemology encouraged the drawing of analogies between classes of artifacts that were only superficially similar, acts that the sociologist Trevor Pinch has referred to as "similarity judgments."[42]

Practical fuel cells should properly have been seen as miniature chemical plants in which the fuel cell—the actual energy converter—was only one component of a complex system. With limited time and resources, however, researchers rarely took a systems approach, concentrating instead on the fuel cell, the heart of the system. This informed a self-reinforcing political economy of expectation. Proceeding stepwise, engineers first developed half cells, then full cells, and then ganged these together in a multicell power unit called a "stack," to distinguish it from the galvanic battery. As engineers developed increasingly larger and more sophisticated stacks, they attracted sponsors.

Naturally, power density was the first criterion by which practical power sources were judged. But durability and cost-effectiveness were equally important. Researchers found, however, that they could boost power faster and with relatively less effort than it took to make stacks cheap and long-lived. As a result, power density, and the ability to increase it, became the chief political capital of the fuel cell engineer. Patronage became linked to this practice. As researchers experienced success with relatively simple stacks and gained confidence, they and their sponsors and managers assumed that more complex designs would work similarly well in more demanding roles. First pressed into practical service in the mid-1960s in highly specialized semimilitary applications using special fuels, fuel cell technology began to be seen by many researchers as having the potential to function in a range of civilian roles.

Perhaps the most important similarity judgment made by practitioners was that fuel cells could convert the chemical energy in hydrogenous fuels to electricity almost as easily as they converted pure hydrogen. Researchers frequently bracketed demonstrations of hydrogen fuel cells with suggestions that a general-purpose multifuel cell might be possible. Liberating its operators from the economic and political fetters of the global energy economy by converting the most readily available primary fossil or biomass energy resources with no emissions, this hypothetical device was virtually guaranteed a vast market if it could be developed. In carefully prepared demonstrations that emphasized certain criteria and test results and downplayed or withheld others, fuel cell workers and their sponsors tried to leverage the promise of the super fuel cell in order to fund ongoing research operations, among other rationales. These rituals of presentation or dramaturgy—controlled public performances framing relevant questions and ratifying knowledge claims—were a key source of the idea of the power panacea.[43]

But the dream machine remained a vision for a mix of physical and social reasons. The performance of a laboratory stack, whether using pure hydrogen or carbonaceous fuels, proved a poor indicator of how large integrated systems would behave over hundreds or thousands of hours. And carbonaceous fuel cell systems of all types were far more complex and less reliable than their hydrogen-fueled counterparts. As the fuel cell pioneers George E. Evans and Karl V. Kordesch noted in 1967, the various types of fuel cell had little in common beyond their technical definition as electrochemical cells that electro-oxidized a fuel using an electrode-electrolyte assembly believed to remain essentially invariant over time.[44] But the assumption of invariance did not hold under long-term testing. Engineers discovered that fuel cells were as susceptible to deterioration over time as conventional galvanic batteries. Moreover, as with batteries, the course of fuel cell research and development was strongly influenced by the appliance the power source was designed to serve. Power packs that functioned well in some applications did not always do so in others. Most importantly, fuel cell laboratories were rarely in a position to conduct protracted engineering research as the pace of power breakthroughs slackened and sponsors reconsidered their support. The situation was exacerbated by electrochemistry's relatively low prestige in the United States. By framing the fuel cell as an electrochemical super-engine, researchers were able to broaden their bases of support and buy time for more research. But this tactic inevitably set up crises of expectation when the promise was slow to materialize.

Crafted by the historian Thomas P. Hughes to describe the development of large-scale energy conversion and electric power distribution systems, the oft-cited metaphor of the reverse salient helpfully illustrates this aspect of fuel cell research and development. Hughes saw the progressive development of electric power production and distribution systems as akin to a military front. As in war, the engineering front advanced unevenly. In certain places, pockets of resistance—reverse salients—retarded overall progress and were focused on accordingly. When these areas were mastered, the front could again advance uniformly.[45] Progress in developing fuel cell technology has been similarly uneven. The difference is that the tendency of fuel cell researchers to concentrate on reverse salients within the energy converter and their neglect of crucial auxiliary technologies have more often than not inhibited the development of any kind of effective power system.

Energy, Power, and Ideology

Winner has observed the inadequacy of simple deterministic, cause-and-effect approaches to understanding technology, encouraging instead an exploration of the meaning of technology in civilization. As a characteristically human activity, it is shaped by and shapes people. In laying the basis of a philosophy of technology, Winner draws on Marx and Wittgenstein. As people use technology to alter modes of production, consumption, work, leisure, transportation, war, and healing, they transform themselves materially and ideologically. This, in turn, is reflected in the ways they use language to comprehend reality, construct identity, and develop self-consciousness. As people make physical things work for them, says Winner, our gaze should turn to the worlds people make in the process.[46]

The story of the people who labored to make fuel cells work and the worlds they built in so doing elaborates the relationship between physical and social power and the centrality of energy to both. A relatively neglected idea at the intersection of the history of technology, science and technology studies, and environmental history, the notion that all power, including social power, ultimately derives from energy, note Russell et al., can be a useful way of reconsidering or problematizing conventional historical accounts.[47] It was with this concept in mind that I adopted the electrochemical term *overpotential* as a metaphor for broaching some of the questions of social history I seek to explore. Interpreted figuratively, the word invokes a parallel between the ability of fuel cell researchers to manipulate materials capable of converting chemical energy into useful work and their ability to sustain their own work and reproduce their research communities. Just as engineers sought to manage overpotential in ways that led to high rates of ionization and high power density, they and their backers tried to generate what might be termed social overpotential, promising ever-more powerful fuel cells as a means of attracting funds for personnel and equipment. The stakes were also high for administrators who saw in the technology a means of coping with a range of problems in managing state agencies of innovation. Interpreted literally in a social context, overpotential is suggestive of the construction of expectations that have largely gone unfulfilled.

The sociologist Bruce Podobnik has observed that the scale and technological complexity of the energy conversion chain has the effect of obscuring the totality of the network and the social power asymmetries and systems of hegemony embedded within it.[48] The history of fuel cell research and development helps unveil a certain perspective of this network, revealing the limits of its capacity for change, the relationship between energy policy and energy research and development policy, and the consequences of attempting to use advanced technology as a substitute for politics. Over the years, the activities of fuel cell researchers and their patrons furthered debates on how energy resources should best be employed, although not always intentionally and not always in ways that benefited them. For groups invested in existing energy regimes, fuel cells were attractive as a downstream energy converter with the potential to increase the efficiency of these regimes at or near the point of use. In this manner, high-energy civilization could be made sustainable without having to greatly modify it. Reconciling the contradiction between convenience, sustainability, and environmental lifestyle values, the fuel cell as a power panacea would, ideally, preempt the politics of reform.

As the fortunes of various programs waxed and waned with broader industrial and environmental politics, boosters conflated the qualities of very different fuel cell systems, constructing a transcendent and largely fictional device that could be flexibly interpreted.[49] From the mid-1990s, the fuel cell became increasingly attractive to the middle class and its political representatives as a potent symbol of reconciliation between the traditional fossil fuel–based lifestyle and the nascent green ethos. Others saw it as the means of enabling a decisive break with the old order. The technology occupied a central place in the hydrogen economy, a hypothetical system that, in the words of one visionary commentator, would "fundamentally reconfigure human relationships."[50] In a society conditioned to view technology as the chief agent of progress, the fuel cell as a power panacea made a tremendous impression at various times. In showing how groups with dissimilar interests in the energy conversion chain became enchanted by this idea, this book reveals the sociotechnical mechanisms of a dream machine and its role in perpetuating reveries of technological perfection and limitless economic growth.

1

Device in Search of a Role

Dr. Ellingham remarked that Mr. Bacon was to be congratulated in making an advance which brought the cell into a new field but still emphasized the point that the ERA was sponsoring a piece of fundamental work with no great promise of return.

—Minutes of Electrical Research Association meeting, June 8, 1949

The modern dream of miracle electrochemical energy conversion may be said to have originated with an obscure English inventor with a distinguished pedigree. Working in the leafy environs of the ancient university city of Cambridge, the mechanical engineer Francis Thomas Bacon, a descendent of Sir Nicholas Bacon, the father of the seventeenth-century English lawyer-philosopher Sir Francis Bacon, developed what would become widely acknowledged as the first practical fuel cell.[1] Bacon occupies an unusual place in England's pantheon of illustrious technologists, for he gained fame for inventing a device that by most conventional standards was a failure. Employed in the 1930s by the storied power-equipment maker C. A. Parsons, Bacon learned of a German idea for employing off-peak electricity to electrolyze water and then using the hydrogen so produced in an internal combustion engine. Perhaps, Bacon reasoned, the conversion would be even more efficient if hydrogen and oxygen were combined electrochemically. Aided by a personal inheritance that allowed him to quit Parsons in 1940 and devote himself completely to the project, Bacon sought to build what he termed a "reversible cell." Despite his lack of formal schooling in electrochemistry, he won backing after the Second World War from a succession of sponsors of science and technology who saw his invention as a means of helping revivify Britain's shattered economy, enabling it to compete with the American industrial colossus. In this way, Bacon and his supporters played a key role in triggering the first postwar fuel cell boom.

But fame and success did not come quickly to the gentle-mannered Englishman. For years, he toiled in obscurity in the style of the traditional amateur inventor, constantly searching for new patrons.[2] It was not until the late 1960s, three decades after he had begun his life's work, that his efforts were finally rewarded, although in a way he likely did not anticipate. The inventor had long envisioned the reversible cell as a replacement for the lead-acid secondary battery, first in niche roles and perhaps later in locomotives or the fleets of electric milk-delivery trucks that plied the narrow streets of English towns and cities every morning. With further development, he hoped, the device could be used to power small, short-range electric cars. Instead, the "Bacon cell," as it came to be known, found fleeting first use along the semitropical coast of southeast Florida, a jumping-off point for journeys hundreds of thousands of kilometers long. There, within the confines of the sprawling Kennedy Space Center, Bacon's dream became part of the "American technological sublime."[3] Buried within the stubby cylinder of the Apollo spacecraft topping the 111-meter Saturn V rocket, the most powerful vehicle ever built by human beings, Bacon-type fuel cells would supply the electricity and water that sustained the first people ever to travel to the moon. An advanced version of Bacon's original design later found use in the Space Shuttle, further crystallizing the device's association with space travel.

This indirect triumph of British ingenuity came at a time when the battery and fuel cell community had long since pronounced the Bacon cell a commercial dead end owing to its requirement for pure hydrogen. This was widely viewed as an impractical fuel for most industrial purposes, far too expensive to supply in all but the most specialized applications. Despite its limitations, the Bacon cell was taken by many engineers, scientists, and administrators in Britain and the United States as a sign of things to come in the field of electrochemical energy conversion. For some observers, it was merely a superior kind of electrical storage device. For others, it had far greater potential as a kind of multifuel energy converter. But although it found favor with a series of state-backed impresarios of research and development, the Bacon cell never attracted customers in Britain. In the United States, however, after demonstrations in 1959 that established the technology as the most powerful in its class, the aviation-engine manufacturer Pratt & Whitney licensed it, eventually developing a version that achieved fame powering the Apollo spacecraft.

At first glance, the episode appears to reinforce the claim that Britain was "good at inventing but bad at developing."[4] Articulated repeatedly by administrators and technocrats in the postwar period, this belief helped inform a discourse and subsequently a historiography that held flawed science and technology policy as cause and consequence of national economic decline. The story of the attempt to commercialize the Bacon cell invites comparison with that of penicillin as a declensionist cultural parable in this era, one chronicling "pride over technological prowess, resentment over the loss of opportunity and jealousy of American success." Although penicillin was discovered in England, patriotic science and technology planners believed that U.S. companies exploited the drug's market potential more fully than home industry ever did.[5] Similarly, they feared the Americans might benefit exclusively from the Bacon cell. The main difference, of course, is that the early promise shown by various fuel cell technologies in the postwar period, including the Bacon cell, has gone largely unfulfilled. There has been no lingering sense of loss in British science and engineering communities as a result.

Nevertheless, the story of the Bacon cell is important because it shows how economic pressures, nationalism, and the political economy of research and development in postwar Britain served to construct the idea of the fuel cell—then still an experimental technology—as a power panacea. Like penicillin, the Bacon cell can be seen as constituting a "host of alliances."[6] Bacon's ability to rapidly boost power output in controlled laboratory conditions impressed observers from a wide variety of backgrounds and with different potential uses for the technology. As this occurred, his collaborators and sponsors mused publicly that advanced versions might one day work equally well on carbonaceous fuels. During the Bacon cell's long gestation, its supporters held the device as an example of national technological virtuosity and, when it failed to find customers in its home market, of the arrested potential of British innovation.[7]

It is not just the rhetorical construction of the fuel cell as a power panacea but also the attempts to physically build the artifact that help explain the persistent narrative of unrealized expectations characterizing the technology's history and the emergence of the phenomenon of fuel cell futurism. The story of the Bacon cell sheds light on the problem of applying scientific knowledge in commercial industrial production, a process often impeded by conflicting "research" and "development" institutional cultures. That the Bacon cell's first important champion was the Electrical Research Association, an organization dedicated to supporting incremental rather than radical technological innovation, is symptomatic of the divergence in the ways pre- and postwar industry and government leaders defined research and development and perceived its utility as a cluster of social practices, a shift noted by Sally M. Horrocks.[8] Understanding these relationships and the technopolitics informing the literal and figurative construction of the Bacon cell in turn lends insight into the boom-bust character of postwar fuel cell research.

From "Reversible Cell" to "Fuel Cell"

Language has played a crucial role in informing expectations of fuel cell power. Discovered in the mid-nineteenth century by scientists intrigued by the possibilities of reversing water electrolysis, the phenomenon now known as the fuel cell effect was not originally referred to as such. William Grove termed his apparatus a "gaseous voltaic battery."[9] The phrase "fuel cell" was coined four decades later in the 1880s by Ludwig Mond and Charles Langer in reference to the fossil fuel–derived hydrogen they believed could be used in such an electrochemical device. In the late nineteenth and early twentieth centuries, researchers including W. W. Jacques and Emil Baur attempted to electrochemically convert pure carbon and coal in high-temperature devices.[10]

By the time Bacon began his research in the early 1930s, the electrochemical community was using the term "fuel cell" to denote a hypothetical device that would produce electricity by electro-oxidizing common carbonaceous fuels or "dirty" hydrogen derived from carbonaceous substances, as opposed to pure hydrogen. In theory, fuel cells would consume chemicals in a single irreversible reaction, much like disposable galvanic batteries but with a much longer life span, or so researchers assumed. As they would quickly discover, however, carbonaceous fuels used directly in the cell contaminated catalysts, electrodes, and electrolyte. In contrast, Bacon conceived his apparatus as a reversible electrical storage unit. It would function like a secondary or storage battery, except that it would use pure hydrogen as the storage medium. Electrolyzing the resultant water using electricity produced from an external source would then recharge it.[11] Bacon experimented with the idea throughout the 1930s until, in 1940, he left Parsons to strike out on his own. Obtaining the support of the consulting-engineering firm Merz and McClellan, Bacon subsequently attracted the notice of A. J. Allmand, a professor of chemistry at King's College, London, and an important figure in the world of academic electrochemistry. Allmand provided Bacon space in his laboratory and supervised his work until 1941, when the mechanical engineer was reassigned for war research.[12]

Basing his device on existing electrolyzer technology, Bacon decided against using an acidic electrolyte at low temperature because this required expensive platinum to catalyze the reaction. He instead used an alkaline electrolyte, enabling the use of cheap nickel as a catalyst if the electrolyte was heated above 200°C. At such temperatures, the electrolyte had to be pressurized to prevent it from boiling. Bacon found that increasing pressure beyond what was necessary for stabilizing the electrolyte boosted conductivity and chemical reactivity.[13] Susceptible to degrading reactions with carbon dioxide, the alkaline electrolyte cell required pure hydrogen and oxygen, a serious shortcoming. Although Bacon understood his fuel cell/electrolyzer—the so-called "reversible cell"—as a storage device, it did not have any intrinsic storage capacity but was simply a converter of chemical energy. To complete the fuel cell concept, reactants had to be stored externally. But storing hydrogen posed many challenges. Because of its very low energy density, the gaseous element had to be compressed or liquefied in order to render it practical as a fuel, processes that in turn required considerable energy. Hydrogen is particularly difficult to manage in these states owing to its small molecular diameter and, hence, its propensity to evaporate through uninsulated pipes and vessels. As a result, hydrogen fuel systems had to be built to exacting standards and were very expensive. For these reasons, they had been developed only for experimental applications during the interwar years. But Bacon was not discouraged in this early period, because he did not take a systems approach. He believed instead that the first step was to perfect the reversible cell. Commercial applications would, he thought, become apparent in the future.[14] His first sponsors shared this reductive view.

Postwar Reconstruction and the Bacon Cell

Work on the Bacon cell resumed shortly after the war, a time when scientific research commanded tremendous prestige in Britain and the United States. Industrial and political elites ascribed the Allied victory largely to the marriage of science and industry, a notion influenced by the successful advanced weapons programs. In peacetime, British governments once again promoted science, this time in service of reconstruction. Manufacturers responded enthusiastically. Over the years, however, politicians, technocrats, and the various fractions of industry evolved differing justifications for and expectations of industrial research and development based on their distinct economic interests, as Horrocks has observed. Programs were crucially shaped by the Attlee government's immediate geopolitical and economic goals of protecting the home market, building up export markets, and bolstering the defense sector. These objectives and the policies designed to accomplish them—conscripting scientific and engineering labor for national service, confining trade within the sterling area, and barring import of certain goods—often clashed with and complicated the efforts of certain manufacturers to plan and execute research programs. Control of building permits and goods purchases allowed governments to pick winners, driving many firms to frame their programs in terms of their ability to meet the state's short-term objectives. The result was a concentration of postwar British industrial research and development in advanced defense-related fields such as aviation, nuclear power, and electronics, where government was the chief or only customer, a trend that occurred at the expense of long-term economic growth.[15]

In large measure, the Bacon cell was a product of the movement to mobilize science for industry, one that precipitated a clash between pre- and postwar research cultures that, at root, was a conflict between manufacturing interests. Little more than a conceptual model after the war, the Bacon cell found a home not in the government research establishments or the laboratories of the large manufacturers but in the Electrical Research Association (ERA), an industrial research cooperative. The first such institutions appeared during the First World War. Heavily dependent on products of German science-based industry, Britain was ill prepared to wage a long war of attrition following the severing of trade ties. Among other ad hoc measures, the government encouraged smaller manufacturers to pool their research activities in special associations, establishing the Department of Scientific and Industrial Research (DSIR) to coordinate and help fund them. The first such groups were formed in 1918 and in the early postwar years. In peacetime, however, their work was circumscribed by their cooperative structure in a competitive capitalist economy. Because the various member firms did not necessarily share the same interests, programs were determined consensually and limited to improving existing technologies. Moreover, companies were reluctant to support research that could benefit their competitors. Biased against novelty and possessing limited resources, the research associations played a relatively small role in interwar research and development.[16]

New research associations continued to be organized in the years after the Second World War, but the place of the industrial cooperative in the postwar R&D order was tenuous. As the reconstruction drive gathered steam, the state directed the lion's share of resources to the largest firms with in-house laboratories, although the Attlee government did enlist the cooperative associations in the effort, encouraging them to expand and allocating more resources to the DSIR.[17] Yet their potential was limited by their conservative mandate. Facing increasing marginalization, some cooperatives opted to engage in more speculative research justified by "faith" and "imagination."[18] Not only did such work accord with the prevailing doctrine of science in the service of national industrial recovery, it could also have cohesive effects on associations because it was less likely to trigger commercial conflict among their members, at least in the short term. But this approach could be difficult to sustain, since manufacturers were less likely to support basic research if it did not promise quick practical results.[19]

The ERA's involvement in the Bacon cell originated as such an effort. Founded in 1920, the association concentrated on investigating the physical principles of electric power transmission with the aim of improving efficiency and reliability. It went about its tasks frugally.[20] Not until 1935 did the ERA get its own dedicated laboratory at Perivale, Middlesex. In the mid-1950s, the member industries contributed only a modest annual sum of £100,000 to fund operations, and it took a decade to raise £400,000 to equip a new replacement laboratory at Leatherhead, Surrey. The ERA did not even launch its own journal until 1956.[21] Nevertheless, it grew to become among the largest and most important of the research associations, comparable with some of the laboratories of the big industrial corporations.[22] The Bacon cell project represented a major break with the ERA's traditional agenda, providing an opportunity for the association's managers to reconcile internal and external interests. In the initial flush of postwar technoscientific positivism, the technology offered attractions to a variety of groups. Makers of power-transmitting equipment were intrigued by the Bacon cell's unusual qualities. For ERA leaders, involvement with a potentially revolutionary power source promised to elevate the association's role in the reconstruction effort. The device held similar appeal for state technocrats determined to place Britain in the vanguard of advanced technology. This tacit alliance gave the project momentum.

In early 1946, the ERA assigned a project to the Department of Colloid Science at the University of Cambridge under the leadership of the physical chemist Eric Rideal to see if the reversible cell lay in its field of work. The Cambridge chemists were soon satisfied this was the case.[23] Subcommittee F, the ERA panel responsible for electrical storage, then decided to explore whether the device had potential as a lighter and more efficient replacement for the lead-acid storage battery. One year later, ERA director Stanley Whitehead and a number of members of the subcommittee thought such an application impractical but were reluctant to abandon the project.[24]

It is likely that their attitude was informed by the fuel crisis during the winter of 1946–1947. Neglected during the war as the national government concentrated resources in the armaments sector, Britain's coal-based electrical power system emerged in peacetime weakened by overuse and underinvestment. During one of the coldest winters of the century, it proved unable to cope with an unexpected rise in demand stemming in part from the increased use of domestic electric heaters, a crisis made worse by the Ministry of Fuel and Power's decision early in the year to distribute coal supplies more widely throughout the economy. As a result, power plant reserves were cut from six to four weeks, whereas before the war a supply of 10 to 12 weeks had been considered prudent. In January 1947 the transport network was paralyzed by severe cold, and power plants were reduced to little more than one week's supply of coal, leading to power cuts lasting for 12 days and bringing the electric power system to the verge of collapse. In March the government decreed an emergency program to increase capacity, but the system was not fully able to cope with demand for several more years. The long-term economic and political implications were serious for the Labour government because the lack of electric power capacity forced it to threaten to compel industry to reduce peak demand by staggering the load through work shifts in order to keep the export drive going. This, in turn, risked a confrontation with unions unwilling to compromise after years of wartime discipline and austerity.[25]

It was in this atmosphere that the subcommittee deliberated about what to do next. Here, they deferred to Bacon's mentor Allmand. To him, the reversible cell had more promise as a "fuel cell," and he recommended the ERA continue the project focusing only on the energy conversion unit.[26] Subcommittee F concurred, as did Bacon.[27] This meant abandoning the electrolyzer and using bottled hydrogen instead. Bacon mentions this decision without elaborating in his historical synopsis of the project, but in fact it had important implications for the course of the program.[28] The reversible cell could not be a true fuel cell as it was then configured, because its alkaline electrolyte solidified in the presence of carbon dioxide, whether from the ambient air or from carbonaceous fuels. The power source was therefore restricted to pure hydrogen and oxygen, raising the question of how these substances would be supplied and stored in commercial applications.[29] But the ERA had no mandate to develop hydrogen storage and production technology, moving some subcommittee members to question the rationale for continued involvement. Did the ERA, asked Rideal, seek a "technical" or "industrial" solution to the problem? Did it aim to show that an industrial solution was not feasible? Whitehead responded that the ERA should study areas where knowledge was deficient, noting that the current work was not wholly in the electrical field. With the case at hand, the task had not been to produce something economically feasible but to determine whether the electrochemical technology lay within the ERA's purview. If so, the association wanted to "show that it did work." This was something of a revision of the association's original objective. Having judged, after a year of study, that the Bacon cell had no potential as a replacement for the lead-acid storage battery, members of Subcommittee F advanced new justifications for pushing on. H.J.T. Ellingham, one of more outspoken critics of the technology in the storage role, believed further research could determine whether the apparatus could be converted into a true fuel cell capable of using carbonaceous/hydrocarbon fuels. In the meantime, noted Rideal, Bacon had an excellent record of boosting current density, and further progress could be expected over the next 15 months. Moreover, many on the subcommittee were attracted to the idea of working with a new technology that might yet prove useful. Continuing would be worthwhile, opined one consultant, simply because "no one else seemed to be doing anything on the problem."[30]

These arguments would justify the ERA's support for the Bacon cell over the next eight years. By persisting with the program, the association encouraged the idea that a nonstorage fuel cell was feasible *and* maintained that the technology might also someday serve in the storage role when configured to use hydrogen. Intentionally or not, the ERA made the Bacon cell all things to all people, framing it as a universal energy converter capable of using a variety of substances as fuel. This impression crystallized over time as the association moved beyond exploratory research and committed to developing a breadboard or prototype model.

The Bacon Cell Makes a Splash

In 1949 the Cambridge team settled on the Bacon cell's basic configuration. As the team boosted power output and government supply ministries began to take notice, a debate began over precisely where the Bacon cell might fit in Britain's energy conversion chain, one that ranged widely depending on whether observers assessed existing or potential capabilities. That year, representatives from the DSIR and the powerful Ministry of Supply attended a meeting of Subcommittee F for the first time. Although ERA chief Whitehead supported the program, he admitted that some of the association's members, notably the battery manufacturers, were less interested. Nevertheless, S. F. Follett, the ministry official, was impressed with progress to date and responded that the costs of continued research were "well justified." The technology would certainly interest the Ministry of Fuel and Power, if not the Central Electricity Authority. Indeed, he indicated, the possibilities were "limitless." This reflected the developing interpretation of the Bacon cell as a flexible multifuel system.[31]

A year later, an even larger delegation of government officials representing the Ministry of Supply, the Ministry of Fuel and Power, and the Royal Aircraft Establishment attended a meeting of the subcommittee. Whitehead summarized the program's origins and progress to date. True, the ERA had become involved on the basis of little more than "scientific curiosity," but this phase had now passed. The power and efficiency of the "hydrogen/oxygen" cell had reached levels sufficient to entertain serious consideration of applications. The device could now be compared with the lead-acid storage battery at power levels of more than 30 kilowatts with discharge times of between five and ten hours, opined Whitehead, and the comparison became "progressively better" for higher power levels and longer discharge periods.[32]

By making this claim, Whitehead revived the notion of the cell as a storage device. But what would the device power? Whitehead never mentioned specific types of electrical equipment in this meeting, referring to applications hypothetically, either negatively, in the sense of roles not then filled by existing storage technology, or in the most general descriptive sense, "the direct electro-chemical generation of electrical energy." In neither case did Whitehead specify an application. For him, this question was as much about politics as plausibility. In an earlier meeting in 1949, he had suggested the Bacon cell could serve as an auxiliary power source using hydrogen produced by wind generators. Admittedly, this was a niche role, but it supplied an answer "if someone asked why we were interested in the cell."[33] The chairman of the Merseyside and North Wales Electricity Board saw a "great future" for a hydrogen fuel cell given the impending shortage of liquid hydrocarbon fuels, with a possible near-term role powering electric agricultural tractors. Others were less enthusiastic. The official from the Ministry of Fuel and Power thought the Bacon cell interesting but could not fathom what it might be used for. H. L. Troughton, the senior Ministry of Supply delegate, took a more circumspect view than his colleague Follett. He thought the technology needed more work and in its current form was best suited for powering a navigational aid. The ERA council chair agreed this was a good place to start, for it would show that "something definite had been accomplished." With further development, the cell might prove useful for rail traction and large-scale electricity storage.[34]

Once more, research objectives, expectations, and the ERA's role in the effort had been redefined. Conferees acknowledged that although the practical questions of fuel and applications could not quickly be resolved, this did not mean the project lacked merit. Commercialization remained the ultimate goal, but it could be deferred into an indefinite future. In the meantime, committee members agreed that work would focus on "fundamental difficulties" with the oxygen electrode, which suffered irreversible oxidation over time until it disintegrated completely.[35] In this way, the Bacon cell was made to fit within the ERA's agenda. Left unremarked by the principals was the fact that the unit would have to be a component of a hydrogen fuel supply and storage system if it was to be successfully marketed.

Until the mid-1950s, the research team focused on electrodes—half cells consisting of either an anode or a cathode—and on single cells comprising both an anode and a cathode. As the "fundamental" technology program progressed, it accrued prestige and an ever-higher profile. In early 1951, the Ministry of Fuel and Power joined the ERA in a 50-50 partnership.[36] Then the Admiralty became interested, entering into a contract with the ERA in April 1954 and contributing additional funding.[37] The production of a more durable oxygen electrode in early 1953 garnered international attention. In April, the Office of the Naval Attaché in the U.S. embassy in London asked the ERA for permission to allow an American scientist to visit the Bacon group at Cambridge.[38] Displaying the characteristic British ambivalence to U.S. collaboration in civilian research and development in this period, Whitehead consented on condition the confidentiality agreement in place between the DSIR and the embassy was respected. He warned the Americans to be "very cautious" about the oxygen electrode, "the benefits of which we must secure for this country."[39]

As the ERA struggled to define roles for the Bacon cell, members of the research team developed differing opinions of their objective. In January 1954, the electrochemist R.G.H. Watson indicated that although the initial task was to develop the Bacon cell as a storage device, it might also be possible to modify the technology so that it could consume hydrocarbon-derived fuels. Such a device, he claimed, would constitute "a modern 'philosopher's stone.'"[40] Bacon, on the other hand, thought a hydrogen fuel cell could stand on its own merits.[41] As then configured, however, the Bacon cell could not function in either format outside the laboratory. Unable to accept unreformed carbonaceous fuels, it could not be considered an electricity storage device without the electrolyzer and hydrogen storage components.

In 1954, the research team built its first Bacon cell prototype "battery." Consisting of a six-cell stack of 150 watts, it was exhibited in London. As Bacon prepared to publish his first article in the *British Electrical and Allied Manufacturers Association (BEAMA) Journal* in early 1954, its editor insisted he revise the title so that it referred not to a "Hydrogen-Oxygen Cell" but rather a "Hydrogen-Oxygen Fuel Cell," in order to create more interest.[42] This article introduced Bacon and the Bacon cell to the larger science and technology community. Petroleum concerns had already begun to take notice. In May 1954, Whitehead informed Bacon that a number of companies had inquired about the possibility of running the cell on the waste hydrogen by-product of oil refining.[43] In June, Bacon's article appeared in the *BEAMA Journal*. Four months later, Whitehead asked Bacon to meet with representatives of the Anglo-Iranian Oil Company to discuss the possibility of jointly developing commercial applications of the hydrogen-oxygen cell.[44]

Techno-orphan

As 1955 progressed and fuel and application issues remained unresolved, the laboratory work came to a crossroads. Negotiations between the ERA and the oil companies were inconclusive. In some electrochemical circles there were doubts about fuel cell power.[45] Late in the year, the Bacon cell project was dealt a serious blow. An economic feasibility study completed by the General Electric Company (GEC, also known as British GE) concluded that from the moment hydrogen was produced, by any means, a Bacon cell–based power and transportation network lost money in competition with incumbent systems. The existing infrastructure had the insuperable advantages of mature power technologies that were constantly being improved on an immense scale and, for the time being, a plentiful supply of cheap fossil energy. But while the report rejected the Bacon cell as it was then configured, it did not rule out the possibility that future research might one day make the technology economically viable, perhaps in rail traction.[46]

This left the electrolyzer, long off the ERA's agenda, as the only option for fuel production and, therefore, early commercial development. The problem was that from the start, the ERA had opted to pursue the reversible cell strictly as an energy converter without the hydrogen production and storage equipment. Whitehead had hoped to adapt Bacon cell electrodes for use in an electrolyzer but, as the research team would discover, this was no easy task.[47] A purpose-built electrolyzer, he observed, was too expensive at this stage, even in niche applications, and would throw "everything out of balance."[48]

By late 1955, with the basic physical dynamics of the Bacon cell known, the ERA's work drew to a close. At the end of the year, the Ministry of Fuel and Power let its contract expire, concluding, like the ERA, that the technology was ready for commercial development. But industry remained uninterested and, without a new sponsor, the research team could be kept together at Cambridge for only a few months into the new year. Even the Admiralty, the Bacon team's best hope owing to its possible requirement for a submarine power plant, had lost interest.[49] The prospect of the breakup and loss of the group before a new patron could be found to "adequately exploit for British industry the knowledge they have acquired" was, said Whitehead, a "great pity."[50] In May, the Central Electricity Authority rejected Bacon's request for support.[51] Now out in the cold, he approached John Anthony Hardinge Giffard, the Third Earl of Halsbury and the managing director of the National Research Development Corporation (NRDC).[52]

The National Research Development
Corporation and the Bacon Cell

In a sense, the NRDC was the only remaining institutional avenue through which to pursue the Bacon cell in Britain. The corporation had grown out of the political furor sparked by the shortage of commercial penicillin in Britain in the late 1940s, an event that focused attention on the loss of ownership rights to American pharmaceutical companies. Largely the initiative of technocrats at the Board of Trade determined to never again allow foreigners to exploit British ideas—at least without proper remuneration—the NRDC was established as an independent body through the Development of Inventions Act of 1948.[53] As part of the Board of Trade, it was intended to complement the DSIR, fulfilling the "applied" or "development" aspect of government-backed advanced technology programs by transforming fundamental knowledge produced with public support into hardware for use by industry. So that there was no mistaking the NRDC's mission, the conjunction between "research" and "development" was dropped from its formal title.[54]

Barred from seven military-related government-supported technologies including nuclear reactors and gas turbines, the NRDC supported inventions that were "not being developed or exploited, or sufficiently developed or exploited."[55] The Bacon cell was one of the few innovations that met these criteria. By other considerations, however, the NRDC had good reasons not to invest in it. In December 1954, Lord Halsbury issued a stinging indictment of the ERA's fuel cell effort in the journal *Research*, criticizing the "romantic" desire for technological spectaculars that obscured underlying flaws. It was not that a lack of fundamental knowledge was retarding practical progress, he indicated, but that years of research had revealed no practical solution. The fuel cell simply lacked technical merit. If the best specialists in the field were currently receiving adequate support, little more could be done.[56]

Now, a little more than a year later, Lord Halsbury responded positively to Bacon's inquiry, stating cryptically that he had always wanted to support Bacon's work "if need arose."[57] In fact, the Bacon cell did not meet all of the NRDC's standards. To justify the expenditure of public funds, inventions had to be technically sound, industrially useful, and have commercial potential that was clearly in the "national interest." Industry and the supply ministries had already judged the hydrogen fuel cell deficient in these qualities. Matters were further complicated by an inherent contradiction in the NRDC's mandate. The corporation existed to make money, both for itself and for its clients, by licensing the

technologies it helped develop. Its managers hoped to partner with entrepreneurs prepared to "take the ordinary commercial risks of putting a new product
on the market."[58] But Lord Halsbury had wondered whether industry had missed
enough good inventions to allow the NRDC to be "able to make a living of
developing them."[59] Where the Bacon cell was concerned, there were strong
economic and technical reasons why it was an orphan.

Lord Halsbury's change of heart was likely due to two factors. In the
mid-1950s, the NRDC's portfolio was rather spare, its most promising innovations consisting of cephalosporin antibiotics and computer patents.[60] Rising
American curiosity about the Bacon cell probably catalyzed the NRDC director's
decision to review the project. In early 1956, the Patterson Moos Division of the
Universal Winding Company began talks with the ERA to license the technology.[61] In mid-1956, the NRDC agreed to fund construction of one prototype unit
of up to ten kilowatts. The original plans called for converting another into an
electrolyzer, but this was abandoned following the introduction of a purpose-
built commercial model by the Swiss firm Lonza.[62] Ownership of the Bacon
cell patents was transferred from the ERA to the NRDC, and in mid-1957 the
engineering firm Marshall's of Cambridge was selected to build the device.
Bacon acted as consultant. With British industry as reluctant as ever, the NRDC
believed its sole option was to negotiate a licensing agreement with Patterson
Moos.[63] The American company came to the table with clients with a potential
interest in procurement, an advantage the British side could not match.
Patterson Moos had developed a relationship with the U.S. Air Force, then
experimenting with liquid hydrogen as a jet fuel, and acquired a contract in
early 1957 for a variety of "working models" of hydrogen-oxygen cells, primarily
for exploratory research. The company anticipated similar contracts from all
branches of the U.S. armed services.[64]

Negotiations over the licensing terms were delayed by a dispute over
whether the agreement would cover the exchange of pieces of finished technology, especially electrodes, or a manufacturing process.[65] These issues were
not settled until 1959. At that point, work on a six-kilowatt Bacon cell was well
under way at Cambridge, and the Anglo-American partnership moved into
a new phase, one featuring a player involved in the U.S. Air Force's hydrogen
program. This was Pratt & Whitney, the renowned maker of aircraft engines
and a division of the United Aircraft Corporation. As part of the CL400 project,
a Lockheed-designed supersonic reconnaissance aircraft, Pratt & Whitney had
built a small liquid-hydrogen production plant and a handful of hydrogen-fuel
jet engines. In late 1957 the Air Force canceled this project, but Pratt & Whitney
remained interested in hydrogen fuel technology. In 1959 the company entered
into a joint research and development agreement with Universal Winding,
licensing the Bacon cell from the NRDC with a view to adapting it for aerospace
applications.[66]

The Bacon Cell as Technological Metaphor

The demonstration of the six-kilowatt unit powering an electric welder and a forklift truck in front of assembled visitors in an airport hanger at Cambridge in late August 1959 represented the zenith of the Bacon cell program in Britain. In fact, the event was something of an anticlimax. With the technology seemingly destined to find its first application in an American air- or spacecraft rather than in more prosaic roles in the land of its creation, resentment simmered among some of the guests, who made known their view that the program represented a failure of the British "technological imagination."[67] Precisely how the national interest had been served was, for them, doubtless unclear. Nevertheless, the demonstration caused something of an international stir, raising the profile of fuel cell technology. In September, Bacon was one of the star speakers at the first global conference on fuel cells, staged by the American Chemical Society in Atlantic City. The ensuing panel discussion was dominated by talk of the future of carbonaceous fuels designed for fuel cells and the role petroleum, chemical, and coal companies would play in supplying them.[68] As early as the spring of 1956, U.S. coal researchers had noted that although the Bacon cell was the outstanding example of a fuel cell operating between 200°C and 500°C, such devices had to be capable of using carbonaceous fuels in order to be regarded seriously as a primary power source.[69]

After nearly 14 years of effort, Bacon's team, supported by the ERA, the Ministry of Fuel and Power, the Admiralty, and the NRDC, had developed the world's most advanced power source technology in its class. But only the U.S. National Aeronautics and Space Administration evinced an interest in it. In Britain, the device passed into history.[70] For the NRDC, technology transfer was a satisfactory outcome. Although Bacon appealed for more resources, warning of U.S. domination of a home-grown innovation, the NRDC's new managing director, John Duckworth, was happy to have found a paying customer, even if it was American.[71] The NRDC ended the Bacon cell program in January 1961 and threw its support to H. H. Chambers and his high-temperature molten carbonate fuel cell.[72] In October the NRDC entered a partnership to commercialize this design with British Petroleum, British Ropes, and Guest, Keen and Nettlefolds, a consortium known as Energy Conversion Limited. It also maintained a complicated patent and research arrangement with Patterson Moos (by then a research division of the Leesona Corporation) and United Aircraft in which the parties shared information both on molten carbonate and Bacon cell technology. Far from being perceived as another case of U.S. theft of British ideas, this arrangement was generally well received by the press, likely owing to the belief that fuel cell research would have ended completely in the United Kingdom without American investment.[73]

Conclusion

The Bacon cell grew out of a political culture that understood science-based technology as a means of achieving short-term economic and military objectives. David Edgerton rightly notes the relatively limited role the research associations and the NRDC played in British industrial research and development in this period. Yet it was in these marginal state-backed institutions that the Bacon cell found an audience and the modern idea of the fuel cell as a universal chemical energy converter found purchase.[74] In the early postwar years, a time of electricity shortages and ascendant technoscientific nationalism, the conservative ERA uncharacteristically began to explore a radical concept it believed might trigger a paradigm shift in power source technology. With the reconstruction and rearmament drives essentially complete by the mid-1950s, the NRDC adopted the Bacon cell in an effort to bolster its own fortunes as a fledgling R&D agency and profit in any way possible from American interest in the device. Although the ERA and the NRDC had different agendas, administrators in both agencies believed the device might prove immediately useful in special roles and, perhaps one day, in broader applications employing cheap commercial fuels. It was far easier, and cheaper, for researchers to dramatically boost the power output of Bacon cells using bottled hydrogen than to produce either a reversible cell complete with hydrogen production and storage equipment or a carbonaceous/hydrocarbon fuel cell. This became their primary political capital, fostering expectations that similar rates of progress in durability and fuel flexibility were also possible.

This perception helped drive research year after year. Bacon was not deterred by the problems of producing and storing hydrogen, believing a technological fix lay on the horizon. He sought reassurance from the lessons of history, which, he held, showed that solutions were often found for challenges, such as liquid oxygen storage, previously thought insuperable.[75] This faith in the inevitable progress of technology, a constant theme in Bacon's correspondence and papers, particularly in his later years, echoed prevailing sentiments in both Britain and the United States.[76] It would bolster successive research and development communities over the next four decades as they struggled to bring fuel cells to market.

Rejected at home, the Bacon cell in time became a techno-metaphor, though not in the same sense as other high-profile British successes and failures—penicillin, the V-bombers, and the hovercraft in the former category, or the TSR2 strike aircraft and Blue Streak missile in the latter. Not an inherently military technology, the Bacon cell was nevertheless conceived and brought to maturity at the height of the Cold War, a "supremely transnational" event.[77] As its American sponsors continued its development, the technology took on broader significance. Britain's popular media and scientific community viewed it as a story of domestic success, conflating it with fuel cell technology in general.[78] The Bacon cell's role in the Apollo program also allowed those who spoke on behalf of the national interest to claim a share of the glory of the Space Race.

The Bacon cell was also a potent carrier of the ideal of the universal chemical energy converter, inspiring visions of fuel cell power in the United States. In 1961 Anthony M. Moos, cofounder of one of the companies that had helped bring Bacon's technology to the United States, noted the proliferation of programs since the heyday of that "dedicated pioneer," predicting that the world stood on the brink of an energy conversion revolution based on fuel cells using cheap common fuels such as natural gas and propane.[79] Planners at the U.S. Department of Defense's Advanced Research Projects Agency (ARPA) drew solace from NASA's hydrogen fuel cell program, for to them it seemed to suggest the feasibility of the carbonaceous and hydrocarbon fuel cell technologies they were investigating.[80] Yet speaking strictly in the terminological conventions of the day, the Bacon cell was a hydrogen cell, not a fuel cell, a technology that involved much more complex electrochemistry. Promoters of Bacon and his work had done much to blur these distinctions, encouraging unrealistic hopes. For the remainder of the century and into the next, researchers would frequently promote standards of performance based on hydrogen fuel cells that they were consistently unable to replicate in carbonaceous/hydrocarbon fuel cells.

2

Military Miracle Battery

During the past year a remarkable increase in fuel-cell activities occurred. . . . This upswing in interest is based on the potentials of fuel-cell systems that will operate on fuels and air.

–U.S. Army/ARPA joint report, February 6, 1961

In 1960, eight students at Harvard University's Graduate School of Business published an analysis of the technical and economic feasibility of fuel cells. The preface to their 160-page study featured selections from the 1875 edition of the *Congressional Record* heralding the era of the gasoline-fueled internal combustion engine. This technology, the *Record* declared, would begin a "new era in the history of civilization," one potentially more revolutionary than the "invention of the wheel, the use of metals, or the steam engine." But developing and adopting the new technology would present major cost and technical challenges and introduce considerable socioeconomic dislocation. Gasoline was a dangerous and dirty fuel. Vehicles powered by it could attain high speeds, threatening hapless pedestrians, fouling the air, and displacing horse power on farms, threatening American agriculture as a way of life. But the gasoline engine was a technology that was also "full of promise for the future of man and the peace of the world."[1]

The parallel was unmistakable. Eighty-five years after gasoline engines first challenged and then irrevocably altered the established socio-technical order, suggested the authors, a new energy conversion technology was poised to upset the status quo in transportation. Then, as now, developing and marketing innovations brought risks and rewards. Yet the lessons of history, indicated the authors, were clear.

In its day, this monograph was often cited by fuel cell aficionados. But the historical analogy the Harvard students attempted to draw was not quite precise. The initial path to gasoline automobility had been paved by entrepreneurs with little direct help from the state, notwithstanding the impetus given by government procurement during the First World War.[2] The origins of expectations for a commercial fuel cell, in contrast, can be traced to groups in or linked to the

U.S. defense establishment. The military's interest in fuel cells began in the mid-1950s and persisted through changes in war-fighting doctrine over the ensuing decades, driving the early stages of the first fuel cell boom. The technology first appeared attractive to defense planners as a solution to the special problems arising from the confluence of the nuclear arms and space races and the revolution in transistorized electronics. Of the armed services, the Army was the most interested. Its doctrine of the "tactical atom" held that, on the nuclear battlefield, the traditional order of battle—the multidivision Army group—was obsolete. In such an environment, only small, dispersed, and highly mobile units equipped with the latest communications and surveillance gear could hope to survive. But the miniaturization trend in electronics was not being matched in power sources. Considering existing batteries inadequate and incremental improvements merely an interim solution, the Army sought to develop breakthrough power sources.[3]

Planners in the Advanced Research Projects Agency would become intimately involved with this effort. Since around the turn of the millennium, if not before, ARPA (or DARPA, for Defense, as it was known from 1972 to 1993 and again from 1996) has acquired a sterling reputation among media, policymakers, and some social scientists as a model federal manager of lean and effective cost-shared research and development. The agency's success in certain fields gave rise to the idea of a D/ARPA "system" or mode of general-purpose technology innovation and transfer.[4] Indeed, its very name has taken on almost talismanic significance. In 2009, the Department of Energy created ARPA-E (Energy), a division emulating the D/ARPA approach of using relatively small amounts of money to attract the best and brightest minds in researching high-risk, transformational technological concepts that industry was unable or unwilling to pursue. In this way, the federal government hoped to bridge the gaps between the invention and innovation of energy and power source technologies.[5] To be sure, D/ARPA's reputation stems largely from its role in advancing materials research and especially computer science and technology. But in its early days, the agency engaged with the Army and industry in a now-forgotten program of research in power sources, work that revealed how vastly more difficult it was to succeed in this field than in other technology sectors.

Romantic visions of D/ARPA obscure these troubled early years, a period when planners struggled to define the agency's role in the defense research and development establishment. Instituted by Congress in February 1958 as an initiative of Eisenhower's secretary of defense Neil H. McElroy, ARPA was an ad hoc response to domestic military technopolitics following the launch of the first Sputnik satellite in October 1957. McElroy intended the agency as a corrective to problems in the defense community, particularly the rivalry between the armed services over control of rocketry and their reluctance to sponsor certain promising lines of research. Accordingly, ARPA took charge of large booster programs as planners sorted out military and civilian responsibilities in this realm. In so doing, the institution became America's first space agency. But ARPA also assumed responsibility for exploring radical new concepts the services ignored altogether and that did not clearly fall under the purview of any one of them.[6] When NASA was formed several months later in July, ARPA was relieved of its space technology programs and remade as an administrator of basic research.

In this little-known period of its history, ARPA faced something of an existential crisis. As planners looked for ways of reestablishing the agency as a player in defense R&D, they saw research in the materials and chemistry of advanced power sources including the fuel cell as an appropriate mission. Along with the Institute for Defense Analyses (IDA), a defense policy think tank, ARPA was influential in framing the fuel cell as a universal chemical-energy converter, a technology planners believed could ease the logistics of the Army's fuel-thirsty conventional motorized division. The result was Project Lorraine.

In some ways, the U.S. military fuel cell effort mirrored the British experience. Like the NRDC, ARPA was a fledgling agency with an uncertain future. As in Britain's state science and technology complex, actors used the promise of a power panacea as a form of political currency in order to build alliances and bolster their institutional standing. But the sheer size of the American defense establishment and the crucial dual role played by the federal government as a sponsor of research and development *and* procurer of the advanced hardware it helped spawn ensured that novelties like the fuel cell attracted a much broader support base.

Following the British experience, ARPA planners adopted a linear administrative approach. As with their predecessors, managing the resulting division of labor almost immediately raised definitional problems and jurisdictional disputes that complicated research. On the one hand, ARPA managers initially intended merely to support an investigation of the basic physical principles of fuel cell electrochemistry, leaving technology development to the Army, the main prospective customer.[7] But not all bureaucrats were reconciled to the agency's transformation from overseer of large-scale technological systems to supervisor of basic research. And ARPA continued to rely on a defense advisory system closely linked to industry and, hence, biased toward hardware and prone to self-serving and often erroneous estimates. In the ensuing institutional identity crisis, planners struggled to identify where the science of the fuel cell ended and its engineering began.

This proved a false distinction. Conceived as a relatively short, thrifty exploration of simple apparatuses, Project Lorraine's fuel cell program became protracted and increasingly complex as researchers probed electrochemical phenomena in bigger and more sophisticated devices. As the program progressed and pressure to show results mounted, ARPA planners quite naturally resorted to technological benchmarks. At first, they encouraged contractors to seek the path of least engineering resistance, to work with the most tractable but least logistically valuable fuels in a stepwise effort to understand how the more chemically complex heavier fuels used by the armed forces would behave when electro-oxidized. But when the latter program encountered severe difficulties, the earlier heuristic research became increasingly politically important to program managers. Representing tangible success, it informed an interim technology program that, in turn, engendered hopes for a commercial electrochemical engine that industry was hard-pressed to deliver.

Fuel Cells and the Missile-Age Army

No less than in the United Kingdom, the fuel cell in the United States in 1950s was a technological orphan that found a home in government agencies. The roots of the military's miracle battery lay in the technological requirements spawned by the advent of the nuclear-armed ballistic missile but also in the expansion of the defense research establishment and the class of professional experts employed by it. Their assumptions of the nature of the technology were informed as much by their political interests in this establishment as by their expertise in science and engineering.[8] But in the American context, unlike the British, the idea of the fuel cell as a terrestrial miracle machine was ultimately driven by post-Sputnik bureaucratic politics.

The Army's interest in the fuel cell derived initially from the problem of meeting the power demands of new electronic devices for use in space as well as on earth. Its Signal Engineering Laboratories investigated hydrogen and hydrocarbon systems well before ARPA launched Project Lorraine. Although prohibited by the White House from developing satellites in pre-Sputnik days, the Army was actively engaged in rocket booster development and dreamed of a greater role in missilery.[9] Some of its researchers anticipated that rocket guidance systems and the first space vehicles would have demands for electrical power incapable of being met with existing equipment. Developing suitable power sources for these applications was a task for which they eagerly sought responsibility. In line with the prevailing technophilia in civilian and military circles in the 1950s, Army research and development planners discouraged engineers from seeking incremental improvements. Only exploration of the latest processes and materials, they held, would yield the requisite advances in power source technology.[10]

Others in Army R&D were not so sure. The Voltairean proverb of the best as the enemy of the good, one battery researcher suggested, was especially apt in the field of power sources. Users considered them a "necessary evil" and seemed unaware of the challenges designers faced in balancing the trade-offs of different materials in different applications, noted David Linden at the annual battery conference of the Signal Engineering Laboratories in May 1956. They would not be satisfied, he remarked drily, until they had a weightless, low-volume battery of unlimited capacity that would function without maintenance in all conditions.[11]

In 1956 and 1957, the Army considered advanced alkaline, nickel-cadmium, and zinc-silver-oxide batteries and solar and nuclear devices as the likeliest breakthrough power sources. Fuel cells, on the other hand, attracted relatively little attention as reflected in the battery and power sources conference, an important bellwether of Army and industry thinking. In each of the gatherings in 1956 and 1957, only one presentation was devoted to fuel cells.[12] But in 1957, the Army tested a Union Carbide hydrogen fuel cell powering a small radar unit at Fort Huachuca, Arizona. The next year, fuel cells featured prominently at the conference for the first time. The conferees represented leading U.S. firms with interests in the technology, including Patterson Moos, the National Carbon division of Union Carbide, and General Electric. The tone of their presentations, however, was generally equivocal. All the featured power sources were laboratory models using pure hydrogen and oxygen, and the conferees did not mention specific requirements or applications.[13]

Progress was slow, inhibited in part by electrochemistry's relatively low social standing in the United States, where it had long moldered as an academic discipline. Most of the money spent by the federal government on fuel cell development in the 1960s went to contractors, not to postsecondary or poly-technic institutions. With few career opportunities in American industry or government and little prestige, electrochemical engineering was plagued by a critical shortage of labor. Often forced to improvise, industry retrained physical chemists and chemical engineers and imported European electrochemists.[14]

Work was further hindered by the reluctance of American corporations to engage in collective research.[15] In October 1960, 20 firms—later joined by 26 more—began to support a program of fundamental research on fuel cells at the Battelle Memorial Institute in Columbus, Ohio, agreeing to contribute some $1.725 million over five years. F. T. Bacon believed the institute had competent staff, was well equipped with experimental and construction facilities, and had access to support from other scientific disciplines, but noted that none of the sponsoring companies were prepared to supply the institute with details of their own research on fuel cells. This forced Battelle's researchers to start virtually "from scratch."[16] Corporate competitiveness and secrecy would seriously impede the cost-shared collaborative research ARPA would attempt to promote in Project Lorraine.

Army interest in fuel cell power grew following the demonstrations in 1959 of the Bacon cell and a fuel cell electric tractor. Equipped with a 15-kilowatt alkaline fuel cell using hydrogen-propane/oxygen developed for Allis-Chalmers by Harry Ihrig, this was the first vehicle of any kind to employ such a power source.[17] By 1960, three Army laboratories—the Signal Research and Development Laboratory (the former Signal Engineering Laboratories) at Fort Monmouth, New Jersey, the Harry Diamond Ordnance Fuse Laboratory in Washington, DC, and the Engineer Research and Development Laboratory at Fort Belvoir, Virginia—were engaged in fuel cell work and monitoring the related industry contracts, contributing more than a third of the funds the Department of Defense had expended on fuel cell research and development to date.[18] One of the technologies the Army labs experimented with was the hydrogen-oxygen proton exchange membrane fuel cell (PEMFC). Developed by General Electric researchers L. W. Niedrach and W. T. Grubb in the mid-1950s, the system had a surprisingly prosaic origin. Polymer membrane was then used mainly as a water softener; observing that the material was an excellent conductor of ions, Niedrach and Grubb believed it could serve as an electrolyte in fuel cells operating at 50°C to 100°C.[19] In principle, such a device avoided many of the problems of liquid electrolyte fuel cells. A solid membrane enabled electrode and electrolyte to be combined into a single membrane-electrode assembly, allowing designers to dispense with the heavy porous electrodes of the sort used in the Bacon cell and other liquid electrolyte systems and to build lighter and more compact stacks of cells. Just as important, General Electric's membrane, which cross-linked polystyrene-divinylbenzene sulfonic acid with an inert fluorocarbon matrix, could operate on air instead of pure oxygen, and, in theory, on hydrocarbon fuels. For these reasons, military planners and General Electric believed the PEMFC had potential in many roles, and both ARPA and the Army Signal R&D Laboratory helped the company test the technology. Around the turn of the decade, General Electric won a contract from the Army and Navy for a 200-watt portable demonstration generator for use with portable radars.[20]

But early membrane fuel cells had low current densities and were prone to all sorts of teething troubles including dehydration and high electrical resistance. Moreover, it turned out that the design could not cope with carbonaceous fuels. Partly as a result, the Army worked largely with liquid electrolyte fuel cell systems, both basic and acidic. In 1959 and 1960, the technical problem that most preoccupied Signal R&D Laboratory researchers was the method by which carbonaceous fuels were delivered to the fuel cell. They knew that fuel cells could oxidize alcohol or hydrocarbon vapors at high temperatures but regarded this as impractical. Although gaseous fuels could be stored as liquids, converting them back to gases required special pressure-regulation systems that complicated operational control and added weight and volume to fuel cell systems. The ideal such system was one that could directly use liquid fuels at room temperature.[21]

The chemical composition of these fuels was of crucial importance. Prior to Project Lorraine, the Signal R&D Laboratory had mainly investigated partially oxygenated substances such as alcohols, aldehydes, or fatty acids, mainly in alkaline cells. This work revealed that basic electrolytes were very sensitive to contamination from the reaction products of oxidized carbonaceous fuels. As a result, some researchers favored experimenting with acidic fuel cell systems owing to their insensitivity to carbon dioxide.[22] Encouraging results with partially oxygenated carbonaceous fuels caused some researchers to elide the differences between these substances and hydrocarbons (which are unoxygenated, containing only hydrogen and carbon), including both under the general rubric of common liquid fuels.[23] In fact, these two classes of carbonaceous fuel performed very differently in fuel cells. But success in oxidizing the partially oxygenated liquid fuels likely helped inform the Army's interest in research on the "logistics" fuel cell promoted by ARPA and the IDA for possible stationary and mobile applications.[24]

The Bureaucratic Machine

Project Lorraine was firmly rooted in the rapid expansion of the U.S. defense research base and the subsequent internecine bureaucratic turmoil that followed Sputnik. As administrator of the American rocket program from January 1958 to the late summer of 1959, ARPA had been responsible for fuel cells along with a number of other power source and ancillary aerospace equipment. The technology did not then have an especially high profile in defense research circles.[25] But as planners sought new roles for ARPA following the loss of the space mission, they looked to salvage the aerospace energy conversion program and remake it as one of the agency's principal efforts. This plan posited a much larger and more ambitious role for fuel cell power.

This scheme was actually the brainchild of planners from the Institute for Defense Analyses (IDA). Formed in 1954 by the Massachusetts Institute of Technology at the request of the chairman of the Joint Chiefs of Staff and the secretary of defense, the IDA supplied systems analysis to various military advisory panels, recruiting civilian scientists and managing their activities.[26] In the post-Sputnik era, ARPA and IDA developed an intimate relationship. In the haste to set up a space agency, the White House, Congress, and the Department of Defense arranged for the IDA to take responsibility for organizing and staffing ARPA and provided it sweeping powers to do so. In 1958, the IDA created a special unit to advise ARPA.[27] The institute was given the contract to hire all of ARPA's personnel and then assign them to posts within the Pentagon as directed by ARPA's management. Both ARPA and IDA administrators were allowed to approach private companies that were under contract to the Department of Defense and recruit their staff to work for them for periods of one to two years during which they would receive compensation only from the IDA. In short, industry researchers were paid to advise ARPA on the disbursement of contracts to industry. The opportunities for conflict of interest were considerable, noted Herbert York, then director of defense research and engineering, but were overlooked at a time of national emergency.[28]

Largely recruited from industry, IDA personnel made up the core of ARPA's administrative cadre for the first few years of the new agency's existence. They moved easily between the two bureaus, selecting contractors and greatly influencing the planning of the first projects.[29] They favored technology programs, earning ARPA a dubious reputation in certain scientific circles for a time.[30] The Department of Defense's fuel cell program would grow out of the conflicting interests of these two groups of planners as each worked to foster a new institutional identity for ARPA. Feeling that the new agency engendered too much hostility among the armed services and had too small a budget to effectively guide any advanced technology program from research through development to production, ARPA's director Roy W. Johnson recommended to Defense Secretary McElroy in September 1959 that the agency be remade as a sponsor of basic research. This became its chief role after 1960.[31]

The IDA helped clarify responsibility and the chain of command in ARPA's new mission. In so doing, it linked the aerospace fuel cell to terrestrial applications in a new research program. The basic principles were sketched by Nathan W. Snyder, a power systems specialist who had investigated fuel cells on behalf of ARPA and was working for the IDA in 1960. He recommended ARPA preserve its power source programs on the grounds that knowledge obtained in the course of aerospace research was relevant for terrestrial use; space vehicles and ground vehicles, he claimed, had similar requirements. Much of the work in the space power program, he believed, was of equal or even greater value in terrestrial or marine applications. Following ARPA's new policy line, Snyder argued that there was a need for basic research in energy conversion technology serving all the military services. The work of the Department of Defense in this field, he said, was too heavily weighted toward hardware development. If this continued, scientists would be relegated to improving existing systems rather than making technological breakthroughs. He believed the power requirements of the individual armed services were similar, and as a result ARPA could exploit aerospace power source research to develop fundamental knowledge of value to all. Snyder recommended the agency's energy conversion program be closely linked with its materials science program, its primary effort in basic science. The individual armed services could then fund the expensive demonstrations of energy conversion hardware.[32]

In effect, Snyder proposed to make ARPA the all-services body responsible for basic research in the Defense Department's power source programs. True to its mandate, the agency would be supporting work the defense research establishment was neglecting, not existing hardware programs. Snyder warned that a price in national security would be exacted if science was not adequately supported, invoking linear logic to reinforce the argument.[33] Certain institutions like

ARPA and its industrial contractors, he claimed, were better adapted to promul-
gating basic research. Others—the armed services and their private contractors—
were best suited to applying this knowledge. Of course, Snyder assumed that
there was no engineering involved in the basic phase of research in advanced
energy conversion devices and that fundamental research did not take place
during the engineering phase. Events would confound this assumption.

These principles were adopted by the Department of Defense in its
program of advanced terrestrial power sources. On July 19, 1960, the Office of
the Director of Defense Research and Engineering (DDR&E) initiated Project
Lorraine on IDA's recommendation.[34] This effort encompassed a number of technologies but soon concentrated on the hydrocarbon fuel cell. Although the Army
had a number of ongoing fuel cell projects built around expensive rare fuels like
hydrogen and partially oxygenated fuels such as methanol and ethanol, an additional goal now was to develop fuel cells that could use cheap, common fuels
derived from natural gas and especially petroleum—hydrocarbon "logistics" fuels
including combat gasoline, diesel oil, and kerosene-based jet propellant (JP)-4—
in a wide variety of stationary and mobile applications, both military and
civilian. A conference held in late October between DDR&E and ARPA officials
confirmed that Project Lorraine would concentrate on "physical principles"
and possibly "fundamental engineering principle," but would not administer
hardware development. That was to be left to the various Army laboratories.[35]

In late November, planners from ARPA and the Army Office of Ordnance
Research discussed the program's framework and assumptions, outlined in a
report of February 1961.[36] The group presumed hydrocarbon fuel cells would
"evolve" from existing hydrogen-oxygen designs, although they admitted they
did not expect this to be simple. Their discussions with industry revealed that
little was known about the electrochemical dynamics of operations with carbonaceous fuel, particularly the question of which fuels worked best in various
conditions. But the group was unanimous on one point: alkaline electrolyte was
unsuitable for such fuels and, consequently, acidic designs would have to be
developed. Similarly, little was known about electrodes, then a "black art."
To be sure, the reaction mechanism of the oxygen electrode (cathode) in alkaline
media was understood, but the reaction at the anode, especially where carbonaceous fuels were concerned, was not. Electrode behavior in acid media
presented further unknowns. There were also uncertainties regarding the poisoning effect of sulfurous fuels on electrodes. Given the gaps in the knowledge
base, Ordnance Corps officials cautioned, the "present enthusiasm" for hydrocarbon fuel cells had only a "sketchy" basis. Nevertheless, they recommended
a program of basic research into fuels, electrolytes, and electrodes to be monitored by the Army.[37]

Project Lorraine begs comparison with ARPA's contemporaneous program
in materials science. These two efforts were guided by conflicting management

philosophies indicative of the ongoing struggle to define the agency's role in defense R&D and, more broadly, the federal government's efforts to impose accountability as it enlarged its relationship with industry as a research patron. Before 1959, the Defense Department compensated corporate contractors only for research directly associated with the production of hardware. Under industry pressure, the federal government revised the Armed Services Procurement Regulation XV in November 1959, allowing contractors to charge both undirected basic research and applied research to overhead. As Glen R. Asner notes, the revision represented such a major shift in contracting principles that it spawned a new program known as Independent Research and Development, which spent far more on research than the National Science Foundation. In order to impose accountability and maintain managerial control, the Pentagon moved to strictly define each work practice, linking them in a sequential, linear process: "basic research" increased scientific knowledge, "applied research" expanded the potential of this knowledge, and "development" put this knowledge to work in producing useful hardware. The Pentagon's requirement that contractors separately account for costs in each of these categories led many to structure their programs in discrete divisions responsible for research, development, and manufacturing.[38]

Project Lorraine was organized along similar lines. In contrast, materials science was conceived as an interdisciplinary form of research by federal defense planners, who hoped to emulate what they believed was a new mode of innovation emerging in industry. Impressed by new solid-state science-based technologies flowing from corporate laboratories including the transistor, the solar cell, and polymers, they tried to adapt a similar organizational structure, combining science and engineering under a broad institutional umbrella capable of developing substances suitable for the rigors of spaceflight. Reflecting on this period in 1985, William O. Baker, a former vice president of research at Bell Laboratories and a member of the President's Science Advisory Committee, remarked that materials science was the most promising way to convert new knowledge into "technologic innovation and commercial and public production."[39]

This idea quickly garnered broad support among federal agencies including the Atomic Energy Commission, the Office of Naval Research, the National Science Foundation, the National Academy of Sciences, and the Department of Defense, as well as the White House. York saw materials science research as a premier role for ARPA in its postspace incarnation, and the agency would become the prime mover of what Stuart Leslie termed the "materials revolution."[40] As the lead federal agency in the National Materials Program, ARPA established the Interdisciplinary Laboratories (IDL, later renamed Materials Research Laboratories) at 17 leading U.S. universities between 1960 and 1982.[41]

Although energy conversion was separate from the IDL program, the largest part of ARPA's effort in materials science, it in some ways overlapped it. Both were administered by ARPA's Materials Sciences Office (MSO), and both focused on the basic physical principles of materials with a view to eventual practical application. Here the similarities ended. Although universities were at the center of the IDL, they played a relatively small role in Project Lorraine, particularly in the realm of electrochemistry, with most of the money going to large industrial laboratories. But perhaps the chief differences between ARPA's energy conversion and materials science programs were in expectations and flexibility. The consensus among the defense research community was that the IDL's long-term potential was such that it was unnecessary to link the program with specific requirements. Its timeline, accordingly, was open-ended.[42] It adopted a hands-off approach to management, drafting broad objectives and supplying university laboratories with general work orders relating to the investigation of the theoretical properties of materials. Moreover, IDL let four-year contracts on the grounds that these were much more productive than one-year contracts.[43]

In contrast, Project Lorraine, with no similar broad base of support, was tightly managed along linear lines. Research contracts were let on an annual basis, with renewal contingent on performance. And the science ARPA promoted in this effort was more device-oriented than in the materials science effort. The hope was that research would pay quick dividends in the form of notional technologies that would then be further developed by other parties. Such expectations were stoked in good measure by the revolving-door nature of the defense science and technology advisory apparatus. Pursuing their own technopolitical agendas, Army and IDA planners played a central role in interpreting the relevant science and engineering issues for ARPA managers, who had no real independent counsel. In turn, ARPA managers reinforced the views of their colleagues. The result was a positive feedback loop in which a priori notions of fuel cell technology were reproduced and transmitted throughout the defense research establishment.

Path of Least Engineering Resistance

By fall 1961, ARPA was distributing contracts for Project Lorraine, and fuel cell technology gained a much higher profile within the Army. In his welcoming address at the Power Sources Conference of 1961, Colonel Raymond H. Bates, the Signal R&D Laboratory's commanding officer, observed that thermal energy conversion and fuel cell technology now constituted a major part of the symposium.[44] To be sure, the Pentagon had yet to detail precise roles for fuel cells that would directly use any kind of liquid fuel. Nor had ARPA set milestones before initiating Project Lorraine, and none had been identified as late as June 1961, well after the agency began funneling money to contractors.[45]

Not until the summer of 1962 did ARPA managers begin to draft requirements for a liquid-fuel system. The ultimate goal remained perfecting the oxidation of hydrocarbons, but planners realized that this would take time. As they studied the process by which hydrocarbons were oxidized at the anode, workers at the California Research Corporation (CRC), a Project Lorraine contractor, found unoxygenated substances like propane and ethylene difficult to work with. Employing fuel cells using platinum catalysts and acid electrolytes, they focused on formaldehyde, formic acid, and methanol, the same partially oxygenated organic substances the Army's Signal R&D Laboratory had investigated the previous year. The CRC researchers favored methanol because it was tractable and because they understood the mechanism by which it underwent electro-oxidation. They concluded that this alcohol would produce high current densities when used in a fuel cell, although disposing of the reaction products posed problems.[46]

Project managers at ARPA agreed that research on a methanol fuel cell system might be useful. Accordingly, they developed a plan involving Project Lorraine's two principal contractors. General Electric would concentrate on the complex electrochemistry of the hydrocarbon system, while Esso's Research & Engineering (R&E) Laboratory would develop an experimental methanol fuel cell, employing an acidic electrolyte. The latter now became an important interim goal. To George W. Rathjens, then ARPA's chief scientist and deputy director and a White House science advisor specializing in military technology, the successful development of a laboratory device operating on air and partially oxygenated carbonaceous fuel for an extended period would constitute a breakthrough.[47] He did not, however, specify how this would aid the program of basic hydrocarbon research.

In instituting short- and long-term fuel cell programs—one an effort in basic research, the other oriented toward prototype hardware—ARPA managers redefined their original objectives and views of the utility of fuel cell power. Conceiving the methanol fuel cell as a kind of technological hedge against the hydrocarbon research, they subsequently sought practical justifications for the device. It was true that methanol was not a logistics fuel, and so such a fuel cell would play a minimal role in the Army's energy conversion chain. Nevertheless, reasoned military planners, high power and silent operation were desirable attributes of a special-purpose fuel cell.[48] And they believed a methanol fuel cell could be modified to give it a flexible fuel capability and, hence, broader applicability. Since the technology would function even better with hydrogen, they held, it could be made a component of an "indirect" system, consuming hydrogen-rich gas converted from hydrocarbons by a reformer.[49] True, the reformer was an untried auxiliary device without which the possible applications of methanol fuel cells were limited. But that was a problem for the future. For the moment, the point of experimenting with the methanol system lay in the practical experience it would provide in preparing the ground for research on the more complicated hydrocarbon systems to bear fruit.

As in the Bacon cell program, the question of what fuel cells—methanol or otherwise—would actually power once they were fully developed was a secondary consideration.

Shaping Expectations for the Super Fuel Cell

As Project Lorraine's agenda in terrestrial fuel cell power was revised, Esso R&E and General Electric began to work in earnest in a field in which neither had much experience.[50] Each had strengths that planners doubtless calculated would complement the other. Like Shell, Esso R&E began investigating fuel cells in the late 1950s in order to prepare suitable fuels should the power source become commercialized. Accordingly, the technical parameter it was most interested in was the rate of fuel consumption.[51] General Electric, on the other hand, was a leading developer of fuel cell hardware and one of the biggest promoters of the technology in the 1960s. It hoped to adapt military fuel cells for the civilian market, although its workers knew little about hydrocarbon electro-oxidation and relied on oil companies for expertise in this field.[52]

But Esso R&E and GE did not formally collaborate, working instead in concert with a number of subcontractors. General Electric had two fuel cell facilities, the Research Laboratory at Schenectady, New York, and the Direct Energy Conversion Operation (DECO) manufacturing plant at Lynn, Massachusetts. Assigning GE the task of constructing a breadboard fuel cell with an acidic liquid electrolyte as a model of a practical hydrocarbon-fueled military power plant, ARPA itself contracted with the American Oil Company and the California Research Corporation to investigate the physical principles of low-temperature hydrocarbon electro-oxidation.[53] In 1961, Grubb and Niedrach were detailed to the hydrocarbon fuel cell program and began working with a small team at the Research Laboratory on highly reactive, relatively expensive light hydrocarbons— propane, propylene, and cyclopropane—in liquid sulfuric acid and alkaline electrolytes using a variety of catalysts.[54] Of these fuels, propane was cheapest, but cyclopropane proved the most reactive and was the only one of the three substances researchers were able to fully electro-oxidize by the summer of 1962. Charles F. Yost, ARPA's assistant director for materials science, regarded this as a success. Acknowledging that cyclopropane was not a logistics fuel—it was then most widely used as an anesthetic—Yost held that experimenting with it was useful in learning how to directly electro-oxidize common liquid fuels.[55] In September, ARPA extended the Research Laboratory's contract for another year, charging it with understanding the unique activity of cyclopropane and achieving equal or better performance with liquid hydrocarbons.[56]

The advisory feedback loop that linked ARPA with the IDA and the industrial contractors was the main mechanism informing planners' estimates of progress. The question of whether the fuel cell was suitable for military electric drive was a particular subject of controversy. In the late summer of 1962, two IDA analysts got into a dispute over this issue that sent conflicting signals back to ARPA. The incident centered on a draft report that was to have been jointly prepared by the analysts. What instead happened was that the paper had been completed by one analyst, who then forwarded it to the DDR&E without informing his colleague. Authored by a scientist named Robert Hamilton, this report was strongly critical of ARPA's management of the fuel cell program. Citing poor project selection and information distribution and nebulous task orders, Hamilton called for assigning an administrative aide to Project Lorraine's manager. More importantly, he thought there was no justification for replacing internal combustion power with fuel cell electric power because internal combustion costs were so low, only two cents per mile to operate an Army truck. Hamilton recommended cutting funding for the fuel cell portion of the program by 41 percent.[57]

Annoyed, Hamilton's colleague G. C. Szego wrote a dissenting report in the form of a memorandum to S. S. Penner, his IDA superior. It was imperative, said Szego, that Hamilton's views not be presented to ARPA, for it might well lead to the cancellation of a valuable program. He took issue with Hamilton's cost estimates, citing sources including Ernst M. Cohn of the Army Research Office Durham (the former Office of Ordnance Research, renamed in February 1961) among other informants that indicated that each gallon of gasoline supplied to U.S. forces in Korea during the 1950–1953 war had cost $10. Szego calculated that fuel cell electric trucks employing regenerative braking would save about $1,650 per battlefield day per 200-horsepower truck. Citing figures supplied by General Electric, he claimed fuel cells would cost between $300 and $500 per kilowatt by 1970, a figure he considered conservative. The company forecast that costs might be even lower by the mid-1970s, less than $100 per kilowatt. Szego also believed fuel cell technology lent itself to cheap methods of manufacture including parts stamping. The most serious unavoidable expense was the platinum catalyst, which represented about 20 percent of the total capital cost, but he thought the noble metal could easily be recycled from spent fuel cells. Szego recommended increasing ARPA's fuel cell budget for fiscal year 1963 by $700,000 to a total of $2.4 million and accelerating and expanding the hydro-carbon program, a position paralleling the view of the Army Engineer Research and Development Laboratory, which was interested in the fuel cell in large stationary and automobile applications. This meant paying more attention to the two ways hydrocarbons could, in principle, be used in fuel cells: injecting and oxidizing them directly in the device, or reforming them into a hydrogen-rich gas prior to injection. The former was more desirable, wrote Szego, but the Engelhard Corporation's new hydrocarbon reformer made the indirect method more viable in the near term.[58]

This episode is striking for what it reveals about the construction of assump-tions of fuel cell power and the larger decision-making process within the defense advisory apparatus. Szego's view of fuel cell electric drive was influenced at least in part by workers at Allis-Chalmers, the company that had built the first such vehicle.[59] But the cost estimate of $300–$500 per kilowatt, obtained from General Electric's sales representative, was a projection that assumed major technological and manufacturing advances. True, the sales representative had noted that it was difficult to predict trends in the field, particularly since the fuel cells General Electric was currently developing were for special applications and would bear only a "limited resemblance" to the types the company planned to manufacture in quantity.[60] But the high current densities Szego had cited as evidence of progress were produced by alkaline hydrogen fuel cells, precisely the technology military and industry planners were working to replace because of its susceptibility to contamination. Szego also reinforced ARPA's predilection for methanol. He believed this fuel would not place excessive demands on the existing industrial infrastructure, being a cheap chemical already produced in quantity as a by-product of petrochemical refining and wood processing and easily stored at room temperature. Yet he acknowledged that introducing methanol into the Army supply chain was "not especially desirable."[61]

Personal politics played an important role in informing expert opinion. The two IDA analysts shared management responsibilities, with Hamilton studying magnetohydrodynamic, thermionic, and photovoltaic systems and Szego concentrating on electrochemical and thermoelectric technologies. Szego's indignation stemmed not only from the fact that his colleague had failed to consult him on a major issue of policy but also that he had judged technical issues not within his purview.

Szego's views prevailed within the IDA and likely influenced thinking in ARPA. In October, an ARPA internal memorandum reaffirmed that Project Lorraine's ultimate goal was to develop devices capable of electro-oxidizing logistics fuels. Another memo in December couched expectations even more explicitly, noting that the hydrocarbon fuel cell was to be made available for "widespread terrestrial military application," especially in vehicles. Developing fuel cells with high power densities capable of using common liquid fuels, the anonymous author claimed, would halve the fuel requirements of an armored division. The timeline was reset to five years, an extension of previous estimates.[62] Researchers had quickly learned that liquid logistics hydrocarbons could not easily be directly oxidized in a fuel cell. But the program was not canceled. Instead, ARPA, guided by the IDA, funded interim hardware. In addition to supporting Esso R&E's direct methanol fuel cell, ARPA also decided to invest in a reformer/fuel cell system, a "one-shot" effort to use field-grade fuels.[63] For this program, the agency would not fund research but instead purchase a complete reformer built by the Engelhard Corporation, which was to be integrated with an Allis-Chalmers five-kilowatt alkaline fuel cell. The whole system was then to be transferred to the Army Engineer Research and Development

Laboratory.[64] As Project Lorraine wore on, ARPA managers found it increasingly difficult to maintain an organizational division between science and technology.[65] And their adherence to linear management logic would soon induce reverse salients.

Linear Innovation

By early 1963, General Electric's program was showing mixed results. Yost applauded the company's work on cyclopropane, yet ARPA and the Army were dissatisfied with the results of experiments on light hydrocarbons, feeling they did not justify intensified applied research. Accordingly, ARPA fully funded GE's work at the Schenectady Research Laboratory but provided only $200,000 of the $500,000 requested for operations at the DECO manufacturing plant. General Electric's response was to delay the expansion of its test installations.[66] In effect, ARPA's move further compartmentalized the program, impeding interdisciplinary relations between the company's fuel cell laboratories.[67] With durability testing contingent on advances in basic research and with General Electric bearing some of the cost of a test regime ARPA had demoted in priority, the company had little incentive to immediately invest in such facilities. Consequently, it was not prepared to exploit an advance made at the Research Laboratory in late February and early March 1963. Researchers succeeded in electro-oxidizing both propane and n-hexane—a liquid hydrocarbon—an achievement ARPA believed merited intensified applied research.[68] Cleared to use notional hydrocarbon fuel cells as "research vehicles" to further investigate heat transfer and reactivity limitations, the DECO had barely begun preparing test installations by late spring. Only one test stand had been completed, although three more were planned, and a backlog of completed electrodes accumulated. A handful of cells—now using a phosphoric acid electrolyte—had been tested with propane using various catalysts, but very limited data on their lifetime had been compiled.[69]

This event showed how fuel cell laboratory communities developed simplified tests in controlled conditions, revealing the limitations of the stepwise approach to R&D. The oxidation of n-hexane was held by ARPA managers to be of special significance because it was the first time a liquid hydrocarbon, valued for advantages in storage and handling over gaseous fuels, had achieved high power in a fuel cell. To be sure, industry used n-hexane primarily as a solvent, not a fuel. Nevertheless, work proceeded apace. General Electric workers began testing heavier hydrocarbons more representative of Army logistics fuels in phosphoric acid fuel cells employing platinum as a catalyst. By spring 1963, they knew these substances posed a far greater challenge in chemical engineering than the light hydrocarbons, reporting that the performance of n-hexadecane, an important constituent of diesel fuel, was only 20 percent of that of propane and butane.[70] That such a disappointing result was obtained from a component of a logistics fuel using the most active known catalyst must have caused disquiet among researchers and planners. Diesel and JP-4 were complex compounds, and investigators did not yet know how their oxidation by-products would affect fuel cell electrolytes and electrodes over time.

Despite these obstacles, NASA's well-funded and publicized aerospace fuel cell effort provided some measure of comfort. It was true, noted John H. Huth of the materials sciences division in February, that the technology of the hydrocarbon fuel cell lagged five to ten years behind that of the hydrogen fuel cell, but he believed this had to do more with institutional priorities than with physical obstacles. Thanks to NASA's efforts, aerospace fuel cells were nearly ready for use. Conversely, the terrestrial fuel cell effort, ostensibly relevant for the entire Department of Defense, was receiving much less support and would have stalled completely, claimed Huth, were it not for ARPA. More resources for terrestrial fuel cell research, he suggested, might produce results as dramatic as in the space program.[71] This well illustrated the terms by which ARPA managers judged success, revealing a rather shallow understanding of the electrochemical issues at play and, as events would demonstrate, of the nature and pace of progress in this field at NASA as well.

Nevertheless, the public face of the terrestrial fuel cell program was sunny in the spring of 1963. In late April, General Electric declared its breadboard a success in a press conference and demonstration held in New York. Noting that the new fuel cell used "inexpensive fuels"—propane and natural gas—at an efficiency of 40 to 50 percent, the company claimed that the device would also operate on diesel, gasoline, and kerosene, although additives in these fuels reduced efficiency and operating life. The program, said GE, had advanced the state of the art closer to "practical commercial" applications.[72]

In fact, the results were paradoxical. Although the breadboard possessed some features attractive for civil use and others perhaps suitable for the military, as a system it was unfit for any application, civilian or military. Too expensive for commercial operation owing to its reliance on platinum catalyst, the device used fuels that, while cheap in comparison to hydrogen, were not attractive from the perspective of military logistics. Given General Electric's limited work on heavy hydrocarbons up to that point, the company's assertion that these fuels could be used with some lifetime and performance penalties made it a hostage to fortune. Nevertheless, news that the dream of the hydrocarbon fuel cell had been realized was cheered in the electrochemical community. In a congratulatory letter to H. A. Liebhafsky, the leader of General Electric's fuel cell team at the Research Laboratory, Francis T. Bacon lauded the "great achievement," the "ultimate objective" that researchers had been working toward for so long. He had been skeptical that such a feat was possible, wrote the pioneer of the pure hydrogen fuel cell, but "now we know without doubt it can be done."[73]

Faith in the Black Box

General Electric's apparent breakthrough raised the problem of how to transform laboratory fuel cells into advanced research models and, from there, practical power sources. Difficulties along these lines had been anticipated. As early as July 1962, Yost had warned that the basic design of a multicell unit would be dictated by the application it served, hinting of the future complications of adapting demonstration power plants for field service.[74] As 1963 progressed, contractors, systems analysts, and military planners gave conflicting accounts of the state of the hydrocarbon fuel cell effort. The Power Sources Conference in May featured mixed views of the latest results. Ernst M. Cohn, formerly of the Ordnance Corps and the Army Research Office and now in charge of NASA's in-house fuel cell program, delivered a generally positive report on the aerospace cells being developed for the Gemini and Apollo spacecraft.[75] For its part, Esso R&E was bullish about its methanol fuel cell. One year earlier, its engineers had been worried about complications arising from directly injecting fuel into the electrolyte. They interpreted initial tests employing individual half cells as evidence that the carbonaceous fuel-air mixture would work without losing efficiency as a result of their interaction.[76] Now they believed the odds of developing fuel cells capable of 60 percent efficiency in "widespread military application" were excellent.[77]

Conferees from the IDA and the Army were even more optimistic. They invoked the hydrocarbon fuel cell as a kind of miracle battery. Notably, the IDA was represented by Szego, not his skeptical colleague Hamilton. In a presentation that elaborated themes outlined in his report to his supervisor the previous September, Szego sketched a blueprint of the battlefield of the future, one where Army vehicles powered by hydrocarbon fuel cell electric drives would bring about a revolution in logistics. As then, he based his assessment largely on future developments and seemed much more concerned with electric drive than with the fuel cell power source itself, paying little attention to the electrochemical issues attending the most recent laboratory developments.[78]

The Army was similarly bullish. Its fuel cell effort had two divisions, one at Mobility Command's Engineer Research and Development Laboratory (ERDL) at Fort Belvoir, Virginia, responsible for power sources larger than one kilowatt, and the other at the Power Sources Division of the Electronics Research and Development Laboratory at Fort Monmouth, New Jersey, the former Signal R&D Laboratory, charged with developing smaller devices.[79] But it was the promise of the multikilowatt hydrocarbon fuel cell that most fired imaginations. Such devices, claimed the ERDL's B. C. Almaula, had the potential to replace all the Army's internal combustion engines in both stationary and vehicular applications using cheap fuels. Now, he believed, the research produced over the course of Project Lorraine was ready to be applied. The ERDL would, stated

Almaula, develop a five-kilowatt prototype fuel cell using hydrogen reformed from a liquid hydrocarbon, a project it expected to complete by early 1965. As for direct hydrocarbon oxidation in fuel cells, he added, the Army believed that the fundamental principles had been proved and hoped that the doubts expressed by "various scientists and industrial managers" were now dispelled. With this key obstacle overcome, it was time for industry to step in.[80]

But Almaula had hinted that not all was well in the hydrocarbon fuel cell program. And there were also broader indications of trouble. Linden, the circumspect Electronics Laboratory researcher, indicated that a great divide had emerged between what the promoters believed the technology was capable of and what engineers could achieve in practice. Despite the belief of the "sales and public relations people" that the fuel cell was just around the corner, "we engineers," noted Linden, were still having difficulty realizing even the more moderate expectations.[81] Over the summer and into the fall and early winter of 1963, a succession of new reports by ARPA, contractors, and government agencies painted an even bleaker picture, revealing an assortment of basic physical and engineering obstacles. One study by the California Research Corporation (CRC) observed that although cooling was simple where only one fuel cell was concerned, it became much more difficult when many cells were clustered together into a stack. The problem then was no longer two-dimensional but three-dimensional, and a means had to be found of removing heat from the middle of the stack.[82]

There were even more fundamental difficulties. One ARPA manager lamented that "not a single" catalytic or electrocatalytic reaction was fully understood, a major implicit criticism of Project Lorraine given that it had been initiated precisely to illuminate such dynamics. Somewhat wistfully, he speculated that perhaps a neutral electrolyte could be found that would avoid the shortcomings of the strong basic and acidic systems. Particularly troubling was the low-temperature acid cell's requirement for platinum.[83] The CRC noted that replacing all automotive heat engines with electric motors powered by fuel cells using even an extremely thin monatomic layer of the noble metal would require almost two thousand times the then-current annual global production of this strategic element.[84] Another study revealed that the United States produced only a tiny fraction of its annual platinum consumption and was heavily dependent on imported supplies, recommending that platinum be considered only for special applications.[85]

This was no small consideration in a program begun largely on the premise that it would free American military forces from the bonds of fuel logistics. Having begun work on a technology with a major materials reverse salient, ARPA backtracked, letting contracts for basic research on new, cheaper catalysts. The results were disheartening. The Monsanto Research Corporation quickly confirmed that platinum was the most efficient of all known catalysts and that substitutes were unlikely to be found. Interestingly, of the 15 catalysts chosen for investigation, 13 were active not for organic hydrocarbons but for hydrazine, an inorganic (noncarbonaceous) chemical used mainly as a rocket fuel and chosen for study because it, like methanol, was relatively easily electro-oxidized.[86] Monsanto chemists opined that a comparative study of hydrazine might yield insight into hydrocarbon oxidation. After 1965, hydrazine became an increasingly attractive fuel option for the Army as the hydrocarbon project encountered further technical difficulties.

If this was not enough, General Electric's propane fuel cell proved to have poor durability. An acidic system, it was not handicapped by carbon dioxide, but corrosion was a problem, lowering performance and shortening life. This was particularly worrisome since propane was far from the dirtiest fuel researchers hoped to use. Throughout the summer and into the fall, General Electric worked to improve durability, engaging in tests that further blurred the distinction between basic and applied research. On the one hand, the work of the Schenectady laboratory became more applied in nature. Researchers began testing complete fuel cells in order to understand and improve the performance of propane in acid electrolytes. Efforts to make more efficient use of platinum by incorporating it as a powder into various electrode configurations had a technological aspect involving the fabrication of materials and components. Conversely, materials testing now became an additional preoccupation of the Lynn laboratory. The challenge was to seek combinations of cheap substances that were tough as well as reactive. Experiments with a variety of materials in concentrated phosphoric acid at 150°C demonstrated the difficulty of meeting these conflicting requirements. They revealed that not all of the electrodes developed by Schenectady were sufficiently robust to withstand prolonged contact with acid. Materials such as molybdenum, tungsten, and palladium-nickel alloy (platinum black) experienced severe degradation. Electrodes made from noble metals such as platinum and gold slowly disintegrated in the hot corrosive mixture over time. Once these components became active in phosphoric acid fuel cells, a complex mixture of substances was produced that made a quick analysis of their precise composition, and further experimentation, very difficult. Meanwhile, Lynn's test division continued to suffer from the effects of the construction delay triggered by the funding cut of the previous September. By late September 1963, no test stations were fully operational. With two partially completed and three more projected to be ready by November, durability testing was just beginning.[87]

A major federal survey conducted over the course of the year acknowledged these difficulties, linking them with the federal government's conduct of energy and power source research and development. A comprehensive cabinet-level effort, the Interdepartmental Energy Study (IES) found that planning was "inherently complex and imperfectly understood." Part of the problem was that this sort of R&D was shaped by a wide variety of nonconstant variables including natural resources, science, technology, economics, and politics. Accordingly, such work was likened to a "highly nonlinear mathematical problem" and was necessarily highly speculative. It was further complicated by the decentralized nature of federal energy and power source research and development in which a multitude of military and civilian agencies carried out their own programs.[88]

These findings were echoed by the IES fuel cell subcommittee, composed of leading researchers in industry and government. As in the Bacon cell program, actors interpreted the dearth of knowledge in the most favorable light. On the one hand, existing data cast doubt on important claims made for both the aerospace and terrestrial variants.[89] Preliminary tests suggested that fuel cells could attain long-term invariant operation at up to 90 percent efficiency, but little was known of the lifetime and reliability of any of the designs under study, let alone the challenges of scaling up small laboratory devices. For this reason, the subcommittee was understandably reluctant to reject the technology out of hand, feeling that the information gap justified further research. But faced with these unknowns, the committee chose to cling to long-held assumptions. Uncertain whether information relevant to terrestrial designs could be gleaned from the operation of aerospace fuel cells owing to dissimilar parameters— aerospace applications required relatively short lifetimes and pure reactants— the subcommittee nevertheless claimed that all fuel cells including hydrogen-oxygen designs had the potential to consume hydrocarbons if cost-effective reforming technologies could be developed. Of course, this technical fix was far from proven. Nevertheless, the subcommittee believed that the factors limiting the pace of progress were primarily sociopolitical and economic, not inherent to the fuel cell concept, noting that the field of electrochemistry had been crippled by a shortage of skilled researchers resulting from the expansion of more prestigious disciplines such as physics and aerospace engineering.[90]

The IES fuel cell subcommittee's report accurately described the challenges facing Project Lorraine. But it missed the chief contradictions that undermined the program. By late 1963, expectations born of false technological analogies— parallels the subcommittee members themselves subscribed to—clashed against rigidly linear management protocols in a fledgling field that required interdisci-plinarity but had no real industrial or academic constituency. Ostensibly com-mitted strictly to investigating the basic science of electrochemical hydrocarbon energy conversion, ARPA planned to turn the program over to the armed services and industry for development as soon as possible. Instead, science, technology, and engineering commingled in a way that defied the conventional administra-tion of sequential innovation. The inability of researchers to achieve direct hydrocarbon oxidation on notional devices inspired a rush to develop technolo-gies that used chemically simpler fuels but that were necessarily larger and more sophisticated as workers struggled to prove the concept and became ever more deeply absorbed in engineering research. In turn, these artifacts became tangible symbols of progress. But they also raised unexpected new issues of fundamental electrochemistry. As Yost acknowledged, basic questions had emerged on the applied side of the program.[91]

And industry was reluctant to pursue them. As the year drew to a close and prospects for a hydrocarbon fuel cell faded, many firms abandoned the field altogether. This was a real blow, for the Department of Defense relied on manufacturers to underwrite a major portion of the research program, perform most of the engineering, and, it was hoped, one day produce fuel cell stacks in quantity. With companies pulling out, observed Yost, by now director for materials sciences, the military terrestrial hydrocarbon fuel cell program would depend solely on government support.[92] Interestingly, an initial draft of the IES had recommended that an additional $3 million to $4 million be spent on the project, some of which would fund an electrochemical training program, and had advised postponing the methanol fuel cell in favor of more research on hydrocarbon oxidation. Some in the materials sciences division received this advice coolly. John Huth did not dispute that there was a shortage of electrochemists but suggested that an immediate effort to train more workers made little sense given that mastery of low temperature hydrocarbon oxidation lay eight to ten years in the future. Of all fuel cell systems, he wrote, only the methanol fuel cell was "essentially at hand" and could serve a useful purpose.[93]

By the fall of 1963, the agency began moving to end its involvement in an enterprise that had failed to yield quick dividends. Under financial and time constraints and with fundamental problems in fuel cell electrochemistry far from resolved, ARPA decided to phase out Project Lorraine. To accomplish this, planners opted to terminate most of the basic programs and support the hardware projects of the two main contractors for two more years in order to get them into sufficient shape for the military to take them up.[94] The decision was not without its critics, highlighting divisions over the agency's role in the program. General Electric felt that more study was needed to determine how some of the scaled-up hydrocarbon fuel cell systems behaved over time. Some in ARPA agreed. Huth warned that ending basic research would "inadvertently bypass significant factors," leading to a technoscientific dead end.[95] This did not bode well for contractors, whose programs, noted Yost, remained "essentially intuitive" in nature.[96]

Yet the materials sciences division had opposed plans to bolster postsecondary electrochemical training. Some ARPA managers had also become critical of the IDA. Huth slammed one report authored by Hamilton and Szego in late 1963 as a mix of data and opinion that created a misleading impression of what was possible given existing and projected states of the art.[97] By early 1964, however, there was no political will to reform the program. R. L. Sproull, ARPA's new director, wanted an orderly wind-down by 1965.[98] Even news from the aerospace field, long a source of solace to terrestrial fuel cell researchers, was bad. In recommending a final year of support for Esso and General Electric in fiscal year 1965, Huth cautioned against "overoptimism" despite satisfactory progress, noting "developmental difficulties" that had recently arisen in the Gemini and Apollo hydrogen-oxygen fuel cell programs.[99]

In the summer of 1964, one year after he had congratulated Liebhafsky and General Electric for their "breakthrough," Bacon wrote once more. This time, the tone was consolatory: "Are we getting nearer to a really practical direct hydrocarbon fuel cell or is it still a long way off? I wish I knew the answer to this question!"[100] Although Bacon was renowned for his gracious manner, a faint note of satisfaction could be detected in the missive of the old hydrogen fuel cell partisan. For his part, Liebhafsky was keenly aware of the institutional dynamics that had unhinged the terrestrial hydrocarbon fuel cell project. Confiding to Bacon in the summer of 1965, he remarked that General Electric's effort was vulnerable because it relied almost entirely on government support. Continued aid was contingent on the development of "useful" hardware within two to three years, but he did not believe that this was possible. With prospects bleak, Liebhafsky suspected that his best researchers would soon begin to leave for greener pastures. His bitterest complaint was reserved for the culture of corporate secrecy. The worst offender, he believed, was Pratt & Whitney, foreshadowing future clashes between the American company, its collaborators in the British fuel cell consortium Energy Conversion Limited, and the federal government. Liebhafsky held that unless the industrial laboratories started to cooperate, the terrestrial fuel cell program would "flop."[101] He later noted the inadvisability of identifying the individual fuel cell as the province of "research" and stacks of fuel cells as an "engineering assignment." Strictly speaking, research continued in the engineering phase. Sufficient theoretical knowledge of hydrogen-oxygen cells existed to begin arraying them in fairly reliable power plants. But, Liebhafsky noted, the same could not be said of the terrestrial hydrocarbon fuel cell.[102]

Hedge Technology Close-Up: The Methanol Fuel Cell

The efforts of Esso R&E to develop its methanol fuel cell in the final months of Project Lorraine illustrate how design was determined at the intersection of the technopolitics of R&D and the material realities of scaled-up fuel cells. In mid-1963, the methanol system was regarded by many researchers including Ernst Cohn as the most advanced employing an oxygenated fuel. Even so, there were myriad problems. The Army Electronics Laboratory then thought existing designs were too heavy and costly.[103] Another problem was platinum's high reactivity, the very quality that made it so effective as a catalyst in low-temperature fuel cells. Esso R&E's fuel cell was a direct system, which operated by injecting methanol into the sulfuric acid electrolyte. When this occurred, the fuel came in contact not only with the fuel electrode (anode) but the oxygen electrode (cathode) as well. The platinum-laced cathode oxidized some of the methanol before it could be consumed at the anode, greatly reducing efficiency through fuel wastage and increased cathodic overpotential, the fraction of energy produced by the reaction necessary to sustain itself. Moreover, cathodic oxidation produced large amounts of carbon dioxide that physically displaced the electrolyte. As there was no means of preventing this, Esso R&E engineers tried to control the amount of methanol dissolved in the electrolyte.[104]

These issues greatly complicated the construction of full-size prototypes. Through the remainder of 1964 and into 1965, Esso R&E engineers worked to replicate in multicells the high performance they had achieved with individual cells, a key requirement of all fuel cells being long-term chemical invariance with as little recourse to control equipment as possible. These issues loomed large in the case of the methanol fuel cell, noted Esso R&E researchers, because of the additional process complexity demanded by the direct fuel design. The electrolyte had to be circulated and methanol added in response to load demand while maintaining chemical equilibrium, a task complicated by cathodic oxidation.[105] Workers soon discovered the difficulties of operating stacks. Esso R&E's 16-cell unit was wired in series, meaning unlike terminals of each successive cell were connected together (cathode to anode), as opposed to a parallel connection, in which like terminals were linked (anode to anode, cathode to cathode). Series connections produced higher voltages and were more efficient than parallel linkages, an important design consideration for Esso R&E because carbonaceous fuels produced lower voltages and were less efficient in fuel cells than pure hydrogen. This was a serious handicap, since many electronic appliances required high voltages. Wiring stacks of carbonaceous fuel cells in series helped compensate for this shortcoming. But series-linked fuel cells were more prone to failure than those connected in parallel because they were only as robust as the least reliable cell. If the supply of gases to a series-linked cell in a stack was not uniform and hydrogen and oxygen starvation occurred, there was a possibility that undesirable reactions could occur at its electrodes. In certain situations, a weak cell could be "driven" by adjacent cells, reversing the charge of its electrodes and turning it into an electrolysis cell. Instead of being drained by the circulation system, water normally produced at the cathode might be captured and electrolyzed, bringing about the failure of the entire stack.[106]

Esso R&E managers also began to reconsider the economic rationale for the project. It dawned on them that the methanol fuel cell—should it be further developed—was almost certainly destined to remain a niche product. But the primary objective of Esso R&E's parent, Standard Oil of New Jersey, was to sell as much fuel as possible. A draft proposal completed by the laboratory in July 1964 outlined ESSO R&E's view of matters. It called for the continuation of government-funded research, reversing the priorities laid out for it by ARPA in August 1963. Should the program be extended throughout 1965, Esso R&E held, the main goal should be to conduct feasibility studies of a genuine hydrocarbon-air fuel cell "capable of widespread military application." In this report, the laboratory reminded ARPA that the whole point of developing a hydrocarbon fuel cell was to ease fuel logistics. This plan envisioned substituting fuel cells for other power sources in large numbers, a situation that, on paper, better aligned with the interests of Esso R&E's parent company.[107]

Project Lorraine ended well short of its original objectives.[108] Its chief fruit, the methanol fuel cell, had neither been tested nor met doctrinal requirements. Confident that its technology had displayed "realistic performance levels under reasonable operating conditions," Esso R&E rigorously tested half cells and individual single cells—up to 6,000 hours in some cases—but devoted far less effort to multicell assemblies, testing stacks only out to 200 hours.[109] In January 1965, the Power Sources Division of the Electronic Components Laboratory (ECL) took delivery of Esso R&E's experimental 100-watt methanol fuel cell, declaring it a major milestone, the first stack to oxidize methanol directly without relying on external reforming. But the military regarded the reliance on platinum as a major handicap.[110] Esso R&E continued developing the system into 1966, producing another breadboard for the Army. Using platinum and weighing 94 pounds without fuel, it was intended mainly as a research tool, being far too heavy to serve as a portable power plant.[111]

Army Fuel Cell Systems

In 1965, the Army's research and development bureaus considered the state of fuel cell technology. Over a decade their missions, and their corresponding power source requirements, had changed considerably. A new expeditionary war loomed in Southeast Asia, one that had relatively more scope for infantry and less for the massed armor formations built for conflict on the European battlefield. Although some in the Army remained enamored of the concept of a universal chemical energy converter, the inconclusive results of Project Lorraine and the need for equipment that could quickly be pressed into service in Vietnam made power sources for "manpacked" electronics equipment relatively more important after 1965. Emulating their ARPA colleagues, Army managers sought the swiftest possible solutions. Among the options were two types of fuel cell, both based on alkaline operating systems: the reformer/fuel cell combination, initially considered by ARPA planners as the simplest way to use hydrocarbons in fuel cells, and the direct system using hydrazine, an exotic nitrogenous fuel.

Allis-Chalmers and Engelhard had built reformer/fuel cell plants using ARPA funds as part of Project Lorraine. Mirroring earlier justifications for the methanol fuel cell, researchers believed hybrid systems could supply valuable lessons relevant for direct hydrocarbon operations and provide useful service in the field. By the mid-1960s, each Army laboratory had an indirect system under prototype development. With support from ARPA, the ERDL began work with Allis-Chalmers and Engelhard on a five-kilowatt unit in 1963, while the ECL contracted with Pratt & Whitney for a 500-watt reformer/fuel cell in 1964.[112] Conceiving the Allis-Chalmers/Engelhard system as a way to allow engineers to exploit existing reformer and alkaline fuel cell hardware, ARPA planners also wanted to use it as a platform to test components they hoped to employ in the direct hydrocarbon fuel cell. As with the methanol fuel cell, they claimed that the reformer/fuel cell would be able to stand on its own merits.[113]

But the fuel purification process merely externalized the electrochemical complexity of the direct hydrocarbon system without minimizing it. A 1967 Mobility Command analysis ranked reforming, hitherto widely regarded as the path of least engineering resistance, as the most difficult of all the various fuel cell systems.[114] Researchers made two key assumptions: first, that integrating a reformer with a fuel cell represented a tolerable engineering challenge; and second, that reformers could easily convert oily logistics fuels to a hydrogen-rich gas that could be used in alkaline fuel cells. Neither estimate held under the tight fiscal and time constraints imposed by the ERDL. Begun in the summer of 1963, the reformer/fuel cell had not yet been operated as a complete unit by the spring of 1965, with only individual components subjected to testing.[115] Contrary to expectations, Allis-Chalmers proved incapable of integrating the components of the fuel cell, let alone the integrated system. The former was plagued by poor electrode quality control, a reverse salient that undermined the whole system. No difficulty was encountered running individual cells, but they would not operate uniformly when arranged in stacks. Consisting of four modules linked in series, the device suffered from uneven performance and occasional cell reversal (whereby weak cells became electrolysis cells). On the other hand, the reformer was reported to have worked relatively well by itself.[116] As a system, however, the reformer/fuel cell was a failure. The pumps, the control equipment, and the scrubber required to clean the alkaline electrolyte turned out to exert a considerable parasitic power demand. Bulky, noisy, overweight, complex, and inefficient, the unit required 40 minutes to start and was unable to operate at full output for more than a few minutes.[117] In what was becoming a typical refrain in the terrestrial fuel cell field, Army engineers claimed the exercise had "proved the feasibility" of the reformer/fuel cell concept, providing valuable knowledge for future programs.[118]

But the Army's hydrocarbon fuel cell program was doomed by a problem that transcended the various combinations of electrolytes and direct and indirect approaches attempted throughout the 1960s. The military's logistics fuels, particularly JP-4, were simply too chemically complex for reformer technology to adequately process. Originally developed by the U.S. Air Force for wide availability and maximum performance in aircraft, this jet fuel contained more than 30 different substances including anti-icing agents, antioxidants, and corrosion inhibiters, as well as sulfur, a major electrode contaminant.[119] Not only did JP-4 corrode reformer catalysts during the conversion process, the resulting hydrogen-rich fuel gas still retained sufficient sulfur to poison fuel cell electrodes.[120] As in Project Lorraine, Army engineers reverted to a simpler fuel, in this case naphtha, a sulfur-free light fraction of refined petroleum.[121] And as in previous fuel experiments, this expensive fuel was hardly representative of the substances in the Army's supply chain. Nor were researchers nearer to solving the puzzle of direct hydrocarbon oxidation. The heavier and more complex the fuel, the greater the quantity of platinum required to catalyze the reaction.[122] The path of least engineering resistance led once more into a complex maze of compromises. Unable to develop suitable direct or indirect hydrocarbon fuel cells capable of meeting the immediate requirements of the land forces, engineers looked to the hydrazine fuel cell after 1965 as the best hope for advanced electrochemical energy conversion.

Humbler Hopes: The Hydrazine Fuel Cell

The Army's doctrinal justifications for the hydrazine system were similar to the ones ARPA advanced for the methanol fuel cell and reformer/fuel cell combination. Investigating this technology independently of Project Lorraine since 1962,[123] the Army saw it as a research tool with some potential practical utility, a means of reconciling its long- and short-term goals for power sources. An expensive and volatile inorganic (noncarbonaceous) substance most notably used for power and propulsion by NASA as a rocket fuel, hydrazine, like methanol, had no logistics value. Army planners knew hydrazine was too costly to use in large multikilowatt fuel cells. But they were impressed by its operating properties, for hydrazine could be relatively easily electro-oxidized compared with the logistics fuels, producing high current density. Believing hydrazine would present fewer obstacles for fuel cell operating systems than hydrocarbons and alcohols, researchers conceived it, like the methanol design, as a direct system. The technology first appeared desirable to the ERDL as a test power source to help perfect automotive electric drive, simulating the performance characteristics of direct hydrocarbon fuel cells under electric motor loads. At such time as direct hydrocarbon fuel cells were perfected, the Army reasoned, they could be coupled to electric motors with minimal difficulty.[124]

With the dream of a general-purpose hydrocarbon fuel cell indefinitely postponed, although not abandoned, planners sought other roles for the hydrazine fuel cell. Its value, they held, was that it was a silent power source near the prototype stage that might prove superior to conventional batteries in soldier-portable communications and surveillance appliances as well as electric vehicles.[125] By 1966, Army engineers were confident that the hydrazine system's simplicity and practicality would allow it to be the first fuel cell deployed in combat in Vietnam.[126] They soon discovered, however, not only that hydrazine was no more manageable than hydrocarbons in fuel cells but also that it presented wholly new challenges. A toxic nitrogenous compound, hydrazine is very sensitive to temperature change and is easily decomposed into ammonia. Because the chemical reacts with a wide variety of materials, fuel cells using it had to be kept spotlessly clean to prevent the substance from decomposing prematurely. Moreover, the materials comprising a hydrazine fuel cell meant for service in arctic or temperate climates might not be suitable for a device intended for operation in the tropics.[127]

Hydrazine's volatility posed unique challenges for the design of fuel cell control equipment. Like all direct fuel cell systems, the hydrazine variant needed sophisticated auxiliary systems to inject fuel, remove water, and cleanse the electrolyte. These, in turn, had to cope with the duty cycle or load demand of a particular appliance. As Army researchers gained more experience testing scaled-up hydrazine fuel cells in the latter half of the 1960s, they discovered that these finer points of engineering research could be game-breakers. A major complication in designing control systems for a general-purpose fuel cell was the fact that different appliances had different duty cycles. Radars, for example, imposed a more or less steady load, while radios used power intermittently. The load profile of a specific appliance had unique consequences for the chemical balance of a fuel cell and, thus, its service life, dynamics that further varied according to the type of fuel, electrolyte, and electrode materials used in the fuel cell reactor. Control equipment had to be designed to take these changing electrochemical conditions into account. Only after all the various components had been assembled into the power source and the unit coupled to an appliance could a fuel cell be considered complete, and accurate judgments regarding reliability and durability be made. But researchers typically considered control equipment only in the later stages of development. It was as crucial a component as any other but, with research funding increasingly restricted as the war in Vietnam escalated, represented a price in time and resources that the Army laboratories were increasingly unable—or unwilling—to pay.

The most important Army hydrazine fuel cell program of the period was Mobility Command's Union Carbide 300-watt generator, the only fuel cell to see service in Vietnam. The life cycle of this technology illustrates how linear and reductive R&D imperatives helped frame the fuel cell as a discrete power source in and of itself rather than a component in a complex energy conversion and utilization system. This device originated as a response to an Army requirement developed in 1966 for a light, silent power source for communications equipment, particularly the AN/PRC-47.[128] Built by the Collins Radio Company, this large, powerful radio was used mainly by the U.S. Navy and Marine Corps in vehicles, fixed installations, and as a soldier-portable unit. During the Vietnam War, forward air controllers favored it for its long range. But high power came with costs. The radio quickly drained its power source and the overall system was very heavy.[129] At 45 pounds (20 kilograms), not including its 17-pound (7.7 kilogram) silver-zinc BB-451/U battery, the device was a heavy burden for an individual in jungle terrain, although it was considered a "manpack" radio. In fact, the radio typically required two soldiers, one carrying the receiver/transmitter and the other packing the batteries.[130] So important did the Army regard light and long-lived power sources that it made the Union Carbide 300-watt fuel cell prototype a crash project, giving it the high-priority ENSURE (Expedite Non-Standard Urgently Required Equipment) status.[131] Produced in only five months, the earliest version impressed Army engineers. They cited successful trials of the 33-pound (15 kilogram) unit in audible noise, high temperature, humidity, altitude, and inclined operation tests. Proclaiming it the most advanced fuel cell of any type then nearing operation, the Army deemed it robust enough to be placed in a limited production run of 100 units and pressed into service by late 1966.[132]

Like all types of fuel cell, however, the full-size hydrazine stack displayed adverse phenomena under long-term tests that did not manifest in earlier test phases and that were serious enough to cause major delays. When these power units were coupled with an appliance and subjected to simulated operating conditions, the fuel constantly underwent side reactions, even when the stack was not under load.[133] In conditions of high temperature and high fuel concentration, the electrical current released by the chemical reaction could decompose unspent hydrazine into ammonia and hydrogen, resulting in efficiency losses. This could also cause the hydrazine itself to evaporate, posing an inhalation hazard for the operator.[134]

These factors impinged on the design of control equipment. Union Carbide was determined to develop a general-purpose fuel cell that could cope with a variety of load profiles, meeting every possible operating requirement in all climates.[135] Its engineers first tried using a mechanical feed device of sufficient sensitivity and robustness to cope with the hydrazine system's very low fuel consumption and the rough handling and conditions it was likely to experience in the field. But the viscosity of hydrazine varies with temperature, and this made it difficult to develop a standard, all-weather fuel feed system.[136] The duty cycle of two-way radios further complicated the design of fuel control equipment because these appliances drew more power to transmit than to receive and their operators tended to use them periodically. A mechanical fuel control system was adequate for a fixed power level but supplied too much fuel under conditions of partial load.[137]

By 1969, Union Carbide realized the idiosyncrasies of the hydrazine system ruled it out as a general-purpose fuel cell. In April, some three years behind schedule, 20 300-watt units were shipped to Vietnam for trials. Engineering compromises had yielded a technology that did not possess the qualities of light weight and quiet operation that designers had assumed were intrinsic to it and which they believed made it useful for portable tactical roles. Noisy and smelly, the fuel cell weighed 40 pounds, seven pounds more than the 1966 version and much heavier than the battery it was designed to replace. As such, it was unsuited for soldiers operating near enemy positions. After years of experience, moreover, the Army had learned that no individual piece of equipment weighing more than 20 pounds could be considered "man-packable" in tropical conditions. The Union Carbide unit was instead classed as "man-manageable," meaning that it could be hauled by an individual soldier intermittently over distances of 50 to 100 meters. It was used on boats and helicopters and in base areas powering large appliances such as command-post radios and ground surveillance radars, environments where quiet operation was of little consequence. One Army researcher admitted that the value of silent operation, for long an important if not the prime justification for special-purpose fuel cells using special fuels, was difficult to quantify and was more likely psychological than anything else.[138] The point was moot, in any case, as this power source was anything but stealthy.

The Army Electronic Components Laboratory and Monsanto, its chief hydrazine fuel cell contractor, experienced similar problems. But because their fuel cell was considerably smaller, they faced even greater challenges than Union Carbide and the ERDL. As the Union Carbide researcher Karl Kordesch noted, the technical issues of fuel control were much more difficult in smaller systems because their consumption was so low. A 100-watt unit required only one milliliter of fuel per minute, and the Monsanto design was 60 watts. Metering such minuscule flows would be further complicated by the rough handling and nonhorizontal orientation the power unit could be expected to undergo while strapped to a soldier's back.[139] These issues led the ECL and Monsanto to consider a different approach from ERDL's and Union Carbide's. They decided to develop a hybrid battery/fuel cell that shaved the demand peak of the radio: the fuel cell charged the battery during periods of low demand and then combined with it to meet the heavy wattage when the radio was used as transmitter. This prevented the fuel cell from being subjected to the stress of sudden and irregular demand that played havoc with overall chemical equilibrium. Conversely, such hybrid power sources offered few advantages when an application steadily and continuously drew power.[140] In the battery/fuel cell hybrid, Army and industry researchers in essence declared the concept of a general-purpose fuel cell impracticable.

Even with this innovation, ECL and industry engineers could not overcome the inherent shortcomings of the hydrazine fuel system. Unable to control destructive chemical side reactions, researchers opted for a radical solution in one of the final designs. This transformed the fuel cell into a disposable power block with a useful life of 750 to 1,000 hours, to be replaced as necessary. The durable component of the technology was instead the power-processing and control module, which had a lifetime of around 5,000 hours.[141] In a sense, then, this particular fuel cell ended up as a kind of disposable primary battery, negating the assumed principal advantage of the fuel cell as an energy converter that would work as long as it was supplied with fuel.[142] Although an Army evaluation team appreciated the "intended flexibility" of the hybrid device, the system left much to be desired in terms of cost-effectiveness.[143]

Conclusion

The story of the Army's involvement in hydrazine fuel cell technology reveals both how judgments of fuel cell power varied according to the context in which they were made and the crucial role of dramaturgy in perpetuating work along lines that might otherwise have been abandoned. As the hydrocarbon programs stalled, the hydrazine fuel cell provided the Army laboratories with badly needed successes. In an environment where resources were increasingly at a premium, workers in the Electronic Components Laboratory considered the hydrazine fuel cell technically sweet, valuing high power over the drawbacks of volatility, toxicity, and cost.[144] Soldiers unfortunate enough to have used these systems undoubtedly would have felt differently. Reflecting on the era in 2000, one veteran researcher quipped that although the power density of hydrazine fuel cells compared favorably with the latest contemporary designs, "lifetime, both of the fuel cell and its operators," was another matter.[145]

In some quarters, the hydrazine fuel cell was regarded as little more than an effort in public relations. Writing in 1966, D. P. Gregory of the British fuel cell group Energy Conversion Limited questioned the use of the technology as a means of gaining experience with direct fuel cell systems. He believed this application was designed to show that continuously fed devices in general were conceptually valid. The relatively advanced state of development of the hydrazine fuel cell, suggested Gregory, allowed designers to claim that the goal of a direct hydrocarbon fuel cell was near. In short, the technology had been exploited for political purposes.[146]

But the hydrazine fuel cell was a false analogue, for it possessed significantly different electrochemical qualities from the hydrocarbon fuel cell. Moreover, it was a retrograde solution, for it employed an alkaline electrolyte. In a sense, research had come full circle. In large measure, the fuel cell boom of the late 1950s had been triggered by the successful demonstration of alkaline fuel cells using pure hydrogen. In turn, the deficiencies of this technology helped stimulate Project Lorraine and other efforts to develop acidic hydrocarbon fuel cells. Problems with that design led researchers back to alkaline electrolyte systems in the form of the hydrazine fuel cell.

During the Power Sources Conference of 1972, Mobility Command's John B. O'Sullivan reflected on the history of the military's terrestrial fuel cell program. In seeking an explanation for the waxing and waning of expectations over the previous decade, he offered no single theme but noted the dramaturgy that had underpinned the boom. Employing metaphors of pseudoscience, O'Sullivan described how the fuel cell had been sold as a "technological elixir" in the manner of the quack physician prescribing "patent medicines." Part of the problem in the early days, he wrote, had been an "inverse relationship" between the variety of research approaches and the sum of knowledge. Fuel cell research and development had always been conducted with the expectation of a breakthrough "just over the horizon." But every advance brought a bewildering array of trade-offs.[147]

The larger question was why ambitions for fuel cell power were so much greater in the United States than in Europe, where researchers concentrated on the modest goal of developing specialized fuel cells using exotic fuels. Although the dream of a general-purpose electrochemical engine was not uniquely American, it galvanized a broad spectrum of actors in a society where, as the historian David F. Noble notes, popular enchantment with innovation was unrivaled.[148] Military-industrial planners who prized performance over durability and cost were particularly besotted by it. Rapid, ill-planned expansion of the defense research complex and the cascading requirements and bureaucratic rivalries this engendered made managers, scientists, and engineers receptive to the idea of a power panacea. Defense research intellectuals saw the fuel cell as a means of justifying ARPA's existence as the agency made the difficult and doctrinally incoherent shift from managing technology programs to managing science programs. Drawing false analogies from the early stages of research, credulous officials in ARPA, IDA, the Army, and industry believed a breakthrough technology could be thriftily and speedily developed by incrementally improving existing designs. Under an inflexible and inefficient system of administration that had little place for engineering research, assumptions for fuel cell power gradually crumbled. Only the new civilian space agency was prepared to commit the resources required to put the technology into practical service.

3

Fuel Cells and the Final Frontier

> Like so many revolutionary technical concepts, the fuel cell is the direct
> result of the extraordinary demands of the space age.
>
> —Wernher von Braun, August 1964

Of the postwar fuel cell enterprises, the programs sponsored by the National Aeronautics and Space Administration in the 1960s and early 1970s were the largest, best funded, and had the highest profile until the automobile-centered boom of the 1990s and 2000s. From 1962, NASA served as a sponsor and customer of hydrogen fuel cells, paying contractors tens of millions of dollars to build systems for the Gemini, Apollo, and Shuttle spacecraft. More than any previous patron, the space agency introduced fuel cell technology to the public and informed broader expectations of its potential. As with the Bacon cell in Britain, American aerospace fuel cells became symbols of advanced technological prowess in NASA's campaign to justify the elaborate set-piece pageantry of the Space Race.

Little is known of how this project contributed to the fuel cell and alternative energy technology fields as well as energy research and development policy in general. Indeed, NASA's broader economic legacy is largely terra incognita. In the 1970s, aerospace history scholars including Vernon Van Dyke, John M. Logsdon, and the official NASA historian Eugene M. Emme interpreted NASA's social influence mainly in political terms. To Logsdon, the key question was how science and technology were employed for national goals, rather than how space policy shaped science, technology, and the economy.[1] Joan Lisa Bromberg's study of NASA's relationship with the aerospace industry was a notable effort to address this lacuna.[2] But Bromberg revealed little of how America's adventure in space influenced non-aerospace science, technology, and commerce.

Fuel cell researchers A. J. Appleby and F. R. Foulkes hold that NASA was responsible for making the fuel cell concept a practical reality, contributing "enormously" to the fields of electrochemistry and electrochemical engineering science in the process.[3] It is true that NASA did do much to advance the development of terrestrial fuel cell power, although not directly. Fuel cell technology was important to the agency in two ways. Planners regarded it as suitably advanced but essentially proven space-age hardware capable of meeting certain demanding aerospace requirements. In a program laden with superlatives, the fuel cell also became an important marker of engineering virtuosity, not least because NASA managers claimed the device would benefit the civilian marketplace. The space spin-off was an important political consideration for NASA as the 1960s progressed and the agency faced increasing criticism that the science and engineering it supported were irrelevant to social realities.

These expectations were outlined in a promotional pamphlet prepared for the NASA-sponsored Second National Conference on Peaceful Uses of Space at the Seattle World's Fair in May 1962. Among the range of space techniques and technologies that promised broader terrestrial use were "amazing new sources of power" including the fuel cell. With further development, indicated the pamphlet, the technology would allow homes and businesses to generate their own heat and power, reducing dependence on the centralized gas and electricity grid and presenting a business opportunity as potentially lucrative as space communications.[4]

The reality was that NASA had neither the capabilities nor the mandate to directly develop a terrestrial fuel cell. Still, as we have seen in the case of Project Lorraine, the space fuel cell program was of crucial ideological significance to managers of terrestrial fuel cell programs because it established the precedent of practicality. With the Pentagon's fuel cell effort faltering by the mid-1960s, NASA's aerospace variants became the leading technological exemplars. Like ARPA, the agency helped popularize the notion of a general-purpose electrochemical engine but to a much larger audience owing to the highly public nature of the space program. Far from spinning off a terrestrial fuel cell variant, however, NASA and its contractors sought to adapt technology originally developed for terrestrial purposes for use in spacecraft, a costly process fraught with difficulty and not at all a good indicator of the odds of successfully developing an efficient and affordable commercial fuel cell. As applied in spacecraft, fuel cells operated in conditions more akin to a controlled laboratory environment than those they would encounter in daily use on earth.

Powering the Space Age

Of the government agencies interested in fuel cells in the postwar period, NASA embraced the technology like none other. Forged in a national security emergency, NASA's institutional culture developed what the historian Howard E. McCurdy has referred to as a "frontier mentality." Dominated by engineers and scientists, NASA's workforce was steeped in a professional ethos that encouraged the pursuit of the latest new technologies. When new systems were put into service, engineers hoped to shift their energies to wholly new challenges.

Routine operation of existing systems, on the other hand, was often derided as mere "management."[5] This philosophy paradoxically made fuel cells attractive, yet complicated their development because it afforded no tolerance for the tedium of engineering research. In his consultations with "space experts," Vice President Lyndon B. Johnson (then chair of the White House's National Aeronautics and Space Council) had been advised that a moon mission required no technological breakthroughs but rather a development of "existing capabilities."[6] Although this may have accurately described the state of certain propulsion technologies at the time, fuel cells were hardly proven hardware when NASA planners first began investigating them for use in spacecraft in 1959. As in the Department of Defense's terrestrial fuel cell program, building an aerospace fuel cell was not simply a matter of applying known scientific principles. And the operation of the first such devices was anything but routine. When contractors encountered severe problems, NASA provided them with considerable resources in an effort to press the technology into service as quickly as possible.

The possibility of using fuel cells as a lightweight replacement for batteries in spacecraft was first broached during the meeting of the Goett Committee, a group struck by NASA headquarters in April 1959 to study the technical options for a variety of space missions.[7] After assessing a number of power sources, the Space Task Group (STG), the agency's nascent chief planning bureau, concurred, deciding that fuel cells were best suited for the moon mission on the grounds of weight and reliability.[8] In principle, the technology's main limitation was the weight of fuel. For short missions, conventional batteries had the advantage. Radioisotope thermoelectric generators, which produce electricity through the natural decay of radioactive material, could provide power for very long periods but required prohibitively heavy shielding if they were to be used in human-carrying spacecraft. For missions of up to two to three weeks, however, STG/Manned Spacecraft Center planners believed fuel cells offered the ideal balance between power and weight, as well as a number of other advantages for spaceflight. They allowed spacecraft designers to dispense with arrays of externally mounted photovoltaic panels, which they believed might obstruct docking procedures. Powered thus, they reasoned, vehicles would be more maneuverable, allowing non-solar-oriented attitudes for tracking stars.[9] Finally, planners hoped crews could consume the water produced by hydrogen-oxygen fuel cells, allowing further weight reductions. As NASA's chief of electrochemical systems Ernst Cohn put it, fuel cells seemed to represent a "maximum in utility and a minimum of problems."[10]

Such enthusiasm was not universally shared within the agency. At least one deputy associate administrator warned D. Brainerd Holmes, the head of the agency's human spaceflight program, that fuel cells were untried and advised the STG to carefully review developments in the field. True, major strides had been made since 1957, noted Thomas F. Dixon, but the technology was not yet near operational status.[11] But his was a voice in the wilderness. Even before President Kennedy formally announced the moon mission and as the STG drafted criteria for the Apollo auxiliary power unit, Pratt & Whitney had impressed many with its fuel cell program. In an emerging technology field that featured few real competitors, the renowned maker of aerospace engines had the insuperable advantage of owning a license for the Bacon cell, the world's most powerful and proven fuel cell design. Possessing the same procurement powers as the armed services by virtue of the National Aeronautics and Space Act of 1958, NASA dispensed with the traditional practice of competitive bidding.[12] It agreed that Pratt & Whitney would build a 250-watt pilot fuel cell that would serve as the basis for a full-size stack of between two and three kilowatts for the moon mission spacecraft.[13] The company's successful development of this prototype led NASA to award it the Apollo fuel cell contract in March 1962.[14]

Bacon Cell in Space

The Bacon cell was indeed among the most practical fuel cell designs in the early 1960s. But adapting this terrestrial power source for use in space raised a whole host of unanticipated complications and trade-offs between high performance and reliability stemming from variations in pressure and electrolyte concentration. Designing the technology with a civilian market in mind, Bacon had opted for cheap nickel over platinum as the catalyst. However, the only way nickel became sufficiently reactive was by immersing it in potassium hydroxide at temperatures above 100°C. As he experimented further, Bacon learned that raising the temperature to around 200°C in an electrolyte concentration of between 27 and 50 percent increased performance without requiring more catalyst. Increasing the pressure on the electrolyte to prevent it from boiling had led him to discover a correlation between higher pressure and higher current density. He had then raised pressure well beyond what was needed to stabilize the electrolyte, reaching a maximum of 600 pounds per square inch (psi) or 41 atmospheres (atm), later reduced to 400 psi (27 atm).[15]

The result was a relatively cheap device that produced high power. Pratt & Whitney initially tried to adopt this formula. But in 1961, a test unit exploded at the Leesona laboratories, and NASA and its contractor began to view the high-pressure cell as too dangerous for aerospace applications.[16] They wanted to use the Bacon operating system at lower pressure, but that meant lower power. To compensate, Pratt & Whitney engineers increased the concentration of the electrolyte. This, however, brought fresh problems. Experiments on two British-built Bacon cells in 1960 had revealed that the electrolyte concentration had important implications for the physical integrity of the stack. At high concentrations, the electrolyte solidified as it cooled after deactivation, warping the bracing structure containing the cells, which had to be periodically readjusted.[17]

As a result of these early tests, the Pratt & Whitney researcher in charge recommended lowering the electrolyte concentration. But the decision to proceed with a low-pressure device meant that a 30 percent aqueous solution of potassium hydroxide would boil at 204°C. To compensate, Pratt & Whitney increased the electrolyte concentration to between 75 and 85 percent and raised the operating temperature to 260°C. The resultant technology worked at a pressure of only 3.4 atm.[18] With safety and power requirements seemingly reconciled, this operating system was selected for the production model, the PC-3A-2.

The trade-off was a power source with a highly corrosive operating system, one that was much less durable than the original Bacon cell. The higher operating temperature and potassium hydroxide concentration exacerbated the problem of thermal expansion that Pratt & Whitney had first observed in 1960. Start-up and deactivation were lengthy procedures to avoid damaging the electrodes as the concentrated electrolyte changed from a solid at room temperature to a molten liquid at operating temperature.[19] One hazard arising from rapid deactivation and depressurization was a phenomenon known as "cold-popping." This occurred when reactant gases became trapped in the cooling, hardening electrolyte. When the fuel cell was reactivated, these gases formed a bubble between the electrode and the melting electrolyte, reducing the active electrode area and overall performance.[20] High temperature and high electrolyte concentration also created an altogether different, and more difficult, problem. When the fuel cell was producing power, nickel ions leached from the cathode to the anode, where they accumulated and created an elongating, branching encrustation known as a "dendrite" for its treelike pattern. This nickel growth eventually bridged the electrolyte gap, creating a short circuit that damaged or destroyed the cell.[21] In short, the Bacon cell as adapted for spaceflight was more difficult to operate and less reliable than its terrestrial counterpart.

Emulating trends in the terrestrial fuel cell efforts, the aerospace fuel cell program featured a linear mode of management that prevented engineers from fully understanding these problems. As with the Manned Spacecraft Center's other units, NASA's Propulsion and Power Division left most of the actual work to contractors.[22] Like their counterparts in Project Lorraine, engineers took a piecemeal empirical approach to fuel cell research and development, working quickly and without considering how the power source would function as part of a chemical energy conversion and electrical supply system. The results would be similarly unpleasant. In part, this was a consequence of the political passions driving the Space Race. Robert B. Hotz, editor in chief of the trade publication *Aviation Week & Space Technology*, provided a caustic example of how ideology interfaced with engineering in this milieu. In the April 29, 1963, issue he excoriated the "Eisenhower-type" detractor who argued for a slower, more cautious pace in developing space technology, framing such criticism as virtually treasonous in a period when the Soviet Union had inflicted the worst national humiliation since British men-of-war abducted American sailors on the high seas.[23]

But haste was folly in aerospace engineering, especially in the case of fuel cell research and development. Pratt & Whitney engineers executed the first tranche of tests of the PC-3A-2 rather perfunctorily, seemingly intent on confirming existing assumptions of durability. They accepted the prevailing wisdom that fuel cells did not deteriorate over time, unlike batteries. As researchers in Project Lorraine were discovering at around the same time, however, this was false. The company rushed the unit into production before engineers were fully aware of how it would respond under sustained use in realistic operating conditions. Although the early tests of 1960 had revealed some of the complications of lowering pressure and increasing temperature and electrolyte concentration, the dendrite phenomenon took researchers by surprise, manifesting well after the basic design had been locked in.

The result was a something of a fiasco. In early 1964, Pratt & Whitney delivered a batch of three prototype units to North American Aviation, the prime contractor for the Apollo spacecraft. Heralding the event as a "major milestone," the Manned Spacecraft Center claimed that the fuel cells had been shipped after completing successful acceptance tests under simulated launch conditions.[24] Then rated for at least 400 hours or nearly 17 days of operation, all three prototypes suffered dendrite growth and quickly failed. One lasted 112 hours, while the other two were severely damaged after being shut down and restarted. North American Aviation returned all three to Pratt & Whitney for rebuilding.[25]

It was a considerable setback. With many units already "in the pipeline" when the failures began to occur, noted William E. Rice, then head of NASA's Power Generation Branch, costs skyrocketed.[26] The space agency played a greater role in a more rigorous and realistic testing program begun in early 1966. Conducted at the Manned Space Center's White Sands facility in New Mexico, this involved integrating fuel cells into the Apollo spacecraft's propulsion and power system.[27] Embodying NASA's conservative test ethos of using full-size prototypes to analyze components, this extensive regimen finally debugged the PC-3A-2.[28] Progress came at a price, however. As one engineer in the Power and Propulsion Division recalled, the sensitive power source had to be "babied." And with the moon program on a crash footing, NASA was willing to sacrifice economy for speed. Engineers discarded units worth several hundred thousand dollars apiece like a "flashlight battery" after they had accumulated 400 hours of operation, consuming up to 30 production copies before the power source was deemed fit for service.[29]

The fruit of almost 35 years of research, the Bacon cell in the form of the PC-3A-2 eventually performed well, providing electricity and water for astronauts in eleven Apollo flights. In this configuration, the technology contradicted the original assumptions that distinguished fuel cells from conventional batteries in the first place: electrode invariance and long life. In immediate technological terms, Pratt & Whitney and NASA transformed a device designed for the commercial market in Britain into a highly specialized, costly, and short-lived technology that was tricky to operate. Like the Army Electronic Components Laboratory's hydrazine fuel cell, the final version of the Bacon cell became a sort of expensive disposable primary battery that, outside NASA's celestial sphere of patronage, was as much a technological orphan as the original device had been in Britain. Of the various organizations that had sponsored the Bacon cell over the years, only NASA was prepared to subsidize the costs of supplying hydrogen. Despite the program's difficult development, the fuel question was relatively simple where only a handful of spacecraft were concerned.

Not surprisingly, the fuel cell community, and Pratt & Whitney above all, viewed the PC-3A-2 as a dead letter in energy conversion well before the device was first used in a human-piloted mission in *Apollo 7* in October 1968. Its success in space validated Bacon's lifework, yet the company saw no future for his invention. Its historical arc was unintentionally metaphorized in a congratulatory letter from Pratt & Whitney on the occasion of the successful flight of *Apollo 8*, informing Bacon that his "grandchildren performed flawlessly . . . and went to their fiery Valhalla as heroes."[30]

Making Space Pay: NASA, General Electric, and the Dual-Use Fuel Cell

In marked contrast to Pratt & Whitney, General Electric was a proponent of the space spin-off and one of the most confident voices of fuel cell commercialization in the 1960s. It was true the company failed to complete development of the direct propane/methane hydrocarbon breadboard begun with ARPA support and demonstrated to some fanfare in April 1963. It instead placed great store in the proton exchange membrane fuel cell (PEMFC), a power source the company was preparing for the Gemini spacecraft and believed had much greater potential in the civilian market. Unlike its competitor, General Electric developed the most elaborate fuel cell dramaturgy of any firm in the 1960s, becoming the leading promoter of the concept of the dual-use power source. In other respects, however, the company's experience with aerospace fuel cells would strikingly resemble that of Pratt & Whitney.

Approved in December 1961, six months after the announcement of the Apollo project, Gemini had been conceived by NASA as a means of gaining experience in spaceflight rendezvous procedures that would be employed in the moon missions as well as an opportunity to test a number of new technologies including fuel cells. The Manned Space Center originally planned to use batteries in the spacecraft, but as mission length, and the weight of batteries, was increased, it selected the membrane fuel cell as the primary power source on the grounds of light weight and simplicity.[31] As impatient and confident as in its terrestrial fuel cell program, General Electric saw little need for engineering research on aerospace fuel cells. The company completed its production plant at Lynn in 1962 but only fully activated the facility's test station in early 1964, a legacy in part of Project Lorraine.[32] General Electric felt the PEMFC concept had been essentially proved in preliminary tests, announcing in May 1962 that single cells had demonstrated a lifetime of 2,000 hours or 83 days of continual operation, well above the two weeks of the longest planned Gemini mission.[33] McDonnell Aircraft, the Gemini spacecraft prime contractor, was already convinced. In March it let a $9 million subcontract for the system.[34] In the summer and fall of 1963, without subjecting the two-kilowatt power plant to long-term tests, General Electric started production.

Like Pratt & Whitney, the company had made an expensive gamble, with similar consequences. The chief shortcoming of the proton exchange membrane fuel cell was its polymer membrane, its core feature. This material was prone to dehydration and cracking, especially at higher temperatures, allowing hydrogen and oxygen to directly react and, potentially, combust or explode. Good quality control was essential because the system was highly susceptible to contamination from other materials used in different parts of the cell.[35] This, however, was not forthcoming. Failures struck production models, and repairs were slowed by faulty parts and materials supplied by subcontractors and by delays in their replacement.[36] In late November, persistent engineering and manufacturing problems forced General Electric to shut down the production line.[37] As the program fell behind schedule, NASA considered abandoning it altogether. The Gemini Project Office asked McDonnell to study the possibility of replacing the PEMFC with conventional batteries on the longer rendezvous missions. The problem was not simply the reliability of the fuel cell stack itself but also its functionality when integrated into the spacecraft's electrical and cooling system. Although laboratory tests in 1963 indicated the power unit had a lifetime of around 600 hours, modifications to the spacecraft's cooling system increased the stack's operating temperature, drastically reducing useful life to between 150 and 200 hours.[38] In response, General Electric reorganized its Lynn plant and brought in a new manager, Roy Mushrush, with sweeping new powers to use whatever resources were necessary to turn the operation around.[39]

Even as General Electric struggled to ready the membrane fuel cell for space operations, its public relations bureau made bold claims. Marketing literature published in 1964 associated the technology's projected performance in spacecraft with potential terrestrial roles. Acknowledging that such applications could take years to develop, General Electric nevertheless claimed that fuel cells were capable of using "readily available fuels," a gross oversimplification.[40]

Later that year, the company ratcheted up its promotional efforts, planning a media extravaganza around a temporary exhibit of its first 1,000-hour production unit at the National Air Museum in Washington, DC. In a measure of the importance it accorded the event, General Electric asked Massachusetts senator Edward M. Kennedy to officiate the handover as a representative of a state in which the firm had major investments.[41] But Frederick C. Durant III, the museum's assistant director of astronautics, demurred, believing it imprudent to display the technology before it had been proven in a practical application. He advised General Electric to delay the exhibition until after the fuel cell's first flight in space.[42]

Indeed, the device had not accrued a distinguished service record. In January, it had malfunctioned during its first operational test in the remotely controlled *Gemini 2* spacecraft and was shut down before the rocket even lifted off.[43] McDonnell and NASA deemed the fuel cell too unreliable to be used in the first two piloted Gemini spacecraft, opting instead for conventional batteries in launches in March and June. Fortunately for General Electric, Mushrush accepted Durant's advice. But the manager did see fit to remark that the 1,000-hour ground test was significant in and of itself because it represented a major advance in the development of "practical fuel cell power" in general.[44]

The flight of a membrane fuel cell system on the eight-day piloted flight of *Gemini 5*, which lifted off on August 21, finally provided General Electric with the practical success it craved. In a press conference held less than two weeks later, officials portrayed the event as a landmark in the history of electric power. Arthur M. Bueche, a GE vice president for research, declared fuel cells the first "practical major power source to be developed since atomic energy." The *New York Times* quoted him promising that the power source would be commercially available in early 1966 for remote television cameras, golf carts, forklifts, and scuba diver sleds.[45]

To be sure, such applications were hardly commensurate with Bueche's grandiloquent rhetoric. What was more, the fuel cell's performance in *Gemini 5* had been far from ideal. The prime contractor's report admitted the power system had displayed "unusual modes of operation" but did not elaborate.[46] A story in the *Houston Post* provided more detailed information, pointing to flaws not in the fuel cell stacks themselves but rather in their reactant storage components, underscoring the sensitive connectivity of the system. The cryogenic hydrogen and oxygen tanks contained probes that produced heat to maintain pressure so that the stacks could continue to draw fuel as they drained the vessels. During the flight, one of the system's oxygen tank heaters failed, although planners never learned exactly why, as the utility portion of the spacecraft was destroyed during reentry.[47] They decided to install conventional batteries in *Gemini 6* for its two-day journey before reverting to the General Electric fuel cell for the remaining six flights.

In these missions, the power source performed tolerably well, perhaps indicative more of the degree of redundancy that had been built into the system than the unit's intrinsic reliability. Although individual stacks failed in *Geminis 11* and *12*, the remaining ones were able to supply sufficient power. But project managers never gained sufficient confidence in the fuel cell to use it as the spacecraft's sole power source, instead integrating it with silver-zinc storage batteries to ensure reliability.[48]

This, in a sense, was a repudiation of the technology, for it had been rationalized as a battery replacement to save weight. Nevertheless, in January 1966, after much delay, the Smithsonian Institution finally staged a 10-day exhibit of a 1,100-hour production unit "identical" to the one that had served on *Gemini 7*. Its official press release repeated Bueche's claim that General Electric had achieved a landmark in power source innovation.[49] Robert Cohen, the NASA official the historians Hacker and Grimwood believed was responsible for recommending the membrane fuel cell for Gemini, was more circumspect. Reflecting on the period, he opined that batteries were far more reliable and easier to use and had fuel cells "beat hands down."[50]

Still, General Electric was committed to the technology and worked to develop it for years after the conclusion of the Gemini program. Toward the end of the decade, with help from NASA's Manned Spacecraft Center and the Office of Advanced Research and Technology (renamed the Office of Aeronautics and Space Technology), GE researchers finally solved the problem of membrane dehydration and cracking, developing a method of humidifying the hydrogen fuel and oxidant before they were pumped into the cells.[51] An even more important advance was Du Pont's Nafion, a perfluorosulfonic acid polymer membrane. Originally developed in the early 1960s for chlor-alkali production, this fully fluorinated membrane delivered higher power and proved more robust than General Electric's old membrane when used in a fuel cell because its carbon-fluorine bonds proved more stable than the latter's carbon-hydrogen bonds. In the mid-1960s, the two companies began collaborating in this application of the new material.[52]

But these dramatic improvements came too late pay a dividend. By the end of the decade, the already limited aerospace fuel cell market shrank further. In 1971, General Electric's fuel cell division was dealt a blow when NASA selected Pratt & Whitney to supply a handful of fuel cells for the Space Shuttle. It must have been a bitter pill, for agency insiders rated the company's latest membrane device as very good.[53] With the conclusion of the moon program, NASA downsized its human spaceflight activities, eliminating its in-house aerospace fuel cell program in 1972.[54] Deprived of government support, General Electric worked on membrane fuel cell technology until early 1984, when, after almost 30 years of effort, it sold this line of research to the United Technologies Corporation (UTC), its longtime rival.

The Fuel Cell as a Dividend from Space

To an extent perhaps unique among U.S. federal institutions, NASA relied on evocative imagery as a means of generating support for its programs.[55] Politicians and bureaucrats keen to demonstrate the agency's broader social relevance highlighted the fuel cell as one of a number of technologies spawned by the adventure in space that symbolized American pride, pluck, and ingenuity. Vice President Johnson's famous address at the Goddard Memorial Award Dinner in Washington, DC, in March 1962 laid out the logic of the space spin-off. Appealing to patriotism and material self-interest, he announced that the true purpose of the space program was not to "peek into the windows of heaven or to preen our national pride" but to endow humanity with new inventions and wealth. For Johnson, "space research" combined basic research with a Second World War–like mobilization of national industrial resources that would yield new knowledge, advanced materials, and a bonanza of new products. He alternately shamed and tempted consumers, noting that the space program cost less annually than the $5 billion Americans spent each year on "face powder, lipstick, and nail polish" and promising that "for every nickel we put in, we get a dime back."[56] As NASA increasingly justified its work in pragmatic terms, the fuel cell assumed an important place in the agency's public relations drive. For years, planners suggested that the space agency was contributing to the development of commercial fuel cell technology, both indirectly, through the production of goods and knowledge that could be adapted for broader use, and directly, in the sense of dedicated work to this end. Blurring distinctions between space and terrestrial applications, the spin-off campaign made an important impression on government contractors engaged in fuel cell R&D.

As early as 1962, the power source was framed as a fruit of the Space Race. That year, the editors of *Time* magazine published a book claiming that the technology offered "probably the broadest range" of application of the various spacecraft power sources then under development. Indeed, they claimed, the fuel cell was "already being carried over from the space industry" by General Electric and Westinghouse, where it promised to revolutionize the production and distribution of electricity on earth.[57] In October 1962, John E. Condon, the assistant director of NASA's Office of Reliability and Quality Assurance, told a gathering of the Mount Vernon, Ohio, Chamber of Commerce that the space program was expected to give rise to whole new industries, not simply individual pieces of technology, including some devoted to new power sources, notably fuel cells.[58] Five months later, James Dennison of NASA's Office of Technology Utilization addressed the annual meeting of the National Association of Business Economists in Cleveland. Elaborating a standard NASA theme, Dennison framed the space program as an effort in basic research in which the specific outcomes and products could not be known in advance. The goal of the

Technology Utilization Program was to identify "incidental knowledge"—new materials, processes, and techniques with direct terrestrial relevance produced over the course of the program—that NASA was prepared to offer "at no charge to all comers, no strings attached." Among them was the fuel cell, "one of the most promising developments" in power sources. Along with the Department of Defense and industry, claimed Dennison, NASA had done extensive research to make the technology commercially feasible. Experiments then under way to develop cheap, long-lived fuel cells capable of using piped natural gas promised a "revolution" in decentralized electricity production.[59]

The Gemini and Apollo projects sparked a rash of newspaper stories suggesting or predicting future terrestrial applications of fuel cell technology.[60] No less a personage than Wernher von Braun engaged in the boosterism. A ceaseless promoter of the moon program, the rocket builder and former Waffen-SS *Sturmbannführer* (Nazi Party rank equivalent to major) extolled the virtues of fuel cell power throughout the 1960s. In a 1964 interview published in *Popular Science*, he spoke at length about the technology, repeating the canonical assumption that it was "not tied" to the Carnot cycle and also the NASA-inspired idea that it was a by-product of the space age likely to find "increasing acceptance on earth."[61] He was even more rhapsodic in a 1967 address to members of the Foreign Investors Council at the Kennedy Space Center. Invoking a vision of domesticated fuel cell power, he claimed the technology was an "almost classic" example of a space spin-off, one that would interest many industries outside the aerospace sector. One company had already built a fuel cell electric tractor, and others were exploring the idea of a reversible fuel cell for home use. At night, off-peak electricity would electrolyze water into pure hydrogen. In the morning, home owners could start up the fuel cell and enjoy silent power all day long. One might even be able to plug in one's fuel cell electric automobile at night in the same way. Not only would this be an economical way to drive, held von Braun, it might also be a solution to the growing problem of air pollution, an early reference to the environmental benefits of the technology.[62]

Despite such rhetoric, NASA was not directly involved in terrestrial fuel cell research and development. In the early 1960s, the space agency was one of the three main federal sponsors of research and development along with the Department of Defense and the Atomic Energy Commission (AEC), which together accounted for 93 percent of federal R&D expenditure in 1963. However, there were important differences in how these agencies conducted such activities and in the objectives these activities were intended to serve, as Bruce L. R. Smith has observed.[63] Like the AEC, NASA encouraged the devolution of advanced technologies into the civilian economy. Indeed, an unparalleled portion of the space agency's budget went directly to its contractors.[64] But NASA was less heavily involved than its counterparts in supporting basic research at universities. Perhaps uniquely, it was a "pure technology agency," designed to advance certain kinds of hardware for certain goals.[65] In the early 1960s, the immediate objective of U.S. space policy, observed Johnson, was to remedy the disparity in rocket boosters between the United States and the Soviet Union. Propulsion, he held, was the most important advantage the Soviets then had in space.[66] Given the short-term time frame of the Space Race, rocket superiority had to be achieved with relatively tried technology, largely an engineering task, albeit on an unprecedented scale.

In sum, the space agency's R&D programs were optimized for a singular mission and managed in a way intended to maximally benefit the aerospace industry. No less than its other technology initiatives, NASA's fuel cell program was shaped by these imperatives. The agency's role was limited to funding contractors and providing support for testing and diagnostics, work coordinated by the Manned Space Center. To be sure, NASA's Office of Advanced Research and Technology (OART) did have an in-house fuel cell effort that was part of the Lewis Research Center's space power program. But it was small and lacked a clear mission, reflecting the contradiction in the agency's desire to be a leader in science and technology on the one hand and its desire to bolster manufacturers on the other. Part of the former National Advisory Committee for Aeronautics (NACA) before it was absorbed into the new national space agency in 1958, Lewis, like the other NACA centers, had been founded to conduct research in service of the aviation industry. After consolidation with NASA, the centers were supposed to continue this role in an expanded capacity, supporting civilian and military aerospace research and future space missions, and adapting space technologies for terrestrial use.[67] But such fuel cell work as was done at Lewis was confined to aerospace applications, and even here there was little for engineers to do. A review in April 1963 found the program adrift, mainly because NASA was preoccupied with its near-term agenda in human spaceflight. Future power requirements were unclear.[68] Over the following 15 months, Lewis would work more closely with the main fuel cell contractors, but fuel cells remained the center's lowest priority in space power technology.[69]

The heart of the OART's fuel cell effort was not at Lewis but at the Marshall Space Flight Center in Huntsville, Alabama. Begun in 1962 by the Office of Manned Space Flight, this program was wholly devoted to a Bacon-type cell developed by Allis-Chalmers as a backup to Pratt & Whitney's own troubled Bacon cell.[70] Such was NASA's concern about the reliability of the PC-3A-2 that as late as December 1968 it awarded Allis-Chalmers a $3.5 million contract to produce four stacks for qualification testing.[71] Here, then, lay the scope of the space agency's interest in fuel cell power. It was willing to do whatever was necessary in order to secure reliable systems for the Apollo project, including sponsoring concurrent lines of development. Moreover, planners were satisfied that the Bacon cell design, with origins dating back to the 1940s, could serve the agency's needs well into the future. Thus, in 1966, George Mueller, the associate administrator of Manned Space Flight, vetoed plans for a new 90-day fuel cell system, believing the potential of the Pratt & Whitney and Allis-Chalmers systems for human missions lasting between 28 and 56 days had not yet been fully exploited.[72]

The fact was that NASA committed few resources to researching and developing either terrestrial or advanced aerospace fuel cell technology. How did the agency reconcile this reality with its claim that its fuel cell program would pay terrestrial dividends? The answer is that NASA officials often crafted different messages for different science and technology communities. In this regard, the activities of Ernst M. Cohn are instructive. As the director of electrochemical systems at the Space Power and Electrical Division of OART, Cohn was in a unique position. A former member of the Army's Ordnance Corps and Research Office, he had played an important role in justifying the hydrocarbon fuel cell and launching Project Lorraine. Cohn's move to NASA in the early 1960s placed him at the administrative head of a relatively small and unimportant department that had little technical involvement in an effort dominated by the Office of Manned Space Flight. But Cohn did much to promote the gospel of spin-off. Liaising with the larger science and technology community, he offered contrasting visions of NASA's fuel cell program and its relevance for terrestrial applications in his addresses and internal reports. For example, the tone of Cohn's presentation at the Army-sponsored 17th Annual Power Sources Conference in May 1963 was temperate. Concentrating mainly on the power-to-weight trade-offs of future space power sources including the various regenerative fuel cell types, he added in passing that the concept of a biological fuel cell consuming gases produced by human waste or metabolism then under consideration for deep-space journeys had also been proposed for use with a pacemaker.[73]

Cohn made much bolder claims in a presentation that September to the American Chemical Society, one of the oldest and most important scientific societies. This time, he explicitly framed the biofuel cell as a space spin-off, holding that the technology had captured the public's imagination as a solution to water pollution by harnessing the hydrogen produced by microbes consuming raw sewage. Cohn also articulated the canonical assumption that information produced in NASA's aerospace fuel cell programs would be equally valid for terrestrial fuel cell research. In solving the problems of fuel cells in the space power role, he concluded, the space agency hoped to advance the technology in ways that would benefit the economy. These assertions were amplified both by virtue of their placement as the first item in the published proceedings as well as by the fact that Cohn appeared to speak as a representative of NASA, not just the OART and its Space Power Division.[74]

But in his contribution to the Kennedy administration's Interdepartmental Energy Study drafted several weeks earlier, Cohn had painted a complex and pessimistic picture of the state of fuel cell technology. Repeating his view that the power source would have many terrestrial applications, he added important caveats. Although the device had attracted great interest in the United States, he wrote, it was in an embryonic stage of development. The current enthusiasm, he observed, stemmed from the belief that the technology was not subject to the limitations on heat-engine efficiency, even under ideal conditions, as dictated by the Carnot cycle. In theory, fuel cells could attain efficiencies as high as 60 percent, compared to the 40 percent typical of commercial power plants. But the problem with such assumptions, Cohn indicated, was a gap between theory and the engineering of the day. By late 1963, researchers had only just begun building multicell prototypes. Because most testing had hitherto occurred on single cells, maintenance costs were largely unknown, as were details on longevity and reliability. Systems engineering of full-scale stacks had "barely begun," but it was likely that failure modes would prove far different than for single cells. Aerospace fuel cells would certainly provide valuable experience, Cohn wrote, but he doubted this would prove relevant for terrestrial applications considering that such designs used pure hydrogen and oxygen over a period of only two weeks at most. Nowhere in the report did Cohn indicate that he believed a space fuel cell could be spun off for use on earth.

As for biofuel cells, he wrote, power densities would have to be increased "100-fold" before they could compete with existing electrochemical systems. Indeed, biocells were not expected to play a major role where alternate forms of power were available. Ameliorating water pollution could be accomplished more effectively with other technologies, he added. Crucially, noted Cohn, fuel cell systems using conventional fuels were currently too expensive, although costs would drop when demand warranted mass production, echoing the ARPA and Army line. He cast doubt on the estimates cited in *Fuel Cells: Power for the Future*, the influential report produced by the Harvard Business School students, believing their claim of capital costs of $7.50 to $10 per kilowatt was far too low. The gasoline internal combustion engines of the time produced about 12 kilowatts per cubic foot and cost about $3 per kilowatt. In contrast, fuel cell stacks then typically produced between one and three kilowatts per cubic foot. Actual costs for commercial fuel cells, Cohn noted, were likely to range from around $50 per kilowatt for short-lived devices to between $150 and $300 per kilowatt for more durable equipment. In comparison, General Electric's Gemini fuel cell then cost a staggering $47,500 per kilowatt. And external hydrocarbon reforming, he noted, was a complicated chemical process that would add cost and detract from efficiency.[75] Probably the frankest assessment of the terrestrial fuel cell project to date, the report demolished many of the myths that had propelled the fuel cell boom of the early 1960s. Cohn's self-censorship during his address to the American Chemical Society in the fall of 1963 was emblematic of NASA's role in sustaining the idea of the power panacea as a dividend from space at a time when the dream of terrestrial fuel cell power was fading.

NASA and Commercial Fuel Cell Development

What, then, was the legacy of NASA's involvement in fuel cell research and development? There is no question that the fuel cell was not a genuine space spin-off, having existed in conceptual form since the mid-nineteenth century. If anything, a kind of reverse spin-off had occurred, with existing terrestrial designs being adapted for aerospace purposes.[76] Only aerospace fuel cells can be said to be a direct product of the space program. But even the claim that fuel cells were first practically applied in spacecraft is suspect, for it turned on the assumption that the technology was routinely used in this role. To a large degree, the aerospace fuel cell program was as experimental, and nonroutine, as previous trials of fuel cells in prosaic terrestrial applications. To be sure, NASA did help solve one of the membrane fuel cell's main reverse salients. But this did not have immediate commercial implications. And subsequent improvements to the technology after the late 1960s would owe much more to efficient new polymer membranes developed by Du Pont and later Dow.

Yet for manufacturers of all kinds in the late 1950s and early 1960s, space had an irresistible cachet. Putting technology into it—or, much more commonly, applying aerospace motifs in the design of everyday consumer goods— conferred not only prestige but also a literal sense of being on the frontier of science and technology.[77] Of NASA's contractors, General Electric had one of the keenest understandings of the power of space symbolism, playing up its membrane fuel cell as a product of the Space Race despite the fact that the company first demonstrated it in the form of a terrestrial 200-watt hydrogen-fueled portable generator for the Army and Navy.

The idea of the space spin-off was criticized almost as soon as it appeared in public discourse. Raymond A. Bauer, chair of a panel organized by the American Academy of Arts and Sciences in April 1962 with help from NASA to study the social effects of the space program, observed that people tended to develop a belief in the potential applicability of space technology in terrestrial roles mainly through analogy rather than demonstrable evidence of similarity.[78] This well described the Department of Defense's interpretation of the aerospace fuel cell program. The space agency did not itself directly advance terrestrial fuel cell technology, but the space program's material and ideological legacies were another matter. Through its patronage of Pratt & Whitney, NASA would make a major contribution to the basis of a commercial fuel cell industry, one that found opportunities for growth in the 1970s.

4

Dawn of the Commercial Fuel Cell

The natural gas fuel cell would allow every straight gas utility company to provide a competitive service with every straight electric utility company in the nation.

–Robert H. Willis, president of the Connecticut
Natural Gas Corporation, 1971

The path to commercial terrestrial fuel cell power was far longer and more convoluted than even the more clear-eyed of its early champions might have guessed. Only in the early 1990s did a model—the PC-25, a phosphoric acid fuel cell (PAFC)—appear on the market. Its provenance can be traced back to the height the Cold War in the early 1960s, a period when Pratt & Whitney first entered the fuel cell field, ostensibly to develop a commercial terrestrial variant. Instead, the company became wholly absorbed in the aerospace fuel cell project, one governed by the technopolitical imperative of performance at any price, where patient engineering research was sacrificed for hasty empiricism made possible by generous infusions of federal cash. With the moon project nearing fruition, the company switched focus to the PAFC, a technology knowledgeable observers expected to be the first terrestrial fuel cell to market. But Pratt & Whitney's fuel cell division had enormous difficulty adapting to the realities of the post–Space Race era, not least because the chief desideratum for the phosphoric acid system was cost-effectiveness. Notoriously difficult to achieve in fuel cell systems, it was an especially tall order in a period when the federal government was curtailing its involvement in the field. As a result, Pratt & Whitney and its parent United Aircraft Corporation (UAC; renamed the United Technologies Corporation in 1975) regarded its own terrestrial hydrocarbon fuel cell program with an odd ambivalence during its formative years in the late 1960s and early 1970s. Why, then, did the company persist with this long, costly technological gestation?

The answer lies in the complex relationship between Pratt & Whitney/UAC and a succession of federal fuel cell sponsors. Utterly dependent on government patronage, as truncated and inconsistent as this was over the years, the company's fuel cell division sought to parlay its considerable experience in the field into commercial domination. But there was only a small market for PAFC technology in the late 1960s and early 1970s, and the federal government was not among those groups interested in it. Consequently, Pratt & Whitney/UAC regarded the government as a dangerously capricious force that had distorted the free play of the market, giving rise to a product nobody wanted. There was an element of truth to this, one that raised the question of precisely what role the space program had played in advancing fuel cell technology.

But the company was too smitten with the dream of electrochemical energy conversion to abandon it. To be sure, several factors worked in its favor. In a period convulsed by great change—economic and political turmoil, shifting cultural values, and evolving federal energy and industrial priorities—Pratt & Whitney/UAC was able to invest its PAFC with political multivalence. Like all fuel cells, this power source could be scaled according to demand and produced no noxious emissions, unlike traditional thermal plants, or radioactive waste, unlike nuclear plants. Owing to its resistance to carbon dioxide, the device was also capable of using cheap carbonaceous fuels cleanly and efficiently, at least in principle. By appealing to certain or all of these qualities as circumstances warranted, Pratt & Whitney/UAC was able to extract investment from a series of patrons, recalling the experience of the Bacon cell program.

Such interpretive flexibility helped make Pratt & Whitney/UAC's PAFC program survivable. As the federal government withdrew from the fuel cell field in the late 1960s, the company took advantage of challenges facing utilities in uncertain times. It first enlisted natural gas companies, which hoped to use fuel cell power to compensate for the vagaries of the natural gas market by allowing them to diversify into the electricity sector. For their part, electric utilities were initially attracted by the possibility of using the phosphoric acid fuel cell as a means of incrementally meeting demand at a time of rising energy costs. But it was the move by gas utilities that convinced electric companies to join in what became something of a race in electrochemical energy conversion technology. Throughout this period, PAFC partisans of all stripes emphasized the fact that the technology was virtually emissions-free. Indeed, the confluence of the environmental and energy crises eventually compelled the federal government to revisit terrestrial fuel cell power. Under strong popular pressure, some political and manufacturing elites began considering conservation and alternative energy and power sources while remaining committed to maintaining the petroleum and fossil fuel paradigm.[1] For some, the phosphoric acid fuel cell was the ideal tool to reconcile these goals.

In this way, fuel cell technology found a niche in U.S. postwar energy policy. In the decentralized and adversarial world of energy technopolitics, where politicians confronted demands for cheap but environmentally sustainable power, Pratt & Whitney/UAC promoted the terrestrial fuel cell as a machine whose time had come. It could, executives claimed, efficiently and cleanly consume fossil fuels without compromising performance and, hence, be seamlessly integrated into the existing fossil fuel system. To federal policymakers, supporting this program was a wise hedge. Although few believed it would yield a true power panacea in the near term, such work was politically defensible in the new Department of Energy, created by the Carter administration in 1977 as a cabinet-level energy bureau intended to centralize federal energy research and development. As with the NRDC and ARPA, underwriting fuel cell work presented an opportunity for a new agency to help fulfill a mandate of supporting technology for national goals. In a period when industry's will to commercialize terrestrial fuel cell technology wilted under protracted engineering research with no certain payoff, the federal government would eventually intervene to preserve the dream of an electrochemical engine at its lowest ebb.

Space Seed: The Genesis of PAFC Technology

Like its alkaline aerospace cousin, the phosphoric acid fuel cell was born of the permanent national security emergency. It emerged not as a response to demands for more efficient energy conversion technology from the civilian market but was instead a product of the multifarious interests of militarized industry and the decentralized U.S. federal science and engineering establishment during the Cold War. Like General Electric, Pratt & Whitney had entered the fuel cell field with dreams of military and civilian sales. Unlike GE, it did not regard its technologies as suitable for dual use, even on paper. To be sure, the company did develop a small 500-watt experimental fuel cell that was installed at a pumping station owned by the Columbia Gas System in 1962.[2] But during the late 1950s and early 1960s, Pratt & Whitney's most developed fuel cell design was the Bacon cell, and adapting this technology for use in space consumed the energies of the company's fuel cell division for most of the decade. Because NASA's specifications put that device "right outside any commercial use," Pratt & Whitney intended to use the proceeds of sales to the federal government as the springboard to a commercial product.[3]

To ensure that the federal government had no proprietary claim on its technology, Pratt & Whitney tried to insulate the money streams flowing into its private and public projects. By mid-1964, two of the company's three fuel cell departments were devoted to the Apollo project and, hence, were government funded, while a third, known as Advanced Power Systems and responsible for basic research, was supposed to draw only from in-house resources. But the work of this group overlapped the government-backed aerospace and terrestrial military projects and the ostensibly private carbonaceous fuel cell program.[4] So it was impossible to prevent the pooling of knowledge and hardware produced with government and private funds. In March 1968, Pratt & Whitney fuel cell officials admitted as much, confirming to British researchers that their work with alkaline systems had been useful in developing acidic fuel cell technology.[5] In April, Pratt & Whitney president B. A. Schmickrath stated that experience gained in the space program had been "invaluable" in enabling the firm to develop fuel cell designs with commercial potential. Seven years of service for NASA had encouraged Pratt & Whitney to explore technological avenues that it had not considered a decade earlier, promising a "significantly different line of business in the future."[6] The company received the bulk of federal money allocated for aerospace fuel cells, about $100 million of the $170 million NASA had spent by 1971, a figure that in turn comprised more than 70 percent of the total national investment in fuel cell technology to that point. Pratt & Whitney claimed only 4 percent profit, but this was sufficient to pay the Leesona Corporation, which was responsible for conducting most of the company's basic research on hydrocarbon-consuming acid electrolyte fuel cells.[7]

This unparalleled federal patronage helped make Pratt & Whitney preeminent in the field by the late 1960s, enabling it to monopolize the small aerospace fuel cell market and position itself to dominate a potentially much larger commercial market. But the company's fuel cell division had become dependent on the federal lifeline. And with the Space Race incontrovertibly won with moon landings imminent, it faced an existential crisis. The future of human spaceflight was a reusable space shuttle, a project only entering the design stage in the late 1960s and that, in the event, required only a handful of fuel cells.[8] As the aerospace market dwindled, no federal agency stepped in to fill the void. The only potential customer for Pratt & Whitney's phosphoric acid fuel cell was the gas utility industry. For some years, a number of companies had been intrigued by the possibility of using the fuel cell as a means of breaking into the electricity supply market. Reasoning that less energy was lost in gas pipelines than in high-tension power lines over long distances, gas utilities saw such a power plant as a decentralized power generator that could be installed in homes and businesses and, hence, be seamlessly integrated into the pipeline grid. This, they believed, would allow them to compete with the big central stations of the electric utilities. However, many gas utilities had invested not in the phosphoric acid fuel cell but the high-temperature molten carbonate fuel cell, a technology that, in principle, could directly consume any carbonaceous fuel without suffering poisoning. The gas industry had helped pioneer this technology in the United States, supporting work at the Institute of Gas Technology (IGT), its Chicago-based nonprofit research association, since around 1960. But the crippling problems of corrosion and thermal expansion plaguing this class of fuel cell remained unresolved. Six years later, gas utilities reconsidered their plans, adopting a strategy similar to that followed by ARPA and the Army in their fuel cell programs. Remaining committed to the molten carbonate concept as a long-term objective, they now saw Pratt & Whitney's PAFC as an interim system that could soon be deployed. This dynamic would govern the course of commercial terrestrial fuel cell research and development in the 1970s and 1980s.

TARGETing a Market

In 1967, 32 gas utility companies and Pratt & Whitney formed the Team to Advance Research for Gas Energy Transformation (TARGET). A $90 million nine-year program, it aimed to develop several dozen multikilowatt PAFC generators for field testing. Interestingly, this work was not supported by the American Gas Association and involved only individual utilities. Like most fuel cell alliances, TARGET comprised players with diverging expectations of the objectives of research and development, a conceptual divide reflected in the terms of the agreement. The gas utilities understood TARGET as a sort of pilot project demonstrating what they believed were proven technological principles. They provided most of the financing and installed, monitored, and maintained the technology at various on-site locations. As prime contractor, Pratt & Whitney/UAC saw matters somewhat differently. It viewed the arrangement as an opportunity to conduct engineering research, aware from hard experience during the space program that building and testing a practical power plant were only the first stages in a long process of understanding how a stack behaved over time. Indeed, the TARGET fuel cell (designated the PC-11) could best be described as a proof-of-principle demonstrator, not a precommercial prototype. A 12.5-kilowatt stationary generator that used steam-reformed natural gas, the PC-11 had poor durability and very high platinum loadings. The TARGET partners hoped the power source could be produced for $150 per kilowatt in 1967 dollars (about $1,000 per kilowatt in 2011 dollars). At $146,000 per kilowatt (2011 dollars), the unit was far too costly to be commercially viable.[9] For their part, however, gas utilities were under the impression that this figure could be lowered over time. Some believed that Pratt & Whitney would be capable of producing phosphoric acid fuel cells at a rate of 100,000 per year by 1975, claims the company publicly disavowed.[10]

Ironically, Pratt & Whitney bore chief responsibility for stoking such expectations. In research circles, William Podolny, the head of the company's fuel cell division, freely admitted that its partnership with the gas utilities had been a gamble in which the chances of success were slim. Pratt & Whitney still had a research and development relationship with the British fuel cell consortium Energy Conversion Limited dating back to its licensing of the Bacon cell, and in a 1970 visit to its laboratory, Podolny recounted the origins and goals of TARGET. The only way Pratt & Whitney had been able to sustain its experiment in commercial fuel cell power, he intimated to his colleagues, had been to exploit the "peculiar plight" of the gas industry. The desire of the utility presidents to compete in electricity delivery, he said, presented the American manufacturer with an opportunity to assemble funds for exploratory fuel cell development.[11]

Worries about environmental degradation and declining primary energy reserves provided timely additional justifications for a terrestrial fuel cell in the early 1970s. In a period when interest in the technology remained tepid in industry and government, some were intrigued by the device's clean energy qualities. At the World Energy Conference in Washington in September 1971, Environmental Protection Agency administrator William D. Ruckelshaus claimed fuel cells possessed "enormous unrealized potential."[12] Pratt & Whitney/UAC quickly appealed to the emerging green sensibility, describing its fuel cell in 1972 promotional literature as "an efficient, pollution-free source of electricity" that would "conserve our natural resources."[13]

However, the company succeeded in attracting only electric and combination gas/electric utilities. And they were drawn to the phosphoric acid fuel cell not primarily for its environmental qualities but for a combination of reasons stemming from competitive and legal pressures issuing from a broader technopolitical crisis facing the energy and electricity sector. For many years, the electric utility industry was widely regarded as a natural monopoly that could not be exposed to the cutthroat world of laissez-faire competition for two reasons—security and efficiency. As an essential public service, the electricity supply could not be interrupted without catastrophic consequences for the economy and public safety. As the U.S. electrical industry developed a philosophy of mass consumption of cheap power and trended toward increasingly large centralized production and distribution systems, monopoly was further rationalized on the basis of the principle of economy of scale. Nevertheless, Congress never exempted the electric utilities from antitrust legislation, and by the early 1970s, under growing popular pressure, it was increasingly challenging the notion of natural monopoly. This led to a number of investigations and suits by the Justice Department and the Securities and Exchange Commission. Moreover, as *Congressional Quarterly Weekly Report* noted, federal executive agencies were so highly fragmented and congressional lawmaking was so decentralized that representatives often unwittingly altered energy policy and patterns of use, with unintended consequences.[14] In such circumstances, noted law professor James E. Meeks, technological developments such as small solar, nuclear, and fuel cell plants had the potential to alter the optimal structure of the industry in terms of public interest.[15]

These factors helped vault fuel cells into national energy technopolitics for the first time, albeit indirectly. A bill submitted in 1971 to the U.S. Senate Antitrust Subcommittee calling for combination electric/gas utilities to be broken up into "straight" gas or electricity providers on grounds that they were charging ratepayers more than single-service utilities were charging triggered a debate that touched on the role of fuel cell power. Gas utilities were opposed to Senate bill 403, fearing that its definition of an electric company as an entity that generated electricity might preclude them from competing with straight electric utilities since they themselves would not actually generate electricity but instead market an energy conversion device. On the other hand, subcommittee minority leader Roman L. Hruska (R-Nebraska), representing the electric industry, claimed fuel cells threatened to undermine the franchise system that awarded local monopolies to electric utilities, thus upending economies of scale and sparking wasteful competition. One gas utility president remarked that fuel cells would not steal existing market share but simply skim off new customers. At any rate, he added, electric utilities were free to respond by acquiring their own fuel cells.[16]

Indeed, four combined gas/electric utilities joined TARGET in 1971, sparking a new flurry of interest in terrestrial fuel cells that well illustrated the assumptions and contradictions of the culture of research and development that had accreted around the Pratt & Whitney monopoly in the post–Space Race period. In 1972, a group of ten utilities and the Edison Electric Institute paid the company $3 million to study the feasibility of developing a large fuel cell in the multimegawatt class. Nevertheless, Podolny was gloomy. Pratt & Whitney's fuel cell division had been nurtured through its formative years by a rich federal client with an open-ended commitment to research, and this it sorely missed. With renewed support not imminent, Podolny lashed out with characteristic pique. In an interview with *Aviation Week & Space Technology* published in January 1973, he accused NASA of locking the company into a field for which there was no real demand in either the public or private sector. Yet Podolny also denied that the space program had anything to do with the company's terrestrial fuel cell program. Instead, he claimed, the government had given Pratt & Whitney little option but to go the commercial route in an effort to recover investments in personnel and infrastructure made at the height of the Space Race.[17]

Of course, the company had, at worst, broken even as a result of its service to NASA and certainly had the option of ending its sojourn into fuel cell technology. These facts only underscored the utter reliance of its fuel cell division on government support and its vulnerability as federal priorities changed in the 1960s and 1970s. Podolny was trying to have it both ways. He wanted it known that government intervention, through NASA's aerospace fuel cell contracts, had altered the normal operation of the market. Had it not been for that intervention, he claimed, Pratt & Whitney would have abandoned fuel cells altogether. All that Podolny would allow was that the federal government had provided "a meaningful base for exploring commercial activities."[18] In a series of interviews conducted by the Smithsonian Institution in early 1974 as part of a study of the industrial impact of the space program, Podolny took a somewhat different line. His new position was that NASA had simply given Pratt & Whitney the incentive to consider the fuel cell as a commercial possibility, providing little more than encouragement and resources sufficient to train a "few people" in fuel cell engineering. William Lueckel, Pratt & Whitney's program manager for electric utilities, denied even this connection.[19]

The reality was that the space agency played a central role in enabling Pratt & Whitney to develop phosphoric acid fuel cell technology. But NASA did not drive the company down a technological path it might otherwise have avoided. More dependent on military sales after the Second World War than its competitor General Electric, the aerospace-engine maker had entered the field of electro-chemical energy conversion in an attempt to diversify its product line. Pratt & Whitney managers knew that the market for aerospace fuel cells was limited at best and terrestrial fuel cell power was still an infant technoscience. From the outset, they intended to use income from the NASA contracts to develop com-mercial PAFC technology. What made the company so ambivalent about its own terrestrial fuel cell program was that it had been born of federal patrons whose future support was anything but guaranteed. Yet that patronage seemed to threaten private property rights before a cent of profit had been made. No doubt corporate managers grew suspicious when NASA and its political supporters claimed a share of the credit for the terrestrial fuel cell program, fears exacerbated by the fact that the space agency's statutory policy on patents resembled that of the Atomic Energy Commission, in which inventions devel-oped under government contract were public property.[20] In 1971, Ernst M. Cohn criticized Pratt & Whitney for its unwillingness to generate innovations in the aerospace program because they would have less patent protection.[21] Of course, in practice NASA had long since waived this policy in its eagerness to devolve as much space technology as possible for industry's terrestrial use. But Podolny's argument contained at least a grain of truth, for NASA had built up expectations by doing all it could to promote the idea of terrestrial fuel cell power. This, in turn, likely reinforced a belief among Pratt & Whitney managers that the com-pany was well placed to monopolize an emerging field and that it could count on further government aid. But rather than abandon its terrestrial fuel cell program when such support failed to immediately materialize, the power source manufacturer resolved to keep it alive by all possible means.

In 1973, Pratt & Whitney seemed poised to reap the fruits of the competition triggered by TARGET when it convinced the electric utilities to invest in a large fuel cell research project. Funded by the newly formed nonprofit Electric Power Research Institute (EPRI), the counterpart of the IGT, the company developed a proposal for a 26-megawatt on-site plant it claimed could incrementally match load growth cleanly, rapidly, and cheaply and would be suitable both for inter-mediate and peak demand. Nine electric utilities, mostly from the industrial-urban Northeast, signed on. As with TARGET, this arrangement featured unequal terms. The utilities committed $7 million for a demonstration unit and a further $28 million as an initial investment in 56 units that, if delivered, would be worth around $290 million. Pratt & Whitney contributed $14 million. On its face, the agreement appeared to reflect both commercial imperatives and the conventional understanding of linear research and development. In essence, it was another experiment in industrial engineering research. If it was a success and the plants were delivered, the companies would collect royalties on additional sales. If not, they lost their investments. The aerospace giant stood to gain the most by far from the relationship, a fact that did not go unnoticed in the business press.[22]

Publicly, Podolny exuded confidence, suggesting not only that the technology had a multifuel capability but also that it was practically ready for service. He told *Business Week* magazine in January 1974 that the plant was "fast on its feet with respect to fuel supply."[23] With natural gas scarce, held Podolny, the first new units would initially use heavier hydrocarbons developed from coal gasification or even, reported the *Washington Post*, hydrogen produced through hydrolysis; both were capital-intensive processes. The main goal for the time being, Podolny emphasized, was for the utilities to familiarize themselves with the technology.[24]

Of course, such claims were not rooted in extensive engineering research and were expressions of hope more than anything else. The power plant was actually designed to use light hydrocarbons and would prove highly vulnerable to sulfurous fuels, even ones that had been reformed. Podolny seemed aware of the dangers of this game. Sensitive to Francis Bacon's requests for information on the multimegawatt project, he angrily accused him in one October 1973 letter of interpreting previous progress reports too positively and publicly airing claims that returned to haunt Pratt & Whitney. Referring to the project as a "high risk development," Podolny poured cold water on its commercial prospects. He warned Bacon that he, Bacon, could "definitely not say that we are considering marketing fuel cell equipment in the near-term at any power level."[25]

Even as Podolny himself did all he could to frame his company's utility fuel cell program in the most positive light, he regarded the advantageous deal with the electric utilities as a stopgap measure at best. Privately, he was defensive and pessimistic about the future. In November 1973, he confided to a visiting British Energy Conversion Limited researcher that Pratt & Whitney's entry into the field had been an "act of faith," adding that the company was not interested in independent appraisals of its technology program since it was using its own money, a characteristically careless remark. Meanwhile TARGET, Pratt & Whitney's other big fuel cell project, was gulping cash and needed even more, at least $80 million by Podolny's reckoning, funds that were not likely to be supplied by the utilities. Having cut matters too finely on cost and timelines, he admitted, Pratt & Whitney was now counting on government help, asking for $250 million as part of the Nixon administration's Project Independence, the much-touted counterstroke to the energy crisis.[26] Fortunately for the company, the prolongation of this crisis would eventually impel the federal government to reprise its role as the chief patron of fuel cell technology.

Fuel Cell Power Resurgent

As in the 1950s and 1960s, a national security emergency—often traced to the oil embargo of 1973 but ultimately stemming from broader contradictions in the American energy and power system—catalyzed renewed federal interest in advanced technology alternatives including the fuel cell. In a sense, this came as an indirect result of Project Independence, announced in January 1974 as a national crash effort in the style of the Manhattan and Apollo projects to end energy imports by 1980. Assuming that demand would continue to grow at a rate of around 4 percent a year, policymakers called for increasing the production of energy in all forms as well as conserving it and improving conversion efficiency with new technology.[27] In essence, this was a blueprint for modifying the status quo with as little disruption and regulation as possible. In practice, energy policy in the Nixon and Ford administrations consisted of an unhappy blend of voluntary and mandatory conservation measures with existing price controls that for years had fixed the cost of energy below its replacement value and overstimulated demand. The result, noted political scientist Pietro S. Nivola, was an uneven system rife with loopholes, one that offered conservation incentives and subsidies for some but by no means all dwelling and transportation technologies and allowed consumers to spend their energy savings in other sectors. Without a comprehensive regulated efficiency plan, held Nivola, the planned regime was primed for abuse. Not only was it incapable of cutting growth in energy consumption, but it also opened the sluiceway of a "burgeoning pork barrel" in conservation technology development.[28]

Research in terrestrial fuel cell power would eventually flourish in this environment. To be sure, in the mid-1970s the technology had a low profile in a Congress preoccupied with renewable geothermal and solar energy technologies and fuel efficiency standards for new vehicles as part of the 1975 Energy Policy and Conservation Act.[29] But a niche for fuel cell sponsorship gradually opened following the long-term reorganization of the federal energy research and development establishment that began with the breakup of the Atomic Energy Commission in 1974 and the consolidation of resources in the Energy Research and Development Administration (ERDA) and its successor agency, the Department of Energy, a process that spanned three presidential administrations. Initially, however, the prognosis for the field was not good. Playing the role of systems analyst as it attempted to secure a leadership role in energy R&D policymaking during this turbulent period of bureaucratic reform, NASA was sharply critical of Pratt & Whitney's phosphoric acid program. Despite having invested more than $70 million in addition to a similar sum by the gas and electric utilities, noted one NASA manager, the power source maker's effort remained "astonishingly thin." Should trouble arise, the company was ill equipped to deal with it. Estimating the total cost of commercializing utility fuel cells at $500 million over

20 years, he was opposed to the company's request for a government subsidy of nearly $200 million, suggesting instead a $93 million plan for integration and testing, one in which NASA would presumably play a management role.[30]

Government aid on even this limited scale was not yet forthcoming. But very probably NASA's pessimistic report helped set up a justification for state intervention in the struggling fuel cell sector. In 1975, the year of its inauguration, the ERDA committed to terrestrial fuel cell research and development, with the space agency serving in an advisory role. In June, the parties agreed to cooperate in a national program in which NASA was to supply basic and applied technical support and management services for technologies including the solar cell, the gas turbine, and the fuel cell.[31] Support for Pratt & Whitney, however, was not necessarily guaranteed at this point. The space agency remained skeptical of the corporation's efforts, particularly its proposal to verify test data to prove the viability of the multimegawatt demonstrator. Citing a lack of detailed failure and decay analyses, NASA officials suspected the firm was instead trying to wheedle funds for more research and development on unproven principles, paralleling Podolny's privately stated rationale for involvement with the gas utilities.[32] And, some in NASA were hesitant to manage a project they saw as technically weak. At least one internal memo advised against involvement in terrestrial fuel management altogether and recommended pursuing only fuel cell projects of direct relevance for space exploration.[33]

A more fundamental issue was whether the ERDA could reconcile its advanced power sources program with energy conservation, an increasingly popular option. During the agency's first energy conservation conference in December 1975, a broad spectrum of experts including engineers, economists, and sociologists voiced criticism both of short-term technological solutions to the energy crisis and of the commercial potential of fuel cells. Many conferees believed utilities would not accept new hardware such as coal gasification and fuel cells before the 1990–2000 time frame. Yet utility representatives claimed they were encouraged to take this path by government and the banking system. The absence of a federal conservation policy, they held, deterred them from investing their own money in such measures. And they found it easier to secure loans for new technology projects rather than for retrofitting existing facilities to improve efficiency, even though the latter was much cheaper.[34]

Nevertheless, the ERDA pressed forward in 1975 with the National Commercial Fuel Cell program, a research and development effort based on a plan developed by NASA. In some respects this represented an organizational advance over efforts in the 1960s, for it coordinated the fuel cell activities of a number of federal agencies—the Department of Defense, NASA, the ERDA, and subsequently the Department of Energy (DOE)—with those of the private, nonprofit EPRI and the Gas Research Institute (GRI), the former Institute for Gas

Technology. In 1976, these groups formed the National Fuel Cell Coordinating Group.[35] But their basic strategy for developing utility fuel cell power followed the pattern established by the industry players in the 1960s. Ostensibly the plan was devoted primarily to the uncompromising, difficult high-temperature concepts—the so-called second-generation molten carbonate (MCFC) and third-generation solid oxide fuel cells (SOFC). Both had long been lodestones for electrochemical engineers intrigued by the theoretical high efficiency that came with operating at temperatures above 650°C, allowing them to dispense with platinum and directly use unreformed hydrocarbons, particularly coal-derived gases, without poisoning effects. Over the next decade, the ERDA/DOE, NASA, and EPRI would fund MCFC studies and then technology programs at Pratt & Whitney/United Technologies Corporation, General Electric, and Energy Research Corporation, and SOFC work at Westinghouse.[36] As in the 1960s, the advanced technology concepts required major breakthroughs before they could be considered practical, as opposed to merely feasible, a distinction frequently muddied in fuel cell research and development circles. Conversely, the phosphoric acid fuel cell—regarded as "first-generation" technology—was at a considerably more advanced stage of development by the late 1970s, having undergone large-scale field trials and sustained engineering research. As events unfolded, the preponderance of effort would be invested in this interim device.

The federal fuel cell program began to hit its stride during the Carter administration, propelled by an energy policy that was relatively more ambitious and coherent than its predecessors. A mixture of conservative and radical public/private initiatives combining the principles of energy efficiency and energy plenitude, the National Energy Program mixed conservation and research and development with an effort to modify the fossil fuel–based system of internal combustion energy conversion. Although solar energy was given new prominence—Carter famously covered the White House roof with photovoltaic cells—the program was devoted primarily to the massive expansion of coal energy, the centerpiece of which was an $88 billion synthetic fuel program.[37] The plan also attempted to induce industry to produce more efficient heat engines while gradually decontrolling natural gas and petroleum prices. One of the most important new energy policy tools was the Public Utility Regulatory Policies Act (PURPA), a sweeping reform of the electric utility industry. Passed by Congress in 1978, this was designed to foster competition and energy diversity by requiring utilities to purchase power from independent producers if they could produce it more cheaply than the utilities. Encouraging the use of conservation technology and renewable energy resources, PURPA had the potential to decentralize energy production. The short-term technological consequence of this law was the development of cogeneration, the recovery of waste heat from thermal power plants for use in space or water heating.[38]

In this policy environment, federal planners began to perceive Pratt & Whitney's large utility fuel cell program more favorably. The company's fuel cell section, which became the Power Systems Division of the United Technologies Corporation (UTC) around this time, tested its one-megawatt PAFC PC-19, the product of its partnership with the electric utilities, in summer 1977. This first demonstration of a megawatt-class fuel cell convinced the EPRI and the newly formed Department of Energy to support UTC's work in this field. Responsibility devolved to the agency's Office of Fossil Energy, which subsequently directed the federal effort in large stationary fuel cell systems. Additional administrative support was provided by NASA's Lewis Research Center. The increasing cost of natural gas following price decontrol in 1978 further clarified the economic rationale for this technology. United Technologies Corporation followed up with a 4.5-megawatt power plant intended for regular operation at a Consolidated Edison site in Manhattan by 1979. The federal government's share in this $70 million collaboration was 48 percent, the largest single stake. Delayed by regulatory disputes prolonged by the New York Fire Department's unfamiliarity with the fuel storage and processing systems and plagued by a variety of technical problems, the plant was stillborn when it was finally assembled in spring 1984 owing to the short shelf life of its stacks. In storage for nearly seven years, they had been irretrievably damaged when liquid acid electrolyte slowly migrated from its matrices into the porous graphite bipolar plates.[39]

The effort was not in vain, held fuel cell boosters. They noted the New York experiment had set a precedent in that the unit had at least received an operating license and was the only large power plant permitted for urban operation up to that time.[40] An improved version of this same design was developed by UTC and the Tokyo Electric Power Company (TEPCO) and tested between 1981 and 1985, accumulating 2,400 hours of operation on reformed methane. That experiment inspired UTC to produce a design for an 11-megawatt variant in 1983 and team up in 1985 with the Japanese engineering firm Toshiba in a joint venture known as International Fuel Cells. But although TEPCO completed a demonstrator based on this technology in 1990, as fuel cell researcher A. J. Appleby has noted, American utilities declined to order it on the grounds that it lacked flexibility and, at over $7,000 per kilowatt (2011 dollars), was at least $5,500 per kilowatt too expensive.[41]

The U.S. Army had largely abandoned the fuel cell field in the early 1970s, but it monitored ongoing developments in phosphoric acid technology. In 1978 and 1980, it contracted for a batch of 1.5-kilowatt PAFC stationary generators from UTC and Energy Research Corporation. Experiments revealed that the type could not cope with diesel and JP-4 and proved no more able to resolve the historical trade-off between ease of operation and usefulness for logistics than an earlier generation of fuel cells. Once again, the Army concluded that methanol was the only feasible fuel for fuel cells. History seemed to have come full circle.[42]

In the wake of the demise of UTC's multimegawatt demonstrator in New York, the Department of Energy began to play a larger role in utility fuel cell research and development. It bolstered efforts to develop a megawatt-class molten carbonate fuel cell, letting contacts to General Electric and UTC in 1980, and also backed a Westinghouse 7.5-megawatt PAFC system. Most importantly, however, the DOE helped revivify UTC's on-site multikilowatt PAFC, a technology more consonant with the trend toward distributed power. In the late 1970s and early 1980s, the bulk of federal resources for fuel cell research went to this program, making it the most important terrestrial fuel cell effort to date.[43] To be sure, TARGET had failed to yield a cheap durable commercial power plant for the gas utilities, but building and testing 65 PC-II units enabled UTC to gain valuable experience.[44] Just as importantly, that program had kept distributed terrestrial fuel cell power, and the company as its chief interpreter, in the public view. In 1977, the federal government began supporting the field test program of UTC's 40-kilowatt PC-18, the successor of the PC-II. Emulating the management pattern set with the company's megawatt-class fuel cell, this effort was sponsored by the DOE and the GRI and managed by NASA's Lewis Research Center. Unlike its predecessor, the PC-18 was designed to recover waste heat for water and space heating, boosting its rated efficiency from 40 percent in electricity-only mode to 80 percent in cogeneration mode, at least on paper.[45] But potential customers had to account for the costs of retrofitting industrial or residential buildings to accommodate cogeneration if they wanted to take full advantage of the feature.

Government support continued. In 1981, the two federal agencies and the GRI supported UTC in a pilot manufacturing project. A $34 million, 42-month effort, it was designed to evaluate the on-site fuel cell concept and perfect techniques of serial production, distribution, and field support in preparation for introduction into general commercial service in the gas industry.[46] Between 1984 and 1986, 53 PC-18s were manufactured and field-tested by gas utilities and the U.S. military. Believing the 40-kilowatt model too small to be commercially viable, UTC then began work on a 200-kilowatt successor known as the PC-25, a project that received further funding and diagnostic and testing support from the DOE and NASA.[47]

In the mid-1980s, after almost twenty years of effort, UTC was on the verge of commercializing phosphoric acid fuel cells. When the company failed to attract interest in its 11-megawatt design, it opted to produce the PC-25 through International Fuel Cells (the joint venture with Toshiba) in the late 1980s.[48] Introducing the unit for on-site trials in 1988, UTC had sold 250 copies by 2004.[49] After more than a decade of support, the federal government ended direct involvement in the company's phosphoric acid utility fuel cell program in 1992. Together, the ERDA and the DOE had spent at least $334 million, and the utilities and UTC contributed tens of millions more, in addition to NASA's indirect support dating back to the mid-1960s.[50] It had been a long and costly path. After decades of dashed expectations, the day of the commercial fuel cell seemed finally to have arrived.

The Cultural Durability of Fuel Cell Power

The phosphoric acid utility fuel cell was a shining example of the dynamics and paradoxes of the military-industrial mode of innovation as it related to advanced power sources. A by-product of the Space Race, the technology originated in the unique engineering culture that grew out of the national security emergency mission, one that encouraged unrealistic expectations of the space spin-off and the technical and economic challenges of bringing new products to the civilian marketplace. United Technologies Corporation was able to continue its work on the technology through the 1970s, 1980s, and into the 1990s partly because it was able to sell it as an uncompromising super-engine to a few interest groups. As during the Space Race, however, the company's fuel cell program was sustained primarily by the state at a time of national emergency, this time in the form of the energy crisis. Only the federal government proved willing to support the extensive engineering research necessary to realize the PAFC. Yet again, a fledgling government science and technology agency saw value in fuel cell power. Once the oil embargo provided the rationale for federal intervention, the newly formed DOE helped complete what NASA had helped set into motion, supporting a long-term program that aligned as comfortably with the agenda of the Reagan White House as with that of the Carter administration.[51] In a time of spiraling energy costs and pressure for conservation, federal program managers and utility executives perceived fuel cells as a potentially cheap, efficient, and flexible way to produce electricity from fossil fuels in a way that satisfied environmental and quality of life concerns. These assumptions served to give the phosphoric acid fuel cell a political and cultural durability that stood in marked contrast to its actual capabilities.

The results recalled the experiences of fuel cell R&D in the 1950s and 1960s. Researchers ultimately wanted to develop technologies that could electrochemically convert heavy, dirty fuels, but this was as difficult a task as ever. Given the available options, they believed the natural gas–using phosphoric acid fuel cell represented the path of least engineering resistance and could provide useful service until devices capable of using coal-based gases were ready. They were disappointed. Not only did the PAFC require Herculean effort over decades to bring to market; its economic rationale was opaque. The electrochemical conversion of natural gas was hardly the solution for energy independence in the late 1980s even if the technology could be made to work reliably for long periods of time, something that was rarely the case with the first precommercial units.

Nevertheless, the original TARGET project did have unintended ideological consequences in the early 1970s far beyond its limited impact on power delivery at that time. Keeping warm the embers of the idea of terrestrial fuel cell power in a period when interest had cooled, it also helped inspire a vision of a much grander energy order, one where abundance and sustainability would finally be joined. This was the hydrogen economy, an energy techno-utopia in which fuel cells had an important place. Born from the energy crisis, hydrogen fuel cell futurism would eventually play an important role in energy and pollution politics at the turn of the millennium.

5

Fueling Hydrogen Futurism

The electricity will originate from nuclear sources. Electrolysis of water
to give hydrogen will occur at site . . . the hydrogen produced would be
pumped to consumer sites, and converted into work by non-Carnot-
limited devices (i.e., fuel cells).

–J. O'M. Bockris and A. J. Appleby, July 1972

Ever since Grove and Schönbein first succeeded in electro-oxidizing hydrogen
in the mid-nineteenth century, researchers associated hydrogen and fuel cells.
In large measure, the history of fuel cell research and development has been
characterized by a central lexical muddle: the term "fuel cell" was coined by
Mond and Langer to denote a hypothetical device that produced electricity by
electro-oxidizing the hydrogen bound up in carbonaceous fuels. Yet the first
practical devices to be so-referred used pure hydrogen and were more properly
termed hydrogen fuel cells. Over the years, this crucial distinction was elided
by sponsors, pundits, politicians, the media, and, sometimes, by researchers
themselves, for reasons we have examined.

Builders of hydrogen fuel cells, consequently, set standards for performance and ease of operation that were very difficult for builders of carbonaceous
fuel cells to match. In laboratory settings and in specialized aerospace applications, pure hydrogen was the most attractive fuel but also the most uneco-
nomic. As NASA wound down its human spaceflight project in the late 1960s
and early 1970s, the torch of fuel cell power passed from the space agency to the
United Technologies Corporation, gas and electric utilities, and, eventually,
the federal government, all organizations with an interest in the carbonaceous
variants. Curiously, this commercial project, one designed to preserve the
existing fossil fuel production and distribution system, helped reinvigorate the
association between the fuel cell and hydrogen, in turn inspiring advocacy
of a hypothetical revolutionary electrochemical energy conversion order built
around these technologies.

This was known as the hydrogen economy. The brainchild of a small group
of researchers in the early 1970s, the expression connoted a future clean energy
regime. But it was imprecise, for a sort of hydrogen economy was already

embedded in the vast chemical and petrochemical industrial complex built up during the twentieth century. Hydrogen has long been a key industrial input and by-product. When synthesized with nitrogen in the Haber process, the result is ammonia, the chief ingredient of fertilizer and explosives. The petrochemical industry produced and consumed vast quantities of hydrogen. As a storage medium and energy carrier, however, hydrogen had historically generated virtually no interest among technologists outside a handful of futurist thinkers, owing to high cost and technological complexity. More popular were experiments using hydrogen as a fuel for propulsion. The German engineer Rudolf Erren pioneered the use of the element as an additive to improve the performance of fossil fuels in converted internal combustion engines in the 1920s and 1930s. Similar efforts continued sporadically throughout the century, mainly in the United States and Germany and later in Japan. But only the aerospace sector, especially in the United States, ever developed a practical requirement for hydrogen as a propellant.

Not until the early 1970s, with the advent of the energy crisis and popular dissatisfaction with the rising environmental and economic costs of dependency on fossil fuel, did researchers seriously consider the possibility of using hydrogen not simply as a fuel but as an energy carrier and storage medium in an all-encompassing system. This idealized hydrogen economy should properly be understood as a form of hydrogen futurism, a term encompassing a rather amorphous bundle of practices including visionary hypothesizing, scientific modeling, and experimental engineering. Hydrogen futurism is often said to have literary origins that date back to the late nineteenth century. Many hydrogen futurists trace their inspiration to two adventure novels, Jules Verne's *The Mysterious Island* (1874) and Max Pemberton's *The Iron Pirate* (1893). Verne's protagonist Cyrus Harding discusses the possibility of using electricity to produce hydrogen from water, while Pemberton tells the tale of a rogue superbattleship using hydrogen-powered engines to stalk and overcome its prey.[1] Hydrogen boosters paid homage to these two works of fiction so frequently that their influential role became a staple factoid in periodic media summaries of the hydrogen movement over the years.

Hydrogen futurism can be defined as high technoscientific positivism imbued with quasi-mystical undertones, one embracing many scientific and engineering fields. As the most abundant element in the universe, comprising 75 percent of chemical elemental mass, hydrogen has a cosmic holism that many futurists found irresistible. As one noted futurist observed, the substance could serve as the basis of a "totally benign energy metabolism."[2] The simplest element has long captivated the imagination of futurists who, like the fictional Cyrus Harding, saw in Earth's vast water supplies an inexhaustible form of energy waiting to be exploited. Drawing from the technocratic tradition, they believed that widespread use of hydrogen could create a perfect balance between energy supply and demand, leading to sustainable plenitude. Hydrogen's champions frequently framed the substance as a viable alternative *fuel*, a view that in turn helped inform the belief that it was a primary energy resource.

This, of course, is erroneous. The tenth most abundant element on Earth, terrestrial hydrogen is bound up in water and organic matter and requires considerable energy to be liberated. In its pure form, it has the characteristics of a chemical fuel, albeit a quite unique one. The idea of hydrogen as clean energy—as distinct from its actual utility as fuel, catalyst, storage medium, and energy carrier—owes much to the ways futurists conceived of and promoted the visionary hydrogen economy, ideas heavily influenced by fuel cell researchers and developers. The chief stumbling block to the hydrogen techno-utopia was an economic means of producing pure hydrogen on a massive scale. Many experts regarded existing methods as unsatisfactory, seeing electrolysis as too costly and catalytic steam reforming, partial oxidation, and solid gasification processes as inflexible and dependent on increasingly scarce fossil fuels. The ultimate solution, according to some authorities, would involve renewable and nuclear hydrogen.

Relatively little research was done in these areas, however. Most experiments with hydrogen fuel tended to focus downstream in the hypothetical hydrogen energy conversion spectrum, concentrating on the point of use. Many hydrogen fuel demonstrations used internal combustion engines, but over the years enthusiasts came to favor electrochemistry as the most efficient means of converting hydrogen. In the 1970s and 1980s, visionary theorists helped rehabilitate the hydrogen fuel cell, a technology widely seen in the 1960s as a bridge at best and a dead end at worst on the road to the carbonaceous fuel cell, the modern philosopher's stone. In turn, improvements in certain fuel cell technologies in the 1980s and 1990s, and the subsequent research and development bubble, helped legitimize hydrogen futurism, eventually propelling it into the mainstream of the public imagination. In this manner, the fortunes of hydrogen futurism and fuel cell research and development would become inextricably entwined.

Techno-cornucopia

Like fuel cell technology, the idealized hydrogen economy was a marvelously flexible concept. A number of groups agreed on hydrogen's utility but disagreed as to how the substance was best used. As with fuel cells, this divergence in views was informed by economic and professional interests rooted in existing technologies and institutions, governing the ways people believed hydrogen should be employed in existing and hypothetical energy transformation chains. What united hydrogen futurists in industry and government was the certitude that this ubiquitous element would eventually prevail in the energy conversion chain because fossil energy would inevitably be exhausted in the deep future. The only way society could make the remaining primary energy resources transferable and interchangeable in a globalized economy would be to reduce them to hydrogen, their common denominator, or so the argument went.

This belief in inevitability often lent a messianic aspect to the appeals of certain hydrogen futurists. The original adherents, a loose group of scientists, engineers, and bureaucrats, tended to align along particular technological applications of hydrogen, whether as a fuel, energy carrier, or an industrial chemical. Over time, some shifted their emphasis, while others envisioned the element's use in all three roles in a total hydrogen economy. Much as fuel cell researchers elided the distinctions between the various types of fuel cells and fuel systems in their public pronouncements, hydrogen futurists often conflated the real and ideal senses of the expression "hydrogen economy." Partly the result of an insensitivity to the nuances of language, this elision also stemmed from the efforts of some to legitimize their views by associating them with individuals and groups in the chemical, petrochemical, metallurgical, and aerospace industries with existing practical interests in hydrogen. As much as with fuel cells, NASA's long experience with hydrogen fuel supplied a precedent and an analogy that encouraged those interested in hydrogen as an energy carrier and storage medium. Representatives from these groups built alliances, enlisting the support of environmentalists and an array of actors in government, academe, and the energy, power, and transportation industries.

In public discourse, the idealized hydrogen economy consisted of an amalgam of ideas generated from attempts to resolve challenges facing the gas utility and nuclear power industries. These concepts were also strongly influenced by the academic disciplines of energy ecology studies and environmental engineering emerging in the early 1970s. It may be no coincidence that many key theorists of the hydrogen economy were deeply involved in electrochemistry and fuel cell technology in the 1960s. Their preoccupation with the interchangeability of chemicals and electricity helped give rise to the notion of hydrogen as an energy vector or carrier, like electricity, only much more flexible.

The German physicist and fuel cell engineer Eduard Justi was one of the first to develop a formal blueprint of the hydrogen economy. In 1965, he outlined a plan to use solar power in the Mediterranean to produce hydrogen and oxygen, which would then be piped to Germany. Some of the chief standardbearers of the hydrogen economy in the late 1960s and early 1970s were Cesare Marchetti, John O'M. Bockris, Derek P. Gregory, and T. Nejat Veziroglu. Marchetti, an Italian physicist employed by the European Atomic Energy Community, was fascinated by the problem and potential of waste heat from nuclear reactors. He believed it could be stored in the form of hydrogen and used in a wide range of chemical processes for industry and agriculture. Bockris, a South African–born scientist, had one of the highest public profiles of the hydrogen futurists. A noted electrochemist with an interest in fuel cells and energy conversion technology systems, he corresponded with Francis Bacon. A prolific author, Bockris was an esteemed member of the small American electrochemical community in the mid-1960s, performing basic research on hydrocarbon oxidation in support of Pratt & Whitney's commercial fuel cell program as a member of the University of Pennsylvania. Toward the end of the decade, he became a leading promoter of renewable power and energy systems and is sometimes credited with coining the term "hydrogen economy."[3] His views on the subject were less narrow—or focused—than those of other hydrogen futurists. In the early 1970s, Bockris was less interested in short-term primary energy regimes, being concerned mainly with converting different forms of energy by means of battery, fuel cell, and electrolyzer technology. But he assumed that the future of primary energy would be nuclear and, subsequently, solar. In his *Electrochemistry of Cleaner Environments*, published in 1972 as one of the first collections of studies on alternative, environmentally friendly energy technologies, Bockris dwelt on the inevitability of electrochemical energy conversion in a world of dwindling fossil fuels. The storage systems society ultimately chose for automobile transport, he wrote, would be determined by how it managed the transition to new primary energy sources. Fuel cells would play a major role if methanol became an important fuel. Batteries, he mused, made more sense in the deep future, when the nuclear reactor was the dominant terrestrial source of primary energy.[4] Two decades later, this influential and ambitious individual would become a polarizing force in the scientific world, accused by some of engaging in reckless promotion, pseudoscience, and even outright fraud.

Derek P. Gregory had a much more singular vision of the hydrogen economy than Bockris. Like his South African colleague, the British scientist was a familiar figure on the electrochemical scene in the mid-1960s, having dealt with Pratt & Whitney in his capacity as a member of Energy Conversion Limited, the British fuel cell consortium. Gregory subsequently found employment with Chicago's Institute of Gas Technology (IGT), the nonprofit research association of the American gas utility industry. By the end of the decade, the IGT had become a leading proponent of a hydrogen economy, displaying a domestic hydrogen kitchen complete with an advanced hydrogen-fueled cooking range in 1968.

But the IGT had a broader goal stemming from problems in the U.S. natural gas sector generated by a political economy that mixed populist and free-enterprise principles. As the gas utility system expanded massively after the Second World War, the Eisenhower administration worried that consumers were beholden to a handful of suppliers. Concerned with sudden price increases, the Federal Power Commission instituted price controls on interstate shipments of gas in the mid-1950s.[5] The results were paradoxical: consumption soared but prices remained fixed, inhibiting exploration and compelling companies to withhold new intrastate reserves from the interstate market. By the late 1960s, a critical gas shortage had developed.[6]

Speculating that the gigantic pipeline grid built up over three decades might become useless if the natural gas economy collapsed, some researchers saw hydrogen as a means of preserving this vast investment. The IGT drew inspiration from TARGET, a project many gas utilities hoped would allow them to seize market share from the electric utilities. Pipelines, the gas utilities held, were a much more efficient means of moving energy over long distances than high-voltage power lines. In TARGET, the fuel cell functioned as a kind of distributed primary energy converter, allowing home owners and businesses to produce their own electricity by converting methane drawn from the gas grid. In essence, the fuel cell inspired gas utility executives to view the gas grid as a power transmission technology akin to the electricity grid and to analogize natural gas to electricity as an energy carrier. In the IGT's hydrogen economy, hydrogen would replace natural gas as the energy carrier in a fuel cell–studded pipeline system.

Like Bockris, Gregory saw nuclear power as the sole viable source of primary energy in the distant postcarbon future, envisioning its use on a massive scale. In one 1973 study authored for the American Gas Association, Gregory proposed mounting huge reactors generating thousands of megawatts on massive concrete islands. Nuclear-powered reversible fuel cells would electrolyze water during periods of low demand, storing hydrogen in vast underground caverns until needed for peak shaving in times of high demand.[7] The remoteness of these giant plants, in turn, justified the use of hydrogen as the energy carrier, for it made economic sense to convert electricity into hydrogen and then back into electricity, claimed Gregory, only if it was produced hundreds of kilometers from primary markets. In that case, it would be more efficient to convert electricity into hydrogen and pipe or ship it as a compressed gas or liquid than to transmit it in high-tension power lines. For Bockris, such a system had the added benefit of isolating nuclear plants, and the threat of an accident, far from populated areas.[8] Once hydrogen arrived at the point of use, it would become a universal chemical fuel and additive. It would be burned directly in aircraft turbines, domestic cookers, and space heaters and converted to electricity by fuel cells for use in automobiles and buildings, serving as the basis of the chemical industry in a post–fossil fuel world. Gregory and his collaborators assumed that all the requisite technologies to realize this vision were essentially at hand. The existing pipeline grid, they believed, should easily be able to handle the transition from natural gas to hydrogen, and the electricity-hydrogen-electricity conversion cycle could be accomplished efficiently by the hydrogen fuel cell, a device that, although complex and in need of further development, had already been successfully used in spacecraft.[9]

The socioeconomic, cultural, and technological implications of such a system were staggering, so much so that hydrogen futurists often did not fully grasp them. Although many considered the basic technologies already proved, individuals such as Gregory, Marchetti, and Bockris who envisioned hydrogen as an energy carrier were, in effect, proposing a completely new energy conversion culture, one that implied material and social engineering on a vast scale. Their brand of hydrogen futurism was a revolutionary technoscientific and industrial creed, fusing unbridled technophilia with utopian socialism. In 1972, Bockris and A. J. Appleby suggested that entrenched energy interests within the capitalist system produced a sort of political inertia that stultified the inherent creativity of Americans and the innovation necessary to bring about this new world. In contrast, they noted, the socialist bloc countries had the potential to transcend these limitations because they owed no fealty to capital investments in existing technological systems. Unfortunately, wrote Bockris and Appleby, the socialist countries simply adopted the technological systems of the capitalist ones, so as to avoid having to pay initial research costs.[10] Marchetti espoused a sort of secular millennialism, referring to the hydrogen economy as the endpoint in an evolutionary progression toward world government.[11]

Other hydrogen futurists offered a less radical message. Lawrence W. Jones, a physics professor at the University of Michigan, made an explicitly populist appeal for hydrogen. In an 1972 article published in the *Saturday Evening Post*, he outlined his vision of hydrogen fuel as a clean, near-term solution to the problem of dwindling fossil fuel reserves that would entail minimal sociocultural and economic disruption. Like most other hydrogen futurists, Jones believed that nuclear power would one day be the sole practical form of primary energy and the ultimate source of most hydrogen. Unlike them, he acknowledged the complications of replacing the existing energy conversion/automobile transport system. But Jones thought hydrogen could be introduced into current fuel distribution systems and run in internal combustion engines without requiring any technological breakthroughs. Like the supporters of the carbonaceous fuel cell projects of the 1960s and the 1990s, Jones promised a quiet revolution that would achieve energy sustainability and independence while maintaining existing standards of living. Interestingly, he clothed this scheme in explicitly antitechnological rhetoric, noting that his plans might seem "disappointingly simple and anticlimactic" to those with a taste for exotic gadgetry.[12] Yet his system, too, relied on nuclear power, hardly an unsophisticated technology.

One commonality among hydrogen futurists divided by professional and conceptual differences was an affinity for NASA as a pioneer and champion of hydrogen and hydrogen technology. Like the fuel cell boosters of the 1960s, hydrogen futurists frequently pointed to the agency's practical experience with hydrogen to bolster their own arguments. In turn, NASA attempted to exploit this experience to carve out a broader role for itself in federal energy research and development in the post-Apollo era, bolstering the credibility of the hydrogen futurists. Originally attracted to hydrogen as a fuel for propulsion and fuel cell–derived electricity, NASA played a role in developing a series of hydrogen-powered rockets and also helped conduct studies of hydrogen fuel for civilian aircraft. Of course, the space agency's most famous such project was the giant Saturn V, which used liquid hydrogen in its upper stages. But the much smaller Convair-built Centaur upper-stage booster was also important, for this "workhorse" satellite launch vehicle allowed the agency to claim that it had helped routinize the use of hydrogen as a transportation fuel for civilian use.[13] Actually, this program was largely the legacy of U.S. Air Force experiments in the late 1950s. Liquid hydrogen for the Centaur came from hydrocarbons converted through partial oxidation at the "Papa Bear" plant at West Palm Beach, Florida, a facility originally built by Air Products to supply Lockheed's CL400 supersonic reconnaissance jet. Pratt & Whitney's work in developing the 304 engine for that project helped the company win the Air Force contract for a liquid-hydrogen rocket motor in 1958. That venture, in turn, yielded the RL-10, the Centaur power plant, the first such technology to be built in the United States.[14]

Following the Apollo program, former and active NASA workers and leaders closely monitored the growing interest in civilian applications of hydrogen. Claiming unparalleled special expertise in the field, some pursued studies of terrestrial applications of hydrogen. One such figure was William J. D. Escher, a former NASA engineer who had worked on hydrogen rocketry at the Lewis Research Center and was a major figure in the hydrogen futurist community in the late 1960s and early 1970s. Escher's efforts outside NASA paralleled the agency's desire to put its hydrogen experience to work on earthly problems as it attempted to reinvent itself in the post-Apollo era. In 1973, it sponsored hydrogen workshops, one held in spring at the Langley Research Center dedicated to hydrogen propulsion in aviation and a summer study session at the Johnson Space Center devoted to the social implications of hydrogen. Administrator James C. Fletcher had even broader ambitions. He saw hydrogen as a possible means of securing a new role for NASA in a federal energy research and development establishment then being reorganized by the Nixon administration. In response to the Atomic Energy Commission's survey of resources that federal agencies might contribute to what became Project Independence, NASA submitted a proposal citing areas it felt competent to manage including hydrogen fuel and hydrogen fuel cells.[15] Fletcher followed up this submission with a list of appended proposals to the director of the Office of Management and Budget that included a hydrogen-based energy system. He understood hydrogen as a universal fuel attractive for its pollution-free combustion that could be produced from coal and water electrolysis and distributed via a national network of pipelines.[16]

The proposal did not elicit much favor from the National Academy of Engineering. The NAE believed NASA lacked experience and accomplishments in fields requiring an understanding of the civilian marketplace and had devoted little effort to understanding precisely how aerospace technologies would be transferred for commercial use. Vague on cost, timelines, and the role of industry, the visionary scheme, chided the academy, was a departure for NASA, a mission agency designed to execute high-priority, short-term programs.[17] Some fuel cell hands within NASA were similarly unimpressed. Ernst Cohn dismissed the very idea of a hydrogen economy. Although hydrogen remained the only practical fuel cell fuel at that time, Cohn believed it was losing its appeal for pundits amid a new reappraisal of methanol, a substance he saw as much more economically and technologically feasible.[18]

Building a Constituency: Hydrogen's Social Multivalence

Hydrogen futurism blossomed in the 1970s even though the practical technological and economic requirements of the hydrogen economy remained far from resolved. Partly this was because it found fertile terrain in a sociocultural environment receptive to radical solutions to the energy and environmental crises. Equally important was the fact that, like fuel cells, hydrogen had an extremely diverse, although diffuse, array of champions across disciplinary, regional, and national boundaries. A broad swath of the scientific and engineering establishment was in some way connected with hydrogen for the simple fact that the element was practically unavoidable in organic and inorganic chemistry. Whether or not one agreed with the futurist vision, it was impossible to deny hydrogen's importance. As the basic unit of energy, hydrogen was, in a sense, a technopolitical universal. From an early date, hydrogen futurists recognized hydrogen's social multivalence and did their best to exploit it for their own purposes. This enabled them to construct an international network of specialists from almost every field of energy, chemical, and materials study, many of whom had quite narrow interests in hydrogen. Despite the best efforts of futurists to cast the movement they helped raise in the mold of the ideal hydrogen economy, these individuals were not guided by a unifying principle beyond perhaps a shared commitment to research. Given the vast technological complexity of the more all-encompassing versions of the idealized hydrogen economy, hydrogen technology advocates in the 1970s and 1980s concentrated on relatively simple devices, much like the architects of the commercial fuel cell programs of the 1950s and 1960s.

The early years of hydrogen futurism were marked by burgeoning recruitment to the cause and the emergence of hydrogen tropes into public discourse. A spate of articles dealing with hydrogen and the hydrogen economy appeared in popular and technoscientific media around late 1971 and continued over the subsequent two years.[19] In 1973, more than 134 participants attended a hydrogen symposium at Cornell University supported by the National Science Foundation (NSF). The first major international hydrogen conference was held the next year at the University of Miami's School of Engineering and Environmental Design, attracting more than 700 attendees. Designated THEME (The Hydrogen Economy Miami Energy), it received support from the NSF and DARPA (the renamed ARPA), an indication of the "basic" (or speculative) nature of the research on display. One measure of the stature hydrogen futurism had attained in energy science and technology circles was the presence of Edward Teller as banquet keynote speaker. This was a coup for conference organizers, who no doubt calculated that having the inventor of the hydrogen bomb as their marquee attraction would bolster their cause. But Teller's somewhat rambling address must have been a disappointment for the most committed futurists, for in it he evinced little enthusiasm for hydrogen as an energy carrier or even as a fuel, except in aviation. Nevertheless, Teller wholeheartedly endorsed all experimental work in the field.[20]

Such ambivalence underscored the vast divergence in the ways people understood hydrogen in the existing and future energy transformation chains. Only a relatively small group, especially Escher, Gregory, Marchetti, and T. N. Veziroglu, chair of the University of Miami's Department of Mechanical and Industrial Engineering, could be described as true believers with an expansive vision of a hydrogen economy. Many more had more tangential interests. As always, NASA's interests in novel terrestrial applications of hydrogen were restricted to a desire to analyze and manage the programs of other groups. Bockris, a leading luminary within the futurist community, hardly spoke of hydrogen on this occasion, focusing instead on solar electricity. Pride of place in the proceedings was reserved for hydrogen production technology, with advocates of synthetic hydrogen fuel production from fossil fuels by conventional means notably absent.[21] And this discussion was dominated by talk of nuclear power. Within this context, researchers interpreted hydrogen in a variety of ways, some relating to problems with managing existing nuclear power technology and others to the larger question of how to supply cheap hydrogen. The University of Michigan's W. Kerr and D. P. Majumdar approached matters primarily from the perspective of safety. Hydrogen was an inevitable, and dangerous, by-product of nuclear fission in conventional boiling water and pressurized water reactors when cooling water underwent radiolysis. Given that this was an unavoidable reaction, they claimed, it made sense to try to collect the hydrogen and put it to good use. Others including Marchetti, German nuclear researchers, and workers at the Los Alamos Scientific Laboratory of the University of California wanted to exploit waste reactor heat, turning nuclear plants into hybrid electricity/chemical plants.[22]

But under normal conditions, cooling water in conventional reactors is heated only to around 290°C in boiling water designs and around 320°C in pressurized water designs, temperatures insufficient to gasify coal into a methane-rich synthetic gas.[23] For many futurists, water splitting was the ultimate means of sustainably producing hydrogen. Electrolysis was then the dominant means of dissociating water into hydrogen and oxygen but was widely regarded as too costly to be practical on the scale required to service a hydrogen economy. One ERDA-supported study conducted in 1976 at Lawrence Livermore laboratory determined that it would take the equivalent of 1,560 one-gigawatt nuclear reactors to produce hydrogen equivalent to the quantity of natural gas consumed in the United States in 1970, costing $1 trillion, roughly equivalent to the U.S. gross national product of the time.[24] Much more desirable in principle was thermal water splitting, but this occurred only at around 2,500°C, a temperature far beyond the capacity of existing nuclear reactors. As a result, many futurists were attracted to chemically assisted thermal water splitting. From the late 1960s into the 1970s, a number of such processes appeared, beginning with Marchetti and De Beni's "Mark I." Criticized as inefficient, impractical, and highly toxic, relying on mercury and hydrobromic acid, the Mark I process was followed by a number of others including hybrid thermoelectric/electrolytic methods developed by Westinghouse and the Ispra Research Center in Italy. Such processes still required very high temperatures and some sort of power source technology that operated at those temperatures.[25]

Contrary to earlier expectations, the ideal hydrogen economy implied massive complexity, and herein lay its key technopolitical weakness. Futurists set this system in the distant posthydrocarbon future as a means of simplifying the conceptualization process. In 1972, Gregory had written that he and his IGT team had developed their analysis as a means of raising questions and stimulating long-term planning rather than a "clear-cut case" of how the transition from the current energy order to the new one would occur.[26] As he moderated the THEME industry discussion with a handful of utility and power source representatives, Gregory found he often had no ready answers when panelists raised this issue. One way to introduce hydrogen in the near future, he suggested, was in the form of a clean synthetic industrial fuel produced from coal. Some utility executives responded that this brought no advantages over methane. One General Electric researcher pointed out that industrial hydrogen did not remedy the immediate troubles of gas utilities, which were interested in hydrogen strictly as an energy carrier. That, and not the hydrogen infrastructure embedded in existing chemical and petrochemical industries, he noted, would constitute the real "hydrogen economy." Toward the end of the debate, a somewhat exasperated Gregory remarked that there were still some industrial processes that required gas energy, such as the manufacture of lightbulbs. This provoked good-natured laughter among panelists but also underscored the difficulty of identifying hydrogen's place in transitional industrial energy scenarios.[27]

Hydrogen advocacy was still too dispersed across a multitude of institutions of science and technology to possess a coherent identity at this point. Nevertheless, hydrogen advocates became much better organized in the wake of this first international conference. In the months that followed, Veziroglu led a group of conferees in establishing the International Association for Hydrogen Energy (IAHE) and its peer-reviewed scholarly publication *International Journal of Hydrogen Energy*. Meeting biennially in the World Hydrogen Energy Conference (WHEC), the IAHE was primarily a home for academics, although it also became a forum for a variety of nonfuturist researchers in industry and government with disparate interests in hydrogen. The WHEC became an excellent bellwether of aspirations and trends within the growing field of hydrogen studies. The conference also marked a watershed for international hydrogen research, elevating its profile in international energy R&D circles.

Meanwhile, futurists gained important new allies in the quest for hydrogen production technology. In 1977, the International Energy Agency (IEA), an autonomous branch of the Organization for Economic Cooperation and Development, initiated the "Hydrogen Agreement," a loosely organized consortium of 12 organizations representing nine countries and the Commission of the European Communities. It was dedicated to developing advanced hydrogen production technologies including water splitting, particularly by thermochemical means.[28] This work in particular was spearheaded by industry and government energy agencies in the United States, Germany, and Japan. Although planners considered the relatively simple solar concentrator as a heat source, the preponderance of opinion was in favor of high-temperature gas-cooled nuclear reactors (HTGR). Like other high-temperature power devices including the solid oxide and molten carbonate fuel cells, the various HTGR concepts brought trade-offs, especially corrosion. A handful were constructed, including the Peach Bottom facility in Pennsylvania and the Fort St. Vrain plant near the town of Platteville, Colorado, the only two such facilities to be built in the United States. Completed in 1972 by General Atomic, the Fort St. Vrain installation was designed to produce commercial power but routinely broke down and was never mated with a water-splitting process. And thermochemical hydrogen production processes were themselves extremely corrosive, particularly those involving sulfuric acid, creating harsh environments that rapidly degraded commercial metal alloys.[29] Following its involvement in this troubled project, General Atomic gradually withdrew from the field of high-temperature reactor design in the late 1970s, citing lack of demand.[30]

An alternative method was steam electrolysis. Based on the principle that much less electricity was required for high-temperature electrolysis than for low-temperature electrolysis, this process involved dissociating extremely hot steam—around 1,000°C—using a solid oxide fuel cell operated in reverse. First developed by General Electric in the late 1960s, this process gained some attention around the mid-1970s, especially among German researchers. But it was no less vulnerable to the corrosion and thermal expansion that had long plagued high-temperature fuel cells.[31]

Such problems could not be easily or cheaply resolved, and this played a crucial role in characterizing the form and content of hydrogen research and development for the remainder of the 1970s and into the 1980s. As with fuel cells, most futurist applications of hydrogen technology held few attractions for industry unless governments were prepared to underwrite the research. The energy and environmental crises did convince many government planners that it would be wise to investigate advanced hydrogen production and end-use technologies as a hedging strategy. But hydrogen had a low priority for planners preoccupied with immediate problems. Moreover, no government had a strategic vision for hydrogen or a clear understanding of precisely how research would be applied.

The situation was particularly unsettled in the United States. There, hydrogen research was supported by the Electric Power Research Institute, the Gas Research Institute, and the Department of Energy, organizations with distinct interests in energy conversion. By far the most important sponsor, the DOE spent between $20 million and $30 million annually in the late 1970s and early 1980s over ten program areas, mainly basic research, and several million more per annum on HTGR technology, decreasing the emphasis on applied work in this period.[32]

Over time, however, futurist hydrogen communities in the United States and abroad were drawn to end-use and existing production technologies, largely because they were so much simpler and cheaper than thermochemical water splitting. A favorite activity involved modifying internal combustion automobiles to run on bottled hydrogen, recalling the technopolitical calculus of the early years of postwar fuel cell research. There were a number of such projects in the United States in the 1970s and 1980s supported by a variety of local and federal government agencies, most involving single vehicles. They were especially popular in the Southwest and on the West Coast. In the early 1970s, a group of residents of the small town of Perris, California, formed the Smogless Automobile Association, receiving some aid from General Motors, Ford, and the commercial hydrogen manufacturer Linde. Perhaps the best-known American hydrogen entrepreneur in this period was the Utah businessman Roger Billings. He generated headlines with a string of hydrogen automobile conversions, even converting his home in Provo and nicknaming it the "hydrogen homestead." Billings received funding from the Energy Research and Development Agency in support of his converted Winnebago hydrogen-powered bus and became an important supplier of equipment for conversion packages, earning him praise from *Mother Earth Magazine*, which referred to hydrogen as "today's reality."[33] Between 1976 and 1977, the California city of Riverside became the epicenter of a number of hydrogen fuel experiments. Its transportation department operated a hydrogen bus developed with the support of the California Transportation Authority and the Riverside-based Pollution Control Research Institute (PCRI), founded by local engineers, scientists, and physicians. A 1978 project involving the University of Denver, the private company Ergenics, and the Clean Fuel Institute of Riverside, the successor of the PCRI, produced a converted Dodge D-50 light pickup truck that served as a test bed for many years.[34]

More sophisticated were the hydrogen programs of the West German automakers, particularly Daimler-Benz. The company began investigating hydrogen fuel drive in 1972, converting a series of production vehicles to hydrogen power and hydride storage—hydrogen-absorbing metals—over the course of the decade as part of a West German federal government program in alternative fuels. It became the first automaker to develop hydrogen demonstration vehicles for public display, showcasing a pair of vehicles at the 1978 World Hydrogen Energy Conference in Zurich.[35] In 1979, the company embarked on a more ambitious multivehicle demonstration program, and BMW followed with a hydrogen-fueled internal combustion engine research effort of its own a few years later. In the 1980s and 1990s, these corporations would become leaders in experimental automobile hydrogen technology. Refining these programs as insurance against possible shifts in pollution legislation and as an elaborate form of environmental public relations, the German automakers, Daimler-Benz above all, would influence their American counterparts in the political uses of advanced sustainable automobile technology.

Such projects could make for dramatic displays of the apparent practicality of hydrogen fuel for transportation purposes. But they also concealed serious problems with the basic premise of a hydrogen fuel economy, as a series of studies suggested throughout the 1970s. Even as hydrogen proponents of all stripes congregated in the heady atmosphere of the first international hydrogen conference of 1974, industry and government conferees presented research showing that hydrogen could not easily be introduced into the natural gas network. Hydrogen pipelines had to be operated at higher pressure than natural gas pipelines in order to compensate for hydrogen's lower volumetric heating value, only around a third of that of natural gas. And they would have to be larger than natural gas pipelines in order to carry the equivalent energy, requiring a completely new gas-metering system.[36] In addition, there were serious chemical challenges in developing hydrogen pipelines. The element embrittles certain types of steel, particularly high-strength alloys containing carbon and manganese, a problem that worsens as gas pressure is increased. Weld seams were particularly susceptible to this phenomenon. One 1977 study conducted by researchers with Rockwell International and Carnegie-Mellon University criticized the tendency of proponents to cite the 210-kilometer low-pressure hydrogen pipeline in the industrialized Ruhr Valley in West Germany as evidence of the feasibility of long-distance hydrogen transmission, pointing out its use of low-strength, unwelded steel provided little guidance in this regard. Such systems would have to be carefully built and would require considerable materials research.[37]

Of course, few categorically rejected hydrogen as a legitimate subject of research. Hydrogen futurism, on the other hand, was another matter. But the technological requirements of the ideal hydrogen economy raised all kinds of interesting problems to solve. Many in the energy research and development establishment welcomed such work, whether it consisted of specific technoscientific problems or the analysis and assessment of existing hydrogen programs. This was a role NASA relished. In 1976, its Jet Propulsion Laboratory completed the Hydrogen Energy Systems Technology study to assess national needs. The most comprehensive analysis to date, it found that nuclear and solar energy were unlikely to be realistic sources of hydrogen until well into the new millennium. In the interim, the JPL recommended a comprehensive program of research in fossil and nuclear hydrogen production as near- and long-term options respectively.[38]

With no solutions imminent on the production front, work in other fields of the futurist hydrogen economy was irrelevant. This was the message of a 1977 study conducted by the Stanford Research Institute.[39] Echoing other critics, it found that only in aviation did hydrogen offer clear advantages. The report's authors urged that the chief research priority should be hydrogen production technology, with less emphasis on hydrogen pipelines. Of least value were projects involving hydrogen automobiles and energy conversion technologies that required rare and costly catalytic materials. Such work, the authors indicated, should be suspended pending progress in methods of hydrogen production. This was a strong indictment not only of the nascent culture of hydrogen demonstration but of certain types of low-temperature fuel cell technology.

Hydrogen Fuel and the Electrochemical Turn, 1980–1990

Such criticism underscored the essence of the hydrogen project. Government and industry were quite willing to make limited investments in exploring certain downstream components of the hydrogen energy conversion chain. But they were not prepared to put up the billions of dollars necessary for developing and commercializing its aggregate systems. In 1979, the IEA abandoned efforts to integrate high-temperature gas reactors with thermochemical processes. One researcher noted ruefully in 1980 that large-scale demonstrations of key hydrogen systems had yet to occur in the United States.[40] Indeed, political and economic conditions in the 1980s were hostile to the environmental movement and supporters of alternative energy technologies. As conservative governments took power in a host of Western countries and moved to undo the progressive social reforms of the 1960s and 1970s, the world found itself awash in cheap oil thanks to a combination of economic recession, the exploitation of non-OPEC reserves, and OPEC overproduction. Early in the new decade, hydrogen researchers began to acknowledge that the combination of sociocultural, political, and economic forces that had nurtured hydrogen futurism in the early 1970s was no longer in play. Lacking a firm techno-industrial constituency even when conditions were propitious—that is, the era of expensive petroleum—hydrogen futurism weakened with the return of cheap petroleum. As researchers realized that advanced water-splitting technologies would not soon become available, the dream of using hydrogen as an energy carrier began to fade. Instead, futurists increasingly perceived hydrogen as a uniquely flexible form of fuel, the rationale being that such an application would provide an opportunity to demonstrate existing hydrogen production and conversion technologies and hence replicate and test the systems of a hydrogen economy on a small scale.

Researchers invoked this idea as early as the Second World Hydrogen Energy Conference in Zurich in 1978. In his banquet address, the physicist and nuclear engineer Wolf Häfele held that technical barriers ruled out solar and nuclear energy as sources of hydrogen in the immediate future, noting the outstanding problems of how to collect dispersed solar energy and most efficiently use the 15 to 20 million tons of uranium ore of all grades that remained on earth. He instead looked to fossil fuels, remarking that these were far more abundant than commonly thought, particularly reserves of coal, for which there was "no clear upper limit." Converting this into methanol or methane, claimed Häfele, could result in a "50 percent" hydrogen economy, one that would utilize the existing liquid-fuel infrastructure of the transportation sector and ease the transition to the ultimate nuclear-solar-electricity-hydrogen economy.[41]

Workers at the Los Alamos Scientific Laboratory held similar views. In 1981, they wrote that questions of primary energy could be set aside while practical experience with hydrogen fuel in conventional combustion engines and fuel cells was gained. They hoped the military would employ lessons learned in the laboratory's civilian hydrogen automobile demonstration program, which consisted of a truck and Buick sedan converted to liquid hydrogen and an electric golf cart equipped with a fuel cell. Claiming these vehicles employed commercial hydrogen fuel, the researchers framed the experiment as proof that the technology for small-scale demonstration projects already existed.[42]

Of course, hydrogen was commercial in the sense that it was available on the market for industrial use. But it was not intended or priced for regular automotive use. Nevertheless, hydrogen futurists were increasingly drawn to end-use devices. At the Fourth World Hydrogen Energy Conference in Pasadena, California, in 1982, organizers indicated that they would devote particular attention to technologies "in current use," displaying three automobiles converted to liquid hydrogen, including the Los Alamos Buick.[43] This theme was even more explicit at the 1984 conference in Toronto. Conferees acknowledged that the age of hydrogen lay considerably farther off than futurists had estimated in the 1970s. By the mid-1980s, gasoline was once again cheap in the United States, and the chair and secretariat of the IEA Hydrogen executive committee worried about "excessive skepticism." Keynote speaker John Bockris suggested the fortunes of hydrogen futurism were linked with those of the nuclear industry. Rising costs in the operation of conventional reactors and industry's inability to develop high-temperature variants and couple them to thermochemical water-splitting processes had prevented even the pilot demonstration of large-scale hydrogen production. Hydrogen ultimately would save the human race, Bockris suggested, but before that could happen, hydrogen research itself had to saved. The chief problem, noted members of Canada's Hydrogen Industry Council (HIC), was how to make this sort of work sustainable. As a start, researchers had to drop their preoccupation with deep-future scenarios. The "conceptual revolution" was over. As desirable as technological breakthroughs were, they were improbable in the current economic and political climate. If hydrogen research and development was to survive, it had to raise its public profile, and this meant pursuing small-scale applications of existing technologies. These would be a far cry from the grand visionary schemes, noted the HIC researchers, but they had the advantage of being "real."[44]

Sometimes this strategy clashed with the overarching premises of hydrogen futurism. For IEA planners, an important problem was that there was no real international coordination in hydrogen research and no mechanism for drafting priorities, a deficiency that stemmed from the inability of national governments to establish their own coherent energy research and development programs. One ray of good news, however, was the continuing support of IEA governments for fuel cell research and development, totaling $200 million per year. This was only a tenth of what the member nations spent on nuclear work, IEA strategists admitted, but it was still a significant contribution to the end-use agenda since fuel cells could consume hydrogen.[45] Of course, the largest fuel cell programs then under way were devoted to utility generator sets that either internally reformed or directly used hydrocarbons, not pure hydrogen. And although such fuel cells were certainly capable of using pure hydrogen, they were designed to fit seamlessly into the fossil energy conversion chain, the very order some futurists sought ultimately to displace.

The favored demonstration technology of hydrogen futurists remained the automobile conversion, a relatively cheap experiment that grew in popularity in the 1980s. A number of companies including Fiat, Peugeot, and General Motors dabbled in this activity during this period. None, however, was as ambitious as Daimler-Benz. Believing that a hydrogen economy could be realized by renovating existing energy and power infrastructure, the German manufacturing giant launched a scheme backed by the Federal Ministry of Research and Technology in 1979 that in some ways resembled the TARGET project. Designed to showcase the "nonrevolutionary" qualities of hydrogen, the plan aimed to turn homes into chemical plants producing hydrogen using electrolyzers or reformers drawing natural gas. The hydrogen would then be stored in a system of metal hydride containment units, chosen over conventional tankage for use in both home and automobile in order to exploit the propensity of metal hydrides to release heat when absorbing hydrogen and absorb heat when releasing hydrogen. By exploiting the thermal cycle of hydride hydrogen storage and use, planners theorized, home owners could efficiently warm and cool their domiciles. In practice, the project focused much more on the automotive side. Daimler-Benz built five sedans equipped with dual hydrogen/gasoline fuel systems and five pure hydrogen-drive vans, fielding the vehicles in a high-profile demonstration program in Berlin between 1984 and 1988.[46]

Most major U.S. firms were uninterested in demonstration projects as large as this. And there was no coordinated federal program of hydrogen automobility. But a number of federal and nonprofit agencies engaged in on-off hydrogen automobile conversions. By mid-decade, 22 such programs were under way at 13 organizations in the United States, more than in any other country, many receiving support from the Department of Energy's Alternative Fuels Utilization Program. Even the U.S. Postal Service developed a hydrogen jeep demonstrator. The Los Alamos Scientific Laboratory probably had the most elaborate such effort in the United States. It collaborated with the German Test and Research Institute for Aviation and Space Flight, the national aerospace research agency, which provided technical support for the liquid-hydrogen storage tank.[47]

The other key facets of the technoscience of hydrogen futurism in the 1980s were electrochemical materials production and energy conversion technologies. Over the course of the decade, they became increasingly prominent in hydrogen demonstration culture. Electrolyzers and fuel cells are closely related, the chief difference being that the former consumes electricity to make chemical products while the latter consumes chemicals to produce electricity. These devices shared an intimate history. William Grove was inspired by electrolysis to attempt the reverse process in his gaseous voltaic battery. Francis Bacon sought to learn from the experience of electrolyzer manufacturers as he developed his own reversible cell. And both technologies were shaped by a narrow spectrum of materials capable of propagating sustained electrochemical reactions, particularly at high temperatures. As with fuel cells, electrolytic cells employing caustic electrolyte performed best with nickel electrodes, while acidic systems required noble-metal catalysts.

Unlike the fuel cell, however, the industrial electrolyzer had been commercialized around the turn of the nineteenth century. It was used to split water to produce hydrogen for synthesizing ammonia fertilizer and oxygen for metallurgical purposes. Later, the technology was used to produce deuterated or heavy water to moderate nuclear reactors. Electrolyzers were also used in the chloralkali industry to split sodium chloride solution—simple brine—into chlorine and caustic soda. Over time, however, electrolysis was less frequently used to produce hydrogen, losing ground to chemical processes using cheap hydrocarbons around the Second World War.[48] By 1984, electrolysis accounted for only a small fraction of the total annual U.S. hydrogen output of around seven million metric tons. Two-thirds of this total production was produced intentionally, and the balance was a by-product of industrial chemical processes. Of the intentionally produced hydrogen, more than 90 percent was made by steam-reforming methane or propane, 2 percent by partially oxidizing heavy petroleum, and only 1 percent by electrolysis. Almost all this hydrogen was consumed where it was produced, with 48 percent used for ammonia manufacturing, 39 percent for petroleum refining, and 8 percent for methanol production. A mere 1.5 percent was available on the wholesale or retail market, the source of the "commercial" fuel, as the Los Alamos researchers had put it, for hydrogen automobile demonstrations. Like fuel cells, electrolytic cells were a modular technology that operated at the same efficiency regardless of plant size, meaning capital costs were almost proportional to production capacity. Widely perceived as a boon in utility fuel cells, allowing power demand to be satisfied incrementally, this quality was regarded as a handicap in water electrolysis because economies of scale in hydrogen production could never be attained by such means.[49]

Nevertheless, water electrolysis had features attractive to hydrogen futurists and the culture of hydrogen technology demonstration. By the mid-1980s only one thermochemical process, EURATOM's bromine/sulfur cycle, had been made to continually operate at the bench scale. Electrolysis, on the other hand, was an established commercial process. For hydrogen futurists, the technology was a tangible link between the real and ideal hydrogen economies, and it became increasingly central to their vision over the years. Begun in 1974 as one of several alternative energy research initiatives, Japan's "Sunshine Project," for example, initially prioritized thermochemical work but ended up devoting half of its $20 million budget to electrolysis systems. Indeed, the project's single largest component was a pilot electrolysis plant.[50] A review of hydrogen production research at WHEC 1982 focused largely on advances in electrochemical technology. Such was progress in this field, remarked Bockris in the plenary session, that "he almost had to caution against over-enthusiasm."[51] In 1984, researchers at the DOE's Argonne National Laboratory anticipated a key justification for electrolysis within the demonstrative culture of hydrogen futurism. Unable to compete with large conventional hydrogen plants, they noted, electrolytic hydrogen was convenient only in small operations. But its "freedom from fossil feedstock" made it an attractive fuel for the future and a key means of transitioning to a hydrogen economy. Indeed, electrolytic hydrogen was already cheaper than electricity on a delivered-heat basis, researchers claimed, but it could not capture market share because most consumer appliances were powered by electricity, not gas.[52]

For more than a few hydrogen researchers, the fuel cell was an obvious match with the electrolyzer in an energy conversion system. Over the years, an increasing number of papers devoted to these technologies appeared at the world hydrogen energy conferences.[53] Hardly any work on fuel cells was presented in the first international gathering in 1974, and subsequent conference literature on this subject tended to be subsumed under a number of rubrics until 1982, when planners organized a separate fuel cell section for the first time. Around this period, papers dealing explicitly with electrolyzers or fuel cells comprised anywhere from a fifth to a third of those presented.[54] Neglected following the shift to the methane-consuming phosphoric acid fuel cell in the mid-1960s, proponents of alkaline-electrolyte pure hydrogen fuel cell technology, including pioneers like Karl Kordesch and Francis Bacon, were given opportunities to promote their work within the hydrogen futurist community.[55]

By mid-decade, plans were under way to stage the first major demonstration of a total hydrogen fuel production/conversion system employing the key existing components—electrolyzers, hydrogen fuel cells, and hydrogen automotive drive—as well as photovoltaic arrays. Once more, West German enterprise led the way. In late 1986, BMW entered a consortium with Bayernwerk, Bavaria's main electricity supplier, and engineering giants Siemens, MBB, and Linde in a plan to build a solar hydrogen plant in the water-rich Schwandorf district. Backed by the German federal and Bavarian state governments in a fifty-fifty public/private partnership, the 13-year Bavarian Solar Hydrogen project was designed to showcase an entire hydrogen-based energy transformation chain. Unlike Daimler-Benz, BMW preferred to use production sedans equipped with liquid storage in its hydrogen propulsion experiments. But the Schwandorf plant was designed to employ alkaline and phosphoric acid fuel cells drawing hydrogen from electrolyzers powered by banks of photovoltaic cells in an exercise designed to simulate the duty cycle of an electric automobile plant.[56]

However, the traditional liquid electrolyte fuel cell systems would play a minimal role in the conjoined renaissance of fuel cells and hydrogen futurism in coming years compared to the proton exchange membrane system. General Electric's involvement in this technology in the 1980s illustrates the linkage between the fuel cell, the electrolyzer, and hydrogen futurism, foreshadowing events in the 1990s. For three decades after it invented the membrane fuel cell in the mid-1950s, General Electric struggled to commercialize the power source but succeeded only in penetrating the niche aerospace market. The technology's performance in this role left much to be desired. But over the years, General Electric considerably improved the power and reliability of the membrane fuel cell thanks largely to Du Pont's Nafion perfluorosulfonic polymer membrane. Known as an ionomer for its ability to conduct hydrogen ions, this exotic, low-volume material was developed in the late 1960s for a number of industrial electrochemical processes. It was, for example, a safer and more economical replacement for the mercury- and asbestos-based electrolyzers that had been used in the chloralkali industry since the late nineteenth century.

When General Electric failed to attract civilian customers for its membrane fuel cell, it switched emphasis in 1967 and attempted to develop this device as an electrolyzer. The emergence of hydrogen futurism and the world hydrogen conferences in the early 1970s presented the company with a forum in which to promote this technology, in turn lending some prestige to the movement.[57] And as General Electric improved its electrolyzer, with help from the DOE's Brookhaven National Laboratory, the company attempted to leverage this progress in developing the device as a power source. In 1983, it claimed that the successful commercialization of the membrane electrolyzer boded well for the membrane fuel cell in the transportation application. As in the past, General Electric was not averse to stretching claims for its electrochemical technology. The power source manufacturer did not then possess a functional automotive power source but rather a model for a 66-kilowatt methanol reformer/fuel cell it claimed cost $150–200 per kilowatt.[58]

This was a very attractive figure. But it was based on future reductions in the cost of materials, notably platinum catalyst and the Nafion membrane, then worth around $400 per square meter. The episode recalled General Electric's promotional tactics during the 1960s. Long a shortcoming of membrane devices, such costs could not be reduced without further painstaking engineering research. But after nearly 30 years of unrewarded effort, General Electric was no longer willing to continue its commitment to membrane technology. In 1984, the company sold its fuel cell/electrolyzer systems to the Hamilton Standard Division of the United Technologies Corporation, its longtime competitor in the field.[59]

Preoccupied with phosphoric acid fuel cell technology, UTC did not immediately act to develop the proton exchange membrane fuel cell. The concept was instead taken up by an obscure Canadian engineering research firm called Ballard Power Systems, thanks largely to the prompting of the Canadian military research establishment. Founded as Ultra Energy in 1977, the outfit first attempted to commercialize rechargeable lithium batteries, mainly under contract to Amoco.[60] Desperate for funding, the company, renamed Ballard Research in 1979, won a contract from the Canadian Department of National Defence in 1983 to build prototype membrane fuel cells, investigate their potential for military and civilian use, and reduce costs.[61] Nafion was a major obstacle because it was then the only electrolyte material suited for such a fuel cell. Matters changed dramatically in 1986 after Ballard acquired an experimental Dow Chemical perfluorosulfonic acid polymer membrane. This technology had a singular impact on the course of low-temperature fuel cell research and development. Thinner than Nafion and with a higher sulfonic acid concentration and ion exchange capacity, the material produced over four times as much power as Nafion when used in a fuel cell. Ballard's 1986 announcement that it had improved current density by a factor of four astounded J. Byron McCormick, division leader of research and development and deputy division leader of the Los Alamos National

Laboratory's Electronics Division, who blurted that the company had "made the electric vehicle possible." McCormick would go on to manage the development of electronics, batteries, and fuel cells at General Motors.[62]

Ballard's successful experiments inspired new interest in membrane fuel cell technology. In the late 1980s, researchers at Los Alamos National Laboratory and at A. J. Appleby's fuel cell and hydrogen research institute at Texas A&M University began to work to reduce the amount of platinum required by the power source.[63] Over the next 15 years, the membrane fuel cell in the automotive application would become the energy conversion technology the hydrogen futurists had long dreamed of. Nicely complementing the small-scale electrolysis plant, it enabled the development of the fuel cell electric concept vehicle as the most visible and psychologically potent manifestation of hydrogen futurism.

Electrochemical Hydrogen Comes Out of the Cold

The second half of the decade saw growing links between the hydrogen and hydrogen fuel cell communities, the media, and politicians. In 1986, Peter Hoffmann, a journalist and a leading hydrogen authority and promoter, introduced the *Hydrogen & Fuel Cell Letter*.[64] Published with the support of the Hydrogen Research Center at Texas A&M, the Hawaii Natural Energy Institute at the University of Hawaii (HNEI), the Clean Energy Research Institute at the University of Miami, the Clean Fuel Institute in Riverside, California, and Canada's Hydrogen Industry Council, the newsletter became a key resource for hydrogen futurists as an information clearinghouse. It did much to associate hydrogen with fuel cell technology.

Even more important was the rise of a small bloc of hydrogen enthusiasts in Congress. A collegial, bipartisan group with diverse backgrounds and interests, it in some ways resembled the original band of hydrogen futurists. The most committed were Spark M. Matsunaga (D-Hawaii) and Daniel Evans (R-Washington) in the Senate, George E. Brown, Jr. (D-California), Thomas F. Lewis (R-Florida), Robert A. Roe (D-New Jersey), Robert G. Torricelli (D-New Jersey), and Robert S. Walker (R-Pennsylvania) in the House of Representatives, and Daniel K. Akaka (D-Hawaii) in both chambers of Congress. Matsunaga and Brown were perhaps the most ardent hydrogen idealists. A decorated veteran of the Second World War, Matsunaga had a longstanding commitment to clean energy, first introducing legislation supporting hydrogen research and development in Congress in 1981. His advocacy in turn exercised an important influence on the agenda of the Hawaii Natural Energy Institute. Founded at the University of Hawaii with the support of the state government in 1974, the HNEI attempted to lessen the island's dependence on imported oil by harnessing its numerous sources of renewable primary energy including geothermal, ocean thermal, wind, tide, current, solar, and biomass. By the early 1980s, the HNEI had demonstrated the near-term commercial potential of many simpler technologies such as geothermal plants, wind farms, and solar water heating and began to tackle more complex projects, including hydrogen production. Several years after Matsunaga introduced hydrogen legislation in Congress, the state government of Hawaii helped establish the Hawaii Hydrogen Program at HNEI. By the 2000s, this had become the institute's largest single initiative.[65]

Matsunaga's kindred spirit in Congress was Brown. One of the more outspoken progressives in Congress, Brown was a strong advocate of sustainable energy, science education, and planning and investment in advanced energy technologies. He joined Matsunaga in working to expand the government's hydrogen research and development programs through the 1980s. But with gasoline cheaper than ever, there was little pressure on Congress to act. In the late 1980s, U.S. federal funding for hydrogen research and development actually declined to mid-1970s levels.[66]

Fusing Hydrogen Futures

As the decade wound down, a species of hydrogen futurism, one with a strong electrochemical aspect that would in some ways anticipate the fuel cell boom of the 1990s, seized imaginations around the world. In a press conference on March 23, 1989, the electrochemists Martin Fleischmann of the University of Southampton and Stanley Pons, head of the University of Utah's Chemistry Department, revealed the momentous news that they had fused deuterium—an isotope of hydrogen—in an electrochemical cell. Their chief claim was that one such cell had produced four watts of heat using only one watt of electricity in a sustained reaction, yielding neutrons and helium. By-products of a nuclear and not a chemical reaction, the presence of such subatomic particles indicated that fusion had occurred. The scientists' modus operandi was simple: they had electrolyzed deuterium oxide—heavy water—in cells equipped with platinum and palladium catalysts and using lithium deuteroxide electrolyte. Shocked by word that, in such a crude device at low temperature, Fleischmann and Pons had propagated a reaction scientists had previously been able to induce only in expensive and sophisticated magnetic confinement reactors at tens of millions of degrees, and then only fleetingly, researchers scrambled to replicate their results at leading universities around the world. To some, the power of stars now seemed available at the laboratory bench.

The subsequent debate on the reality of cold fusion turned on divisions within the U.S. science community—competition between cold fusion groups at the University of Utah and Brigham Young University; perceptions in certain pro-cold-fusion quarters of the Utahan political, academic, and scientific establishment of sectional prejudice by skeptical East and West Coast elites; and grievances within certain physical science communities.[67] The episode illustrated the insularity and insecurity felt by some chemists and electro-chemists vis-à-vis their physicist colleagues where energy issues were concerned. John R. Huizenga, a professor of chemistry and physics at the University of Rochester and cochair of the Energy Research Advisory Board Cold Fusion Panel, a body created by the Department of Energy to investigate the matter, was scandalized by cold-fusion boosterism at a special session of the 1989 conference of the American Chemical Society. He was taken aback by ACS president Clayton F. Callis's boast that chemists had succeeded where physicists had failed. Regarding cold fusion as a vindication of their field and pleased with the attention they were receiving, electrochemists basked in the limelight. The session featured introductory lectures in electrochemistry delivered by the electrochemists Allen J. Bard of the University of Texas and Case Western Reserve University's Ernest B. Yeager. Involved in battery and fuel cell research, Yeager remarked that "these were exciting times for electrochemists" and exhorted the audience to keep their "fusion fever" high.[68]

Cold-fusion advocates were further encouraged by early support from a variety of sources including the National Science Foundation's Division of Electrical and Communication Systems, the Electric Power Research Institute (EPRI), which had been investigating fuel cell technology since the 1970s, and, briefly, by Edward Teller himself. There were also pro-cold-fusion pockets in a few universities. At the Department of Materials Science and Engineering at Stanford, the physicist Robert A. Huggins worked to reproduce the effect reported by Pons and Fleischmann. A pioneer of solid-state electrochemistry in the early 1970s, Huggins had realized that ions could quickly move and be stored in solids, overturning the prevailing view that the important reactions in a battery happened on electrode surfaces. This work in solid-state ionics would have profound implications for advanced energy storage later in the century, influencing the commercialization of rechargeable lithium ion battery technology. But in mid-April 1989, Huggins reported that his Stanford team had successfully duplicated the cold-fusion experiment.[69]

At Texas A&M, the cold-fusion charge was led by Appleby and Bockris, both esteemed members of the electrochemical and hydrogen communities and long involved in fuel cell and hydrogen research. Then engaged in EPRI-funded fuel cell work, noted the journalist Gary Taubes, Appleby, Bockris, and electrochemist Charles Martin were directed by the institute to shift their efforts toward cold fusion. Founder of the Center for Hydrogen Research at Texas A&M, renamed the Center for Hydrogen Research and Electrochemical Studies when Appleby took the reins, Bockris, Fleischmann's former supervisor at Imperial College London and an old friend, became an especially vociferous advocate of what he termed "nuclear electrochemistry." Cold fusion evoked brief but intense curiosity among federal bureaucrats and politicians. It received enthusiastic support on Capitol Hill from Robert Walker, then the ranking minority member of the House

Science, Space, and Technology Committee. A strong supporter of all things hydrogen and well connected with NASA and the aerospace industry, Walker had to be talked out of earmarking millions of dollars for cold-fusion research by a delegation of leading physicists.[70]

Hydrogen futurism did not suffer from its association with what virtually all authorities came to regard as the pseudoscience of cold fusion because the proponents of the hydrogen economy made no novel physical claims. But the culture of experimentalism in the cold-fusion and hydrogen and fuel cell futurist projects shared some points in common. There was the obvious electrochemical connection and the physical similarities of fuel cells and electrolysis cells. As with hydrogen experiments involving fuel cells, internal combustion engines, and some other downstream technologies, cold-fusion experiments could be done relatively cheaply.[71] A few researchers with a similar psychological makeup also became involved in cold fusion, prompted at least in one instance by their sponsoring institute. Fascinated by the notion of the techno-utopia and possessing distinguished science or engineering credentials, they genuinely believed in the possibility of breakthroughs that would fulfill dreams of limitless energy and power. And both cold fusion and hydrogen and fuel cell futurism were legitimized and bolstered by mainstream sponsors of science and technology.

One must be careful in drawing analogies in this instance. Cold fusion proved to be fantasy. Hydrogen fuel cell power was real. But in the late 1980s, the actual capabilities of this technology on the laboratory bench were a far cry from what it was expected to accomplish in concert with the other components of the idealized hydrogen economy. Accordingly, similar assumptions can be seen at play in these enterprises. On a broader level, faith in the future (that is, the belief in technological progress) had always been a powerful factor in fuel cell R&D. Such work was frequently justified as a search not simply for a better power source but for the ultimate power source. As we have seen, researchers sometimes perceived this project in quasi-mystical terms, as a quest for the philosopher's stone of energy conversion devices. In this sense, cold fusion and hydrogen and fuel cell futurism were driven by similar visionary impulses.

Hydrogen Technopolitics Resurgent

The cold-fusion imbroglio overshadowed the growing respectability of hydrogen and fuel cell futurism in industrial and political circles as the decade drew to a close. The year 1989 witnessed the formation of the National Hydrogen Association (NHA), the first hydrogen lobbying group, based in Washington, DC. To be sure, its founders still constituted a relatively narrow base of support. Foremost among them was NASA, long the chief source of inspiration for hydrogen futurists and still the only organization making routine practical use of hydrogen as a vehicular fuel. The presence of the Department of Energy in the

NHA must have further heartened enthusiasts despite the agency's continuing parsimony in hydrogen and fuel cell research in this period. The HNEI was a logical addition to the NHA. It was well on its way to becoming a leading investigator of renewable hydrogen production technology, reflecting its concern with the unique energy ecology of the Pacific archipelago. Particularly revealing of certain technological and economic assumptions harbored by hydrogen futurists was the NHA's industrial contingent. Consisting of Praxair, Air Liquide, and Air Products, these leading industrial gas companies were rooted firmly midstream in the existing hydrogen energy conversion chain, producing and marketing gases and gas equipment for a variety of manufacturing, research, and medical purposes. They did not, however, sell gases as transportation and heating fuel, the one exception being aerospace hydrogen, a market that Air Products in particular had long dominated in the United States.

More significant from the perspective of public policy was the passage of the Spark M. Matsunaga Hydrogen Research, Development and Demonstration Program Act in November 1990. Originating in a bill introduced by Matsunaga in March 1989, the law satisfied three longstanding concerns of hydrogen researchers. First, it acknowledged the centrality of production technology, declaring it in the national interest to develop an economic means of generating the element for use as a fuel and storage medium in quantities sufficient to reduce dependence on conventional fuels. Second, it recognized the necessity not only of researching and developing these technologies but publicly demonstrating them as well. And the program's five-year timeline implied that results were achievable in the near future.[72]

Still, the law's provisions were modest. It authorized only a small amount of money for research and development—$20 million—spread across fiscal years 1992 to 1995. But it also reflected changing perceptions of hydrogen's political efficacy on Capitol Hill in the late 1980s. Growing environmental awareness, resurgent pollution politics, and rising tensions in the Middle East made Congress increasingly receptive to the precepts of the original hydrogen futurists. In defending HR 4521, a bill similar to Matsunaga's S 693, Brown in essence paid homage to them. As an energy carrier, storage medium, and fuel, he noted, hydrogen was a total solution to the problem of sustainable energy, with a multitude of uses, including as a fuel for the fuel cell electric vehicle. He referred to the "utopian day" when the hydrogen economy the visionaries had spoken of for so long would be realized. Brown admitted the bill's material provisions were humble. It basically coordinated the various hydrogen research and development programs then under way in a number of government departments. But its political effect, he remarked, would be to prod the administration to recognize the importance of hydrogen in energy policy.[73]

As representatives moved to pass the Senate version of the bill in October, they explicitly acknowledged hydrogen's political attractions as a general-purpose energy solution, one capable of reconciling a broad range of socioeconomic and geopolitical objectives and economic interests. Robert A. Roe, chair of the House Committee on Science, Space, and Technology, framed the Matsunaga Act as part of an effort to develop a "long overdue" national energy policy. The timing of the legislation was important, he noted, for it complemented Congress's efforts to pass the Clean Air Act during a period when renewed turmoil in the Middle East precipitated by Iraq's invasion of Kuwait demonstrated yet again the hazards of America's dependence on foreign oil. Clean-burning and in "virtually limitless supply" in the form of water, added Walker, hydrogen addressed heightened concerns for the environment and for the country's geopolitical vulnerability. It was, accordingly, the "perfect fuel for the future."[74]

The Matsunaga Act established a precedent in enabling deepening government/industry involvement in hydrogen and fuel cell systems. Among the law's provisions was the authorization of the formation of a Hydrogen Technical Advisory Panel (HTAP) consisting of representatives from industry, government, and academe appointed by the secretary of energy to assist in program management. Neither entrepreneurs nor technocrats were then especially interested in the panoply of hydrogen production, storage, and conversion technologies. But as lawmakers worked to craft the law, major developments in automobile politics unfolded in California that had important implications for the fortunes of hydrogen and fuel cell futurism. In January 1990, General Motors displayed its "Impact" battery electric concept vehicle at the Los Angeles Automobile Show, suggesting it planned to introduce a commercial variant. Inspired by this seemingly landmark decision, the California Air Resources Board (CARB) introduced clean automotive legislation in the Low Emission Vehicle and Clean Fuels regulation (LEV) of September 1990. This law had revolutionary potential, for one of its provisions required the automobile industry to market zero-emission vehicles in increasing proportions if they wished to do business in the state. The ensuing politics of green automobility would eventually help vault hydrogen and fuel cell R&D, and futurist discourse, into the national technopolitical limelight for the first time.

6

Green Automobile Wars

We're in the position Intel was in 15 years ago with the microprocessor. The fuel cell is a technology that will transform the world.

—Firoz Rasul, president and chief executive officer,
Ballard Power Systems, February 1998

In early 1993, a low-slung, racy sports coupe with a massive, incongruous power plant jutting from its rear deck appeared in the parking lot of an obscure West Palm Beach research and development start-up known as Energy Partners. Expensive, exotic vehicles were not uncommon along a stretch of Florida coastline dotted with affluent communities. Even by these standards, however, the automobile, dubbed the "Green Car" by its inventors, was unique. Although Ballard Power Systems was then emerging as the premier developer of proton exchange membrane fuel cell technology, testing it in an electric bus at around the same time, the Green Car was the world's first full-size electric passenger automobile to use this power source technology, hitherto best known for powering NASA's Gemini spacecraft.

But Energy Partners had no intention of producing commercial fuel cell electric automobiles. John H. Perry, Jr., its millionaire chairman, wanted to use the Green Car to advertise the potential of the fuel cell as a power source for zero-emission vehicles, especially for the California market, where stringent new rules enacted in 1990 promised to radically reshape the way automobiles were built. But the car itself was not really a practical prototype. A custom-built two-seater, it was cramped, underpowered, and unsuited for travel on public roads. And it was pricey, the power plant alone costing over $180,000, and plagued by assorted troubles. Nevertheless, reported the *New York Times*, the path to the commercial electric automobile might be cleared if Energy Partners and other companies investigating fuel cells succeeded in weeding out the bugs and cutting costs.[1]

The implications, suggested the paper of record, were enormous. Perhaps no other consumer product has had as pervasive an effect on society and the environment in the twentieth century as the commercial fossil-fueled, internal combustion automobile. So profound is this influence that scholars use the term "automobility" to refer to the vast complex of technologies, institutions, and associated sociocultural values that enable the act of driving, as well as to the act itself.[2] For decades, the automobile sector functioned as an industrial Atlas, supporting myriad constellations of metallurgical, chemical, petroleum, and parts enterprises. The automobile became a key source of individual identity, a symbol of modernity and an object of desire for the billions of people for whom automobiles promised figurative and literal social mobility. So, too, have the deleterious effects of automobility—accidents, urban sprawl, resource depletion, and environmental degradation—been systemic.[3]

The Green Car did not generate much attention at a time when the automotive media were concentrating on the efforts of the major manufacturers to develop battery electric drive autos. Indeed, in the early 1990s, electrochemical experts believed no existing fuel cell technology met all the economic and technical criteria for the automotive application. Membrane fuel cells could deliver fast load response and high power in a relatively small volume. But despite being continually improved over the years, such devices remained handicapped by high cost, unreliability, and important materials limitations. A number of researchers favored the alkaline fuel cell for all the traditional reasons: it was a proven technology offering relatively high performance using cheap materials. The main shortcoming of this design, of course, was that it required expensive pure hydrogen fuel. Accordingly, predicted A. J. Appleby, the first fuel cell electric automobile to enter the market would likely be a subsidized fleet vehicle that could be readily fueled and serviced from a central depot. For these reasons, he and others expected that the first commercial fuel cell would be a large and expensive multimegawatt central generating station of the kind that had been under active development by United Technologies Corporation and Westinghouse for years, on the grounds that only this type could capture the large markets and exploit the economies of scale that would justify major investments.[4]

Events seemed to confound these expectations. By the late 1990s, several major automakers were working on membrane fuel cell electric drive and were promising commercial production. What had happened in the interim? Progress in alternative automotive technologies, the economic and environmental contradictions of American automobility, attempts to legislate technological solutions to these problems, and a reappraisal of federal science and technology policy combined to create political space for fuel cell power in vehicular applications as the Cold War drew to a close. Fiercely opposed to the government when it intervened to regulate the safety and efficiency of their products (although not when it intervened to shape the fossil fuel automobile paradigm through the construction of highway infrastructure and administration of energy policies that secured cheap oil), American automakers were shocked when California's Air Resources Board (CARB) introduced a Low

Emission Vehicle statute that included a zero-emission vehicle (ZEV) requirement in 1990. In introducing production quotas for the battery electric automobile, then the only practical zero-emission vehicle, the law implied a revolutionary restructuring of traditional industrial relations in the energy and transportation sectors at a time when American manufacturing was experiencing the stiffest foreign competition in generations.

In a way, these events were the culmination of two distinct but linked trends that informed the planning of the renovation of the American energy conversion chain in this period. On the one hand, the overlapping environmental and energy crises of the 1970s inspired Congress to promulgate a series of laws that for a variety of political reasons were premised not on the science and economics of sustainable industry but on the presumed ameliorative qualities of a variety of existing technologies like the industrial exhaust scrubber, the automobile catalytic converter, and smaller, more efficient internal combustion engines. Industry fiercely resisted initiatives like the Corporate Average Fuel Economy (1975), the 1977 and 1990 amendments to the Clean Air Act, and the ZEV mandate.[5]

This era of technology forcing also coincided with the recrudescence of the perennial debate over the economic productivity of basic science, one that had simmered within the federal science and technology establishment for nearly a decade by the time of the appearance of the ZEV mandate. In essence, it turned on the role of government in supporting basic science as a means of stimulating innovation and, hence, general economic growth. Out of that protracted discussion came the idea of the public-private research and development partnership, the brainchild of technocrats, lawmakers, and academic scientists. Sometimes referred to as collaborative research and development, this model of innovation is often dated to the Bayh-Dole Act of 1980, which eased the process of patenting intellectual property produced with public funds. The law heralded a wave of legislation in the 1980s designed to yield an economic dividend from federally funded basic research performed in universities and federal laboratories.[6]

Impressed by the potential of collaborative research and development, the Clinton administration attempted to adapt it as an alternative to the technology-forcing approach to developing sustainable industry. The White House had ambitious goals. Whereas Congress and the state of California aimed to compel industry to adopt what bureaucrats had come to believe was state-of-the-art but feasible technology (often, ironically, thanks to industry futurist dramaturgy), the Clinton administration aimed to enlist the automobile industry in voluntary research programs intended to rejuvenate the industrial base and develop new forms of automobile drive. Industry had complete discretion to dispose of the resulting research as it saw fit. But the White House expected that such programs would help remediate environmental damage, develop sustainable transport, and foster energy independence.[7]

To be sure, collaborative R&D was not new, as the economist David C. Mowery observed, having a very long history dating back to the formation of NACA in 1915.[8] But although the state had played an indispensable role in shaping the post–Second World War system of automobility by subsidizing energy, infrastructure, and the auto-centric suburb, it had no tradition of engaging in cost-shared research and development with the auto industry. Weakened by foreign competition and threatened by new state and federal clean air and fuel efficiency measures, Detroit agreed to participate in the Clinton plan.

Here, then, lay the origins of an important form of "greenwash" or environmental propaganda.[9] It was a kind of technological futurism, originating not so much as a corporate response to government meddling, as it is sometimes interpreted, but as a consequence of the innovation concordat between Detroit and the White House, one spurred by the failing business model of American automakers and science and technology politics in Washington. One result was that research in fuel cell electric drive (already sponsored by the Department of Energy) assumed importance for Detroit in the politics of pollution and innovation in the 1990s. Daimler-Benz and Ballard Power Systems provided another impetus early in the decade when they promised to produce a commercial fuel cell electric car. Meanwhile, foreign and domestic automakers maneuvered for advantage in the U.S. marketplace, competing but also conspiring to undo California's ZEV statute. In the interim, manufacturers grudgingly produced battery electric autos on a limited basis, motivated mainly by a desire to publicly demonstrate the inadequacies of the technology.

But as researchers continually improved the fuel cell electric car through the 1990s, U.S. oil and auto elites began to perceive it a feasible alternative zero-emission vehicle, one potentially superior to the pure battery electric. And with this realization came the use of fuel cell R&D as a bargaining chip in a high-stakes game of pollution politics.[10] Automakers first convinced the CARB to recognize the fuel cell electric as a legitimate ZEV. Then, in exchange for a promise to one day commercialize the technology, they persuaded the agency to roll back its production quota deadlines. In this manner, business leaders, technocrats, and new-economy entrepreneurs, players with otherwise incommensurable short-term agendas, found common ground in fuel cell futurism, altering the course of the technopolitics of sustainable automobility in the 1990s and early 2000s.

Battery versus Fuel Cell Electric Drive

The history of the fuel cell is interwoven with that of the battery, especially in the case of the electric automobile. In important ways, the history of the fuel cell electric auto is but the latest, hitherto unwritten chapter of the history of the battery electric auto, as historians Gijs Mom and David A. Kirsch have noted.

Their relational accounts of electrochemical and internal combustion power are particularly relevant to the history of fuel cell automotive drive.[11] Both question the whiggish tautology of traditional automobile history, the view that the internal combustion vehicle "was the best because it won and won because it was the best."[12] Instead, they hold, the idea of technological superiority is largely grounded in the subjective judgments of groups of engineers, designers, marketers, and industrialists engaging in power relations inherent to the capitalist system. Notions of the inferiority of electric drive derive from the long-standing belief that the lead-acid battery, the simplest and most cost-effective electricity storage medium for most of the twentieth century, was obsolete owing to its weight and limited capacity. This, detractors claimed, made the electric automobile short-legged in comparison with the gasoline automobile. For Mom and Kirsch, the reality was that standards of desirability in automobility were culturally rooted judgments that shifted as certain lines of power source and infrastructure technologies were pursued and others were dropped and as the interest groups that coalesced around these systems competed with one another.

A fundamental characteristic of the history of gasoline and battery electric vehicles, notes Mom, is design and development reciprocity. Builders of each class of automobile tried to embody the best features of the competing technology in their own designs. Both power sources had offsetting pros and cons. In certain circumstances, battery electric vehicles had historically performed well when compared with their gasoline counterparts. Beginning in the late nineteenth century and into the second decade of the twentieth century, battery electric vehicles replaced horse-drawn carriages and elite luxury coaches. The relatively high cost and performance of these vehicles were viewed favorably in relation to the older mode of transport as well as to new gasoline vehicles. As researchers improved battery technology in this period, some European cities found electric taxicabs and municipal vehicles both aesthetically preferable and economically competitive if operated in large fleets served by centralized charging and maintenance facilities.[13]

Early gasoline vehicles, on the other hand, were much less reliable and comfortable than urban electric cars, and less suited for operation in cities. But their greater range and speed were exploited in the United States by automakers that promoted the gasoline car as an adventure vehicle, fostering the bourgeois pastime of "touring," or long-distance country driving. Automakers continued to improve the reliability, comfort, and styling of gasoline vehicles, making them multifunctional, able to compete with electrical vehicles in cities while retaining the long range that facilitated touring. With new stylistic and performance benchmarks thus defined, designers of electric automobiles sought to develop a universal electric car with its own adventure cachet. They emulated the body form of the latest gasoline touring sedans, worked to develop a miracle battery that would even the performance field, and developed the gasoline-electric hybrid. However, the resulting vehicles could still not match the gasoline automobile in range or convenience.[14]

Mom challenges the view that the universal electric passenger sedan failed because of inherent shortcomings in lead-acid battery technology.[15] That such vehicles did not become ubiquitous, he claims, was instead the result of the economic and technological requirements and cultural imperatives of gasoline and electric vehicle production in the context of the historical development of fuel and power infrastructure in the United States. Builders of large centralized electrical power grids such as Samuel Insull saw battery electric vehicles primarily as a tool in their battle against decentralized smaller stations, a means of storing excess electricity and improving load factors. They constructed a network of centralized recharging garages and fleets of electric vehicles optimized for use in large cities, favoring delivery trucks over cars because they had larger batteries and stored and used more power. On the other hand, electric vehicles of all types could not compete in rural markets owing to their relatively short range and, most importantly, the absence of rural recharging infrastructure. The market in the vast unelectrified North American hinterland was instead captured by Ford's Model T, a type well suited for primitive conditions. Whereas batteries were inscrutable black boxes that could be serviced only by specialized technicians, an urban professional class, gasoline motors could be easily maintained by the rural motorist with only a modicum of mechanical skill thanks to their simplicity and to the wide availability of cheap parts and fuel at general stores in small towns throughout America. Its relative technological simplicity made the gasoline automobile amenable to interpretive flexibility, unlike its electric counterpart, with different social groups using and in some cases reconfiguring it in ways that reflected their interests within existing power structures. During the early years of automobility in rural America, farmers adapted automobile gasoline motors as makeshift power sources to serve purposes the original equipment manufacturers had not intended, notably the mechanization of certain household and yard tasks.[16] The success of gasoline automobility, holds Mom, owed much to the unsophisticated, decentralized nature of its supply infrastructure in its early days.[17]

In the years leading up to and shortly after the First World War, however, the fossil fuel–based transportation system rapidly expanded and became at once entrenched, centralized, and ubiquitous in the United States as three powerful groups made common cause. A highly concentrated oil industry, a federal government that made petroleum a national security priority, and a rapidly growing manufacturing sector committed to gasoline power ensured the dominance of this system. In this period, electric drive was increasingly marginalized as American public roads were deluged with cheap, mass-produced heat engine automobiles and, after the war, surplus military trucks. Nonrail electric vehicles were relegated to off-road industrial roles hauling materials in conditions where they were considered to have safety advantages over gasoline vehicles, particularly in enclosed factory spaces. In Europe, however, electric vehicles continued to serve in the short-range delivery role. Following the Second World War, the discourse of alternative automotive technology was reduced to narrow performance comparisons between power sources in circumstances where manufacturers were constantly improving gasoline vehicles, creating new standards that battery technologists had little chance of matching, let alone surpassing. Yet battery researchers never abandoned hope for a miracle battery that would one day narrow the gap and make electric passenger automobility competitive. Defining success in terms of a breakthrough in performance that would match the gasoline vehicle in every way, these researchers tended to perceive even the latest batteries as obsolete. For them, the ideal battery always lay in the future.[18] In this period, fuel cell specialists would periodically take up the challenge of developing an electrochemical holy grail.

Pondering Alternatives

In the decades after the Second World War, the idea of electric drive as an option for passenger automobility, powered either by battery or fuel cell, remained largely beyond the pale in mainstream industry. Only when crises of sustainability wracked the fossil fuel–based transportation system in the second half of the century did opportunities arise for entrepreneurs of electrochemical automotive power in both Europe and the United States. Very often fuel cell researchers had few resources other than their chief article of faith—the potential of the technology to operate as long as it was supplied with fuel—one echoed by journalists and politicians for decades. Demonstrating this capability, however, was then extremely difficult, if not impossible, given the embryonic state of the art and the difficulty of funding sustained engineering research. Moreover, the challenges of developing fuel cell systems powerful and durable enough to cope with the demands of the automotive duty cycle deterred all but the most ardent enthusiasts. Auto companies were uninterested, leaving most of the early demonstrations of fuel cell electric drive in the hands of makers of power sources. As a result, only a handful of demonstrations were carried out before the 1990s, almost all of which were one-off experiments involving converted production vehicles using alkaline fuel cells. The first was Harry Ihrig's electric tractor, demonstrated in October 1959 using a 15-kilowatt Allis-Chalmers fuel cell. As in a number of terrestrial fuel cell programs of the period, several automotive fuel cell experiments in the 1960s employed exotic fuels, mainly to learn more about the electric automotive load profile and its effects on the power source. The U.S. Army's Mobility Command and Shell's Thornton Research Centre in London used hydrazine in electric drive conversions of a three-quarter-ton truck and a DAF 44 compact car in 1966 and 1968 respectively.[19] Shell engineers ruled out the direct use of hydrocarbons as well as onboard fuel reformer technology as too difficult to control and slow to start. Like many others at the time, they looked to methanol as the most attractive fuel, if a cheap catalyst could be found.[20] Around the end of the decade, Union Carbide's battery and fuel cell expert Karl Kordesch converted an Austin A40 compact car to fuel cell power. He claimed that with regular maintenance, his hybrid battery/fuel cell had a lifetime of about 2,000 hours, roughly comparable to that of average internal combustion engines of the time. Purely Kordesch's initiative, not an official project, the vehicle achieved some local notoriety on roads near the Union Carbide facility at Parma, Ohio, in the early 1970s.[21]

The most elaborate fuel cell electric experiment in this period was staged by General Motors in 1964. The first attempt to expressly compare battery and fuel cell electric drive, it consisted of the Electrovair, a Chevrolet Corvair battery electric conversion (an Electrovair II was built in 1966), and the Electrovan, a GM Handivan equipped with an electric motor, a Union Carbide alkaline fuel cell, and hydrogen fuel tanks. Completely filled with equipment, the Electrovan weighed 3,220 kilograms, almost twice as heavy as a standard GM van. But its greatest failing was that it could not be started with a "simple turn of a key." Like Pratt & Whitney's Apollo power plant, the automobile's alkaline fuel cell was extremely sensitive, requiring more than three hours to warm up in order to avoid permanently damaging or destroying it.[22] The project showed that, for all its theoretical advantages, fuel cell electric drive was much more complex and problematic than battery electric drive.

Given such limitations, transportation and power authorities never seriously considered fuel cells in this period. One exception was a special working party of representatives of British government agencies struck by the Ministry of Technology in 1966. Noting that electric drive was already well established in the commercial delivery and industrial sectors in Britain, the Ministry of Transport held that fuel cells in their current form had no obvious advantages over batteries in this role.[23] Researchers and electricity officials held a similar view. One 1966 study conducted for Energy Conversion Limited observed that the advent of automotive emissions legislation provided a rationale for fuel cell electric power, but high weight and a lack of data on capital and operating costs essentially ruled out its use in light-duty vehicles. More feasible, it indicated, were applications where weight and cost were not necessarily handicaps. This narrowed the field to the specialized industrial roles left to battery electric vehicles after internal combustion power swept the field in the interwar years. The report's authors doubted that fuel cell electric drive would be an effective solution for air pollution in itself.[24] Britain's Electricity Council opted for conventional lead-acid batteries for the Enfield 8000, the country's first major postwar experiment with electric cars to be used in cities.[25]

For technologically conservative American automakers, the prospect of commercial electric drive, whether powered by battery or fuel cell, remained anathema. With the emergence of air pollution as a permanent public interest issue in the mid-1950s, researchers and regulators sought to force the industry to adopt the catalytic converter, the simplest means of neutralizing the gasoline engine's most noxious exhaust elements. Detroit bitterly resisted even this modest effort at legislated technological change. It was not until 1975, under pressure from the Environmental Protection Agency, that General Motors embraced this humble remedial device, setting the industry standard.[26] Only when the supply of crude oil was interrupted by political turmoil in the Middle East did the Big Three automakers alter the internal combustion power plant itself. Under the combined pressures of high fuel costs, stiff competition from foreign automakers, and congressional passage of the Corporate Average Fuel Economy (CAFE) regulation in 1975, coupled with the 1970 Clean Air Act, Detroit produced smaller automobiles driven by smaller, cleaner gasoline engines. Industry executives characterized the consequences as "revolutionary," complaining of the vast cost and effort of achieving the 1985 target average of 27.5 miles per gallon for passenger sedans.[27]

Nevertheless, persistent air pollution at home and ongoing political instability in the Middle East in the 1970s justified further experimentation with alternative automotive power technologies. Blue chip industry contributed a handful of one-off concept cars including the Centennial Electric (General Electric), the Postal Van (American Motors Corporation), and the Electrovette (General Motors). A hasty conversion of the Chevrolet Chevette using zinc–nickel oxide batteries, the Electrovette was designed in the late 1970s as a production-ready hedge in case of a catastrophic rise in the price of petroleum. The project was scrapped when $2.50/gallon gasoline did not materialize, but its manager, Kenneth Baker, would go on to work as the director of GM's Impact electric car program in the 1990s. More substantial was Sebring-Vanguard's Citicar program. Several thousand copies of this wedge-shaped two-seater were built by the Florida-based company and by a New Jersey–based successor firm between 1974 and the early 1980s, the largest production run of any post–Second World War battery electric automobile up to that time. Even oil giant Exxon seemed to be preparing for a possible shift to electric automobility, investing in battery research and purchasing a manufacturer of electric motors in 1979.[28]

Electric automobile enthusiasts had even more to cheer about when Congress passed the Electric and Hybrid Vehicle Research, Development and Demonstration Act over President Gerald Ford's veto in 1976. Designed to "determine the commercial feasibility of electric vehicles," the law authorized the Energy Research and Development Agency to survey consumer preferences, draft performance standards, and assess infrastructure requirements of electric automobiles. It also offered incentives to support the manufacture of 7,500 electric vehicles to be purchased by the agency to demonstrate the practicality of electric drive and raise public awareness.[29] About 1,100 vehicles were built under the program's aegis before it was canceled by the Reagan administration.[30] As in the past, this initiative focused on developing a miracle battery that would make the electric auto competitive with the gasoline auto, a standard that created a perpetual cycle where battery performance would always be considered inadequate. This, in turn, served as a pretext not to deploy the technology prematurely lest the public lose confidence in it. Accordingly, the bulk of funds went to basic research, not to infrastructure or demonstration.[31]

With the onset of the 1980s, the shifting fortunes of electoral politics and global economic relations abruptly attenuated government efforts to renovate the energy conversion chain and industry's dalliance with passenger electric automobility. Existing technology-forcing measures achieved modest success. Beginning in 1979, American automakers were able to surpass the CAFE requirement for their automobile fleets by incrementally improving internal combustion engine technology. But the Carter administration's much more ambitious attempt to legislate major modifications to the energy regime (the Energy Security Act of 1980) came to naught. Combining conservation measures, investment in renewable energy programs, and a massive industrial synthetic fuel program designed to preserve the existing fossil fuel–based system of automobility, the project attempted to reconcile energy independence with environmental sustainability but was doomed by the end of the fuel crisis, opposition from the Reagan White House, and lukewarm industry interest.[32] These factors further restricted the scope of experimentation with alternative automotive power sources in the 1980s. By the end of the decade, however, a new phase in the politics of sustainable automobility would emerge, one that turned on conflicting visions of the relationship between the state and the private sector. Reacting in part to technopolitics triggered by technology-forcing legislation in California, the federal government intervened in a way that emphasized public/private cooperation in technological innovation, opening political space for green automobile dramaturgy and fuel cell futurism on a hitherto unprecedented scale.

Regulating the Future: CARB and Detroit's Electric Auto Program

Much has been written on the origins of the California Air Resources Board's (CARB) landmark Zero-Emission Vehicle (ZEV) mandate and its consequences for the politics of automobility in the United States. It was the latest initiative in the state's long-established policy of regulating technological change. The CARB had a special role in the politics of sustainable automobility. Formed in 1967 thanks to a provision of the Federal Air Quality Act (1967) that allowed California a waiver to determine its own emission standards for new vehicles owing to the state's severe air pollution problem, the CARB was the only state agency to have this power. As a consequence of the Clean Air Act of 1970, California gained the right to set emission levels lower than those of the federal government, the only state to arrogate such power. But the CARB shared with Congress and the federal civil service assumptions of the efficacy of technology forcing and the role of technology in liberal democracy. All assumed that environmental benefits would accrue as people exercised their agency as consumers rather than politicized citizens.

The difference, as analyst Sudhir Chella Rajan had suggested, was that the CARB was both more ambitious than the federal government and less capable of bending industry to its will. The state air quality regulator aimed to force automakers to adopt the best available pollution control technology in vehicles earmarked for California in order to satisfy increasingly strict emissions regulations under pain of exclusion from the market. The problem, noted Rajan, was that this did nothing to curb the emissions of millions of older automobiles. The agency compensated by adopting an "actuarial" mode of auto pollution control. Like an insurance company, it would apportion fault in auto pollution among the auto industry and consumers, quantifying the emissions characteristics of the entire light-duty fleet and monitoring individual drivers for compliance with regulations. This approach, Rajan observed, turned on the belief that economic growth would not outpace technological progress. Classifying and monitoring individual automobiles for emissions throughout their lifetime in a massive and growing automobile fleet was an enormously complicated task, one, Rajan holds, the CARB was ill equipped to execute.[33]

This task became increasingly complex after the board approved the Low Emission Vehicle and Clean Fuels emissions (LEV) standards in September 1990. Flowing from the passage of the California Clean Air Act of 1988, which set the basis for the state's air quality management for the next 20 years, the LEV combined conservative and radical technology-forcing provisions. On the one hand, its so-called Phase I requirement for clean-burning gasoline for the existing light-duty fleet was in the gradualist tradition of legislating technological change. But the LEV also set rolling quotas for automakers with annual sales of more than 35,000 light-duty vehicles in California to produce progressively larger numbers of transitional low-emission, low-emission, ultralow-emission, and zero-emission vehicles (TLEV, LEV, ULEV, ZEV). Two percent of their California production had to consist of ZEVs for the 1998–2000 model years, a quota that rose to 5 percent for the 2001–2002 model years and 10 percent in the model year of 2003 and subsequent model years. Companies could trade ZEV credits and were subject to fines for noncompliance.[34]

California law forbade the regulatory agency from specifying the technologies automakers were to use to build the various classes of low-emission vehicles. All, however, were firmly within the fossil fuel paradigm. But the only feasible zero-emission vehicle at that time was the battery electric automobile. And so the ZEV mandate (or simply the "mandate," as it became known colloquially) compelled the seven major automobile companies doing business in California—General

Motors, Ford, Toyota, Chrysler, Honda, Nissan, and Mazda, in descending order of market share—to build a market for this technology. To be sure, the number of ZEVs called for in the plan's early stages was small—only around 22,000 by 1998—in a period when around 1.7 million vehicles were sold annually in the state.

In October, Congress passed the 1990 amendment to the Clean Air Act (CAA), based largely on elements of the California Clean Air Act of 1988. Detroit now faced an apparently concerted federal-state effort to force it develop clean products. But there were important distinctions between the CAA and the LEV standards that highlighted the diverging approaches to legislated technological change adopted by Sacramento and Washington. First, the CAA embodied the gradualist approach. It extended California's Phase I standards to the rest of the country, but compelled the use of "reasonably available" control technologies. And the Bush administration had been able to prevent the adoption of California's more rigorous Phase II standards.[35] In short, the CAA was rooted in a continuum of technology-forcing legislation.

But industry regarded the mandate as a dangerous precedent. Its reaction to it has often been portrayed in conspiratorial terms—antediluvian car and oil barons combining to sink the brainchild of high-minded techno-progressives— a theme developed by policy analyst Jack Doyle in his 2000 book *Taken for a Ride* and by filmmaker Chris Paine, director of the muckraking 2006 documentary *Who Killed the Electric Car?* There is some truth in this image. For many captains of industry, the broader implications of the mandate portended chaos, especially given California's role as a national trendsetter in emissions politics. Commercial electric automobility would change the ways automobiles were built and marketed and significantly alter the automotive supply chain. Old technologies would be rendered obsolete, and markets for new ones would be created. Some interests would be privileged, while others would be punished. And long-established industrial relations and spheres of influence, especially in the energy industry, would be restructured.

The origins of this shift in automobile politics were steeped in irony, highlighting the structural weakness of the world's largest automobile company and the diverging technopolitical strategies of green automobility of the California state and federal governments as the Cold War ended. Many sources agree that the proximate cause of the CARB's decision to draft the mandate was the stunning news that GM planned to market a practical battery electric automobile based on the Impact concept car it displayed at the Los Angeles Automobile Show in January 1990.[36] This vehicle would gain fame, and then infamy, as the EV-1, an automobile that would play a central role in the subsequent dramaturgy of sustainable consumerism. The journalist Michael Shnayerson and the policy analyst Gustavo Oscar Collantes root this program in a public relations effort by GM to rebrand itself as a technology leader at a time when the auto giant faced growing internal and external pressures. Shnayerson, author of the definitive account of the Impact/EV-1, emphasizes the political strife that erupted within the company following Chair and Chief Executive Officer Roger Smith's decision to acquire Hughes Aircraft in the mid-1980s. Widely criticized by the business community, Smith attempted to justify the purchase by commissioning an effort to adapt aerospace technology for use in automobiles. The result was the Sunraycer. Built by the independent California engineering firm AeroVironment and equipped with solar cells and a silver-zinc battery pack developed by Hughes, this was a concept car employing a photovoltaic array, a common power source in spacecraft. The project satisfied Smith's initial goals. But it also inspired Robert Stempel, then an executive vice president, to develop a practical battery electric car. Enlisting Smith's support, Stempel delegated the job to AeroVironment. It crafted the lead-acid battery-powered Impact, an automobile that so impressed Smith that, at its debut, he hinted GM would produce it. Four months later, in a speech delivered at the National Press Club in Washington, DC, just days before Earth Day, Smith reiterated this claim. In July, he retired. Weeks later the CARB, whose members were appointed by the governor but were able to take action without formally reporting their decisions either to the executive or the legislature, passed the LEV statute.[37]

One way of understanding the agency's vehicle emission class hierarchy is to visualize it as a pyramid of narrowing technological options. For the basic transitional standard, automakers planned to modify existing pollution control and fuel efficiency systems. They were less certain about solutions for low-emission and ultralow-emission automobiles, including dual-fuel internal combustion engines that could use gasoline and methanol, because such technologies would require considerable modification to withstand the corrosive effects of alcohol fuels. The ZEV requirement aroused the greatest apprehension of all.[38] The Impact elicited anxiety not only among GM's competitors and the press but also among the leadership circle of its creators. None, writes Shnayerson, were quite sure what Smith and his successors hoped to achieve with this program. From 1990 to 1996, GM's intentions were opaque, largely owing to an ongoing power struggle in the top level of management. In the months after Smith's announcement, the Impact team became a virtual covert unit within the company, a "rogue cell hidden from the corporate immune system."[39] Such secrecy led Chrysler and Ford to regard the Impact as a possible commercial option. They chose to wait and see how market-leading GM would respond, for it had the largest mandate quota. The smaller producers were content to let the automotive giant take the initiative where design was concerned. But no automaker could ignore the threat the mandate posed to their collective interest in the status quo or the potential demand for clean automotive technology.

American automakers simultaneously schemed to do away with the mandate while jockeying for commercial advantage and the favorable publicity that came with green automobility, drawing on industrial alliances old and new.

A willing confederate in the antimandate campaign was the oil industry. Even before the CARB instituted its historic legislation, automobile and petroleum companies had engaged in environmental propaganda.[40] One of the first such initiatives was the Global Climate Coalition. Established in 1989 as a lobby of the largest energy and transportation interests, it was committed to debunking the science of anthropogenic global warming. American automakers had also long made use of the one-off concept automobile, mainly to promote an image of technological virtuosity and whet appetites for more pedestrian commercial products. Increasingly by 1990, such automobiles were also being used to curry favor with management for new ideas.[41] As much as in the commercial market-place, automakers competed in conceptual dramaturgy. Even if the original Impact had been designed simply as an elaborate form of greenwash—and GM had prominently highlighted the vehicle as part of its celebration of Earth Day 1990—Ford and Chrysler were at a disadvantage because they had less sophisti-cated props to deploy in green automobile dramaturgy.[42]

Nevertheless, the Big Three agreed on the chief instruments of their counterthrust against the mandate: legal pressure on the CARB through the American Automobile Manufacturers Association and a campaign to under-mine confidence in existing battery technology. Smith had been careful to place almost equal weight on the Impact's virtues and shortcomings, focusing above all on its lead-acid battery pack. This gave good power, providing sports car–like acceleration, but, claimed Smith, had poor capacity, affording a range of only 193 kilometers. Moreover, the pack would have to be replaced every two years or 32,000 kilometers at a cost of $1,500.[43] Only a superbattery, he implied, would do, one that would surpass the capacity of the best lead-acid units by five or six times, doubling or even tripling the 144–193 kilometer range of an electric auto-mobile using existing lead-acid technology.[44] As Collantes reports, the CARB agreed from the outset. It believed that advanced alternative batteries would be rapidly developed thanks to the battery industry's own confident estimates and growing demand from the electronics industry.[45]

Meanwhile, the Big Three challenged claims about lead-acid technology made by their battery suppliers, moved to constrict in-house lead-acid battery development, and began negotiating with each other on the possibility of col-laborating in developing a superbattery. After more than a year of negotiation, they formed the United States Advanced Battery Consortium (USABC) in January 1991. Coordinated with the help of the Department of Energy, the consortium consolidated all advanced battery research then under way in the auto industry and in federal and federally sponsored laboratories, allowing automakers to award funds to contractors free from antitrust worries because no superbattery had yet been commercialized.[46]

But pollution politics compelled automakers to work with the electric util-ities and the Electric Power Research Institute, which helped fund the USABC and provided technical assistance for the electric vehicle programs. Like their predecessors early in the century, the utilities saw commercial electric drive in the mandate era as an excellent opportunity to market more electricity. By late 1991, the EPRI had the most successful electric vehicle program to date, selling some 150 converted full-size GM vans to utilities.[47] The automakers now had feet in two energy conversion camps with opposing interests, oil companies on the one side and electric utilities on the other.

Advanced battery research was now under one roof. But despite constant consultation among the leaders of their respective electric vehicle programs, the automakers did not cooperate on advanced battery development. They had completely different approaches to the short-term problem of meeting the 1998 LEV deadline. Under Stempel, Smith's successor as both chair and CEO at GM, the Impact team devised a production prototype in 1992, intending to produce the first commercial vehicles using lead-acid battery packs until something better could be devised. Ford and Chrysler, conversely, were far less ambitious. Whereas the Impact had been designed from the ground up as an electric vehicle, GM's competitors planned to convert existing van and truck technology. As Shnayerson notes, their progress was undramatic. Chrysler had a much smaller share of the market, smaller clean-car quotas, and, hence, less of an incentive to commit major resources.[48] And Ford was invested in the highly problematic sodium-sulfur battery. A high-temperature (300–350°C) high-performance system pioneered by the company in the mid-1960s, this power source was plagued by corrosion and the ensuing risk of an explosive recombination of its reactants.[49] Thirty years after it began work on the technology, Ford was still struggling to master it.[50]

Meanwhile, Impact's status remained uncertain. Stempel still hoped for commercial production, but his position within the company steadily weakened as GM lost billions of dollars during the post–Gulf War recession in 1991 and 1992. In April 1992, John Smale replaced him as head of the executive committee of the board of directors. In October, GM's powerful management committee forced Stempel to drastically downgrade the Impact program to a batch of 50 hand-built demonstrators shortly before the board of directors ousted him as chief executive officer. In Stempel's place, a duumvirate arose, Smale occupying the chair and Jack Smith becoming CEO.[51]

For the next two years, the battery electric vehicle programs ran at a low ebb as the Big Three, galvanized by the decision of a bloc of nine northeastern states in October 1991 to adopt the California ZEV mandate, began a counter-offensive against government regulation. To a large degree, this effort blended with an unprecedented experiment in cooperative government/industry research and development launched by the Clinton administration in September 1993. Known as the Partnership for a New Generation of Vehicles (PNGV), the project helped broach the idea of fuel cell electric power as an alternative ZEV technology.

Collaborative Research and Development
and the Search for the Supercar

The PNGV was the first major federal initiative in collaborative research and development in the post–Cold War period. An attempt by the Clinton White House to pool the research and development capabilities of the federal government and Big Three American automakers in radically improving the efficiency of internal combustion technology, the PNGV was an important instrument of the White House's ambitious social policy agenda of sustainable industrial recovery. Formulated at least partly on the model of Japan's Ministry of International Trade and Industry, the PNGV aimed to eliminate the need for government regulation by producing an automobile with desirable social qualities but also uncompromising performance, one consumers would find irresistible. As a Clinton administration official put it, the White House sought to change the adversarial relationship between industry and government by replacing "lawyers with engineers."[52] It hoped that in this way, energy security and industrial recovery could be achieved purely by market means.

Vice President Al Gore touted the PNGV as the automotive equivalent of the Apollo moon project. Its objectives were suitably ambitious: modernize Detroit's industrial base, improve the efficiency, emissions, and performance of conventional internal combustion engine technology, and help manufacturers develop an affordable supercar of triple the fuel efficiency of the average 1994 family passenger sedan (26.6 miles per gallon or 11.3 kilometers per liter) by 2004 without compromising performance or comfort. The lead agency was the Department of Commerce, which coordinated the PNGV-related research with that of other federal bureaus including the Department of Transportation, the Environmental Protection Agency, the National Science Foundation, and the Department of Energy. In turn, these agencies partnered with the United States Council for Automotive Research (USCAR), formed in 1992 as the loosely organized cooperative R&D organization of the American auto industry.

The reality, however, was that the PNGV was not equipped to realize its objectives. The result of a declaration of intent between the Big Three automakers and President Clinton, not authorizing legislation, the partnership received no direct federal funding. Instead, it coordinated previously authorized research. More than half of the federal government's annual expenditure of around $250 million on PNGV-related work came from the DOE's energy conservation budget. Partly as a result, the supercar dominated discussion within the partnership because it was the only one of the program's three goals that could be easily measured.[53]

What was industry's incentive to participate? Writing in 1998, the government's chief technical representative in the partnership believed automakers reckoned their involvement would enable them to dodge CAFE and ZEV obligations.[54] Doyle and Shnayerson were blunter. They saw the PNGV as a form of deception, a Potemkin village born of Detroit's antipathy to regulation and the Clinton administration's eagerness to be seen as a diligent environmental steward.[55] Daniel Sperling, an influential professor of civil engineering and environmental science and policy and the founding director of the Institute of Transportation Studies (ITS) at the University of California, Davis, held similar views. Writing in 1995, he doubted whether collaborative research and development and the national laboratories could contribute much to the project of industrializing electric automobility. Basic science and new ideas were not industry's missing ingredients, Sperling wrote. A far better strategy, he held, would be to improve the cost-effectiveness of existing power source technologies and develop appropriate manufacturing processes. Here, held Sperling, lay the PNGV's fundamental flaw: the program had no mechanism to alter the behavior of the automakers. And in the absence of regulation, Detroit was unlikely to commercialize available alternative technologies, let alone the advanced concepts investigated by the partnership.[56] Six years later, with this prediction borne out, Sperling wrote that the PNGV's major effect had been to accelerate Japanese and German alternative automobile technology programs.[57]

There is much to commend this analysis. Another of the program's collateral effects may have been to idealize the fuel cell electric as a nonbattery electric supercar to an American audience. As in the defense research apparatus in the 1960s, an advisory loop linking academe, industry, and government helped circulate assumptions of the capabilities of industrial-scale fuel cell technology based on news of the latest advances in the laboratory. This dynamic had manifested virtually from the beginning of the federal government's involvement in fuel cell electric drive in the late 1970s. To be sure, planners only became enthusiastic about a fuel cell supercar in the early 1990s, thanks to dramatic improvements in membrane fuel cell technology. A decade earlier, inspired by modest progress in the utility fuel cell programs at a time of rising petroleum costs, they had merely been curious about the possibilities of adapting the phosphoric acid fuel cell for vehicular use. In 1977, the ERDA sponsored a workshop at the Los Alamos National Laboratory to investigate such a project. However, the phosphoric acid design was no better than its alkaline forebears in meeting immediate power demand, taking several minutes to warm up. As in the past, academic and industry researchers looked to the hybrid battery/fuel cell system as a means of compensating for this shortcoming, with the battery providing the initial power boost until the fuel cell was ready for action. Further studies involving the Los Alamos and Brookhaven national laboratories and the U.S. Army yielded "applications scenarios" that found that such a system using methanol and propane fuel was technically feasible in a number of electric automotive applications including cars, buses, trucks, and delivery vans. In 1980, the Los Alamos Scientific Laboratory Electronics division began testing a converted golf cart equipped with a hydrogen-fueled phosphoric acid power plant.[58]

Like all liquid electrolyte systems, however, the phosphoric acid fuel cell was bulky. Near the end of the decade, the Department of Energy began experimenting with the system on a bus chassis, a technically simpler test platform than the passenger automobile or light truck because the power unit did not have to be miniaturized. Begun in 1987 as a partnership between the DOE, the Department of Transportation, and California's South Coast Air Quality District, the Fuel Cells in Transportation Program aimed to build three electric buses equipped with methanol-fueled phosphoric acid fuel cells.[59]

Three years later, the DOE began work on a methanol-fueled membrane fuel cell for light-duty electric vehicles. Managed by the Allison Gas Turbine Division of GM, the project also included Dow Chemical as the membrane supplier and Ballard Power Systems as the stack developer. A team at Los Alamos built the reformer and tested the power unit.[60]

As in previous DOE efforts, this project had a low profile, possessing an annual operating budget of only a few million dollars. In late 1992, however, two academic physicists authored a paper that helped make a case for including government fuel cell research and development in the PNGV.[61] They were only the latest observers to be impressed by the potential of the improved membrane fuel cell as the prime mover of the electric passenger automobile. Henry Kelly, then a senior associate at the congressional Office of Technology Assessment, and Robert H. Williams, a professor at Princeton University's Center for Energy and Environmental Studies, wrote that they doubted whether the American automobile industry could be rehabilitated in a way that was environmentally sustainable and that reduced U.S. dependency on imported fossil fuels if manufacturers remained committed to internal combustion technology. Fuel cell electric power, held the scientists, possessed this transformative potential. Noting recent advances in the latest membrane variants, they understood the fuel cell as a uniquely flexible, efficient, and emission-free power device. An electric car powered by it, they wrote, would be as convenient to refuel (using hydrogen or methanol produced from a wide range of existing feedstocks) and as fun to drive as the gasoline-powered automobile. In short, Kelly and Williams redefined the sort of performance automobile technology should be capable of. For them, the fuel cell was the standard against which the alternatives had to be measured. In a visionary flourish that recalled the musings of an earlier generation of enthusiasts, the scientists claimed that the device moved energy conversion beyond the "age of fire" into the "age of electrochemistry."[62]

Such language strikingly paralleled the rhetorical tenor of the Clinton administration's science and technology policy. Indeed, the proton exchange membrane fuel cell was made part of the PNGV virtually from the outset thanks to a joint decision of the federal government and USCAR representatives.[63]

Kelly likely had a hand in this. Joining the White House's Office of Science and Technology Policy in February 1993 as its assistant director for technology, he became a leading architect of the PNGV and helped point out the advantages of fuel cell power to the administration.[64] As the decade progressed, the tropes that Kelly and Williams had deployed would be gradually taken up by other actors and become a key element in the technopolitics of the ZEV mandate. In 1992 and 1993, however, U.S. automakers did not evince much interest in the technology. As the decade progressed and the first of California's mandate deadlines approached, the federal government and foreign companies took the leading roles in demonstrating how fuel cell electric drive might be a practical commercial ZEV option after all.

Dawn of the Fuel Cell Electric

On its face, the partnership between Daimler-Benz and Ballard Power Systems was an odd one. It paired one of the world's most storied automakers with a tiny, unprofitable engineering enterprise based in the environs of Vancouver, British Columbia, then North America's fifth-largest port but hardly a high-technology hotbed in the late 1980s and early 1990s. These two companies were drawn together by mutual interest even before mandate politics exploded in the United States. The Canadian firm first attracted attention in 1986, when it acquired an experimental Dow Chemical proton exchange membrane and applied it in a fuel cell, achieving a remarkable power density for the time. Shortly afterward, Ballard began concerted work to further develop and test the technology. In turn, these experiments helped bolster the company's moribund business in disposable lithium batteries. Impressed by one of Ballard's early fuel cell power plants, the Vancouver venture capitalist Michael Brown raised $1.3 million in 1987 by selling investors on the company's potential as a leader in membrane fuel cell technology.[65]

This became the basis of the company's efforts to raise investment capital. Gradually, Ballard Power Systems refined an approach that explicitly linked its corporate image with the burgeoning awareness of environmental issues. In 1989, Brown helped broker the resignation of founder and chief executive officer Geoffrey Ballard and the hiring of Firoz Rasul as his replacement. A professional marketer, Rasul had experience in the data collection industry but no background in electrochemistry. In lieu of a commercial product, Ballard Power Systems marketed the hope of fuel cell technology in a world "looking for solutions for both power and environment." And the largest potential market for this product, of course, was electric passenger automobility. Investigating stationary and mobile applications of the membrane fuel cell, the company was increasingly drawn to the latter after Daimler-Benz began to make overtures in 1989.[66] Over time, Brown and Rasul would develop a strategy for corporate growth that edified virtuous automobility as a moral imperative, an ethos reflected in the corporate motto "Power to Change the World."

During the next decade, Ballard would specialize in integrating polymer membrane supplied by Dow, Du Pont, and Asahi Chemical in precommercial fuel cell demonstrators. Costs remained stubbornly high, however, and in the early 1990s Ballard set out to develop its own membrane, an effort led by the chemist Alfred Steck. Meanwhile, Daimler-Benz had become interested in Ballard's Mk 5 stack, one of the most powerful of the day. Producing 150 watts per liter/ kilogram, the unit had some 30 times the power density of General Electric's Gemini fuel cell. At $60,000 per kilowatt, it was similarly expensive.[67] In 1989, one year after concluding its demonstration of hydrogen combustion automobile drive in Berlin, the German automaker leased an Mk 5 for trials, signing a four-year deal with Ballard to build a fuel cell demonstration automobile in 1993. The arrangement was a coup for Ballard. Rasul's early years at the helm had not always been easy. He encountered considerable skepticism from potential investors, one of whom concluded that fuel cells were "another scam like cold fusion" and saw the Ballard chief out of his office.[68] The company scored one of its first successes developing power packs for several electric bus demonstrators in collaboration with the municipal governments of Vancouver and Chicago in the early 1990s.[69] This program did much to raise Ballard's profile, but the company's relationship with Daimler-Benz fixed its attention firmly on the passenger automobile. Over the next seven years, this alliance produced a series of sophisticated fuel cell electric concept automobiles known as the NextCars.

Ballard believed the partnership would allow the company to play a key, if not the leading, role in shaping a whole new paradigm of commercial automobile propulsion. Daimler-Benz's motives in the relationship, on the other hand, were perhaps less obvious. Some observers have claimed that European automakers (and Daimler-Benz in particular) were more responsive to pressure for cleaner products than their American counterparts in the 1990s.[70] But in the United States, the German automotive giant was much less progressive than many of its competitors in developing the technology of fleet fuel efficiency. In the world's largest market, Daimler-Benz and a number of other foreign companies were able to effectively circumvent the regulatory discipline that compelled other automakers to experiment with alternative automobile technologies thanks to their cultivation of the extremely profitable small-volume market for high-performance, polluting vehicles. The federal government's sanctions were far too weak to encourage such companies to improve the efficiency of their relatively small fleets. Daimler-Benz and BMW were prepared to pay stiff fines year after year for violating the CAFE standards, compiling the worst record of all automakers over a period of nearly a quarter century between 1983 and 2007.[71] And neither company was subject to California's ZEV mandate owing to their small market share.

Daimler-Benz's interests in alternative automobility and in its partnership with Ballard probably stemmed instead from a long-standing affinity for advanced technology combined with a rather vague sense of the commercial possibilities arising from the growth of environmentalism and shifts in consumer values and preferences in the 1990s. As befitted a firm whose founders had helped invent internal combustion–powered automobility, the German manufacturing giant perceived itself as a paragon of sophistication, expressed by the unequivocal motto of its Mercedes division—"das Beste oder nichts": the best or nothing. Perhaps more than any other automaker, it was committed to experimenting with alternative propulsion.[72] Perceiving itself to be on the frontier of technology, the company had an ongoing fascination with hydrogen systems dating to the early 1970s. In 1989, Daimler-Benz began investigating fuel cells and, in 1994, became involved with Nicolas Hayek's ultracompact city car project, a venture that eventually resulted in the formation of the company's Smart (car) division. Both initiatives were begun on the watch of Chief Executive Officer Edzard Reuter, a man whose corporate strategy in some ways resembled that of Roger Smith at General Motors. Each executive worked to make his company into a diversified high-technology colossus with interests in the prestigious fields of electronics and aerospace.[73]

Daimler-Benz also had a superbattery for electric drive, the high-temperature (270–350°C) sodium-nickel-chloride combination. Developed in the 1970s and 1980s through a partnership between two government entities—the South African National Physical Research Laboratory (CSIR) and the British Atomic Energy Research Establishment at Harwell—along with the mining giant Anglo American Corporation, the power source technology was acquired by Daimler-Benz in the mid-1980s and used in a high-profile demonstration program in the early 1990s employing Mercedes electric cars and buses.[74] Although exempted from the ZEV mandate, the German company participated in some of the well-publicized electric automobile races that began to be staged in the United States after September 1990, its engineers hastening to point out the high cost and low durability of its battery system.[75]

In adopting the Ballard fuel cell, Daimler-Benz upheld its high-tech traditions. But unlike its work with hydrogen drive in the 1970s and 1980s, the automaker appeared to be thinking more practically about the question of fuel in the early 1990s. It continued to experiment with hydrogen in conventional heat engines, working with Swiss researchers in a project backed by the German Federal Ministry for Research and Technology to explore the use of off-peak hydroelectric power to produce electrolytic hydrogen, not for passenger automobiles but for buses.[76] This was intended for use in a regional power/transport system on the Swiss-German border, an area rich in hydroelectric potential. But by then Daimler-Benz no longer claimed, as it had in the 1970s and 1980s, that a hydrogen fuel system could be relatively easily grafted onto the existing fossil fuel infrastructure. Its officials held that, ultimately, the company would develop the fuel cell electric automobile as a mobile chemical plant, converting carbonaceous fuels to hydrogen.[77]

The debut on April 1994 of the Daimler-Benz and Ballard partnership's first demonstration automobile, the NextCar (NECAR) I, signaled the gradual opening of a new front in the global auto industry's campaign against California's legislated battery electric vehicle. Based on a van chassis and powered by hydrogen fuel cells, the NECAR I was not much more practical than General Motors' 1960s-era Electrovan, its interior being similarly filled with equipment.[78] But Daimler's claim that the Ballard power unit could use methanol strongly suggested that the partners were serious about commercializing fuel cell electric power in the short term. The *Wall Street Journal* reporter Oscar Suris hailed the NECAR I as a breakthrough in "no-battery" electric automobile technology. He suggested that Detroit and Washington would have to respond lest they forfeit "a potential technological edge" to a foreign competitor, noting that the demonstration bolstered voices calling for fuel cell research in the PNGV.[79]

The Road to the Super ZEV

And in 1994 PNGV planners did identify the fuel cell among the candidate 'leapfrog' technologies, along with hybrid diesel and gas turbine–hybrid electric. What did this mean in practical terms? At first, collaborative fuel cell programs were relatively small, in some cases building on existing efforts. For its part, General Motors continued working on the methanol-fueled membrane fuel cell sponsored by the DOE. In September, it moved into the second phase of a 30-month effort intended to yield a 60-kilowatt brassboard (more advanced than a breadboard model) meant for testing outside the laboratory. But in a new development, Ford and Chrysler were assigned responsibility for investigating membrane fuel cell systems using hydrogen. In July, three months after the demonstration of NECAR I, the number two and three U.S. automakers received 30-month contracts from the energy agency worth $13.8 and $15 million respectively to build 30–50 kilowatt power units.[80] The emergence of parallel fuel programs in the PNGV harked back to ARPA's hedging strategy in the 1960s. Once again, planners struggled to reconcile unsatisfactory trade-offs between infrastructure and vehicle performance.

For the moment, fuel cell research and development remained a sideshow in the politics of battery electric automobility. But it grew in importance as GM shaped these politics via its role as industry leader. By 1994, the Impact project had progressed to a limited demonstration phase known as PrEView, designed to place 50 vehicles in the hands of ordinary motorists for field testing with the aid of the electric utilities. Sudden developments in battery technology, however, altered the course of this program and Detroit's approach to relations with the CARB and the ZEV mandate. In January, notes Shnayerson, the Impact team secretly tested a preproduction vehicle using a nickel–metal hydride battery designed by Stanford R. Ovshinsky, an independent inventor and owner of the Ovonic battery company. Although Ovshinsky had developed the technology with the help of a USABC grant, the consortium prohibited him from publicizing the data until extensive engineering research could be performed, pouring cold water on his performance claims. Nevertheless, the test result, over 201 miles (337 kilometers) on one discharge, astounded Impact researchers. Word filtered up the GM chain of command. With popular support for the mandate deepening in California and the northeastern states and GM leaders believing that further lobbying had limitations, Jack Smith upgraded the Impact program. It was now to be run on a precommercial basis of around 20,000 vehicles, produced as quickly as possible in order to claim CARB's early-arrival credits and using Ovonic batteries as these became available.[81]

These events heralded a seeming reversal of fortune for battery electric drive. After years of intransigence, manufacturers abruptly softened their position vis-à-vis the CARB in late 1995. Now they offered to produce a total of 5,000 battery electric vehicles in late 1996 and early 1997, well before the 1998 deadline, and 14,000 in 1998, in exchange for lowering the 1998 quota of 22,000 zero-emission vehicles. It was a curious offer for the auto industry to make because it contradicted its earlier warnings of the risks of a premature market debut.[82] At any rate, the parties reached a compromise. In early 1996, the CARB amended the ZEV portion of the LEV program. Accepting industry's argument that existing battery electric technology could not meet consumer expectations for range, the regulator agreed to eliminate the quotas from 1998 to 2002 altogether and worked out memoranda of agreement with each of the seven largest automakers, committing them to produce 1,800 battery electric vehicles between 1998 and 2000. To compensate for the smaller number of vehicles, the automakers were subjected to stricter emission standards and agreed to continue ZEV research and development.[83] Only the 10 percent quota for 2003 remained unchanged. In January 1996, after years of shrouding its battery electric program in secrecy, GM once again displayed an Impact at the Los Angeles Auto Show, now in the form of the EV-1 precommercial automobile.

For American automakers it was a major victory in their struggle against the mandated zero-emission automobile. General Motors benefited most of all, reaping a double reward. The CARB had always deferred to industry's technical judgment, echoed by the press, that the electric automobile powered by the lead-acid battery was only a temporary solution that actually had the potential to damage the market for electric automobiles by raising performance expectations that could not be satisfied. Now, GM had played a leading role in convincing the emissions regulator that there were still no better short-term options for electric cars than lead-acid batteries and, hence, that further research was necessary.

Of course, the promising Ovonic nickel–metal hydride battery was then in an advanced stage of development, but General Motors was not eager to accelerate that program. And yet it could and did claim to be the first automaker to revive commercial electric car production, evading charges of technological obstructionism without immediately having to produce large numbers of automobiles.[84] All could be delayed until the promise of a miracle battery was fully realized. Other automakers including Ford and Toyota would follow with their own limited-production battery electric automobile programs.

The CARB's retreat also opened the door for the prospect of fuel cell electric power as the basis of an alternative ZEV. As negotiations unfolded in December 1995, some California air quality officials mused that that fuel cell represented the future of electric automobility.[85] And as automakers succeeded in their campaign to discredit lead-acid battery technology, Daimler-Benz and Ballard made fuel cell power seem an ever more credible alternative to the battery-powered ZEV. By the end of 1995, the Canadian company had achieved a power density of 570 watts per liter, well above the benchmark of 400 watts per liter set out in the PNGV's Technical Roadmap, an efficiency goal based on hydrogen fuel.[86] In May 1996, the partners displayed the hydrogen-fueled NECAR II in Berlin.[87] With room for six passengers, this electric van was hailed by the press as a major technical and aesthetic advance over its test-bed predecessor.[88]

For some years, automakers had been favoring the days leading up to Earth Day as an occasion for dramatic revelations of the latest sustainable technology, and April 1997 brought momentous developments. In midmonth, Ferdinand Panik, Daimler-Benz's vice president in charge of fuel cells, declared that the company planned to market a minimum of 100,000 fuel cell vehicles beginning in 2005, having largely solved technical and infrastructure problems. The company punctuated this commitment with a $500 million investment in Ballard, taking a 25 percent stake in its partner.[89] This unprecedented expression of faith in automotive fuel cell power would trigger a research and development boom and a whole new phase of ZEV politics. Several days later, Ford declared that it intended to build an advanced hydrogen fuel cell electric automobile with the support of the PNGV. Environmentalists such as Jason Mark, a transportation analyst with the Union of Concerned Scientists, voiced approval, remarking that the fuel cell—here he meant the *hydrogen* fuel cell—was "perhaps the most promising" sustainable power source.[90] In September, both Daimler-Benz and Toyota unveiled the first methanol fuel cell electric passenger automobiles at an auto show in Frankfurt. Based on the Mercedes-Benz A-class platform, the NECAR III also was the first of Daimler-Benz's fuel cell demonstrators to use a subcompact chassis, although the reformer occupied most of the passenger and cargo space.

There was, of course, a world of difference between pure hydrogen and carbonaceous fuel cell systems, a distinction some researchers and the media often elided.[91] In fact, methanol fuel cell technology faced almost exactly the same challenges as it had in the 1960s. Chrysler's fuel cell chief Christopher Borroni-Bird noted that direct methanol devices (systems that did not use a reformer) required more platinum than hydrogen fuel cells owing to alcohol's low reactivity and still suffered from premature cathodic oxidation.[92] As in the past, integrating fuel cell power units with fuel reformer technology raised a whole new range of difficult technical hurdles.

Fuel Cell Fuel Politics Redux

These sorts of problems compelled PNGV managers to change their approach to electrochemical/electric propulsion, a decision that would have significant implications both for the consortium's research program and for ZEV politics. From 1994, PNGV participants worked simultaneously on hydrogen and alcohol fuel cell systems. But from around 1996 and over the next four years, they began to gradually shift emphasis from methanol to gasoline and then to multifuel reforming, each a successively more difficult and demanding process. By the fall of 1997, Borroni-Bird was speaking of the need to accept the reality that the gasoline infrastructure was here to stay for some time, outlining Chrysler's work in developing reforming technology.[93] Gasoline, of course, was a much more problematic substance than methanol. Its reforming process required a higher temperature (700°C as opposed to 200°C–300°C) and more fuel energy to dissociate hydrogen from the hydrocarbon molecules, bringing with it the problems of thermal expansion and corrosion. And pollutants including carbon monoxide and sulfur had to be removed from the resulting fuel gas stream before it could be used in a fuel cell. All these factors greatly complicated the design of the electric power train.[94]

The new emphasis on hydrocarbon fuel cell systems corresponded with a decision by PNGV managers to prioritize fuel cell technology. Between 1995 and 1998, the DOE's Office of Advanced Automotive Technologies (OAAT) made it the second most important propulsion program after hybrid systems on the government side of the PNGV in terms of research investment, spending an average of about $22 million per year. In contrast, the office devoted an average of around $43 million annually to hybrid technology in this period. But from 1997, federal planners began to increase spending on fuel cell research and development, which climbed to top place among the various technology programs supported by the OAAT in the PNGV by mid-1998. In that year, the office invested around $34 million in the field.[95]

The first results with fuel reformer systems were a good deal more modest than what their promoters claimed. For example, in October 1997, Energy Secretary Federico Peña heralded an experiment in gasoline reformation conducted by the engineering firm Arthur D. Little, the PNGV's main contractor for reformer technology. This, said Peña, was a "terrific breakthrough" on the road to nonbattery electric drive. Jeffrey M. Bentley, the president of the company, was even more ebullient, remarking that the processor/fuel cell system "blows the doors off any battery-powered electric vehicle."[96] The hyperbole dressed up what had been a rather humble test. It had involved only a single Ballard cell over a period of 16 hours, hardly sufficient to give a realistic indication of how a scaled-up gasoline reformer/fuel cell system would behave during thousands of hours of operation in an electric automobile. Criticism was soon forthcoming from Jason Mark of the Union of Concerned Scientists. The federal automotive fuel cell program, he remarked, was "veering off course." Gasoline reforming, Mark noted, made fuel cells more complex, dirty, and costly, undermining their "full potential," which could be attained through the use of "alternative" fuels like hydrogen.[97]

To be sure, the PNGV's fuel cell program had never been dedicated exclusively to hydrogen, having taken a multifuel approach since July 1994. But Mark's views of the technical shortcomings of gasoline reforming were correct. And hydrogen did not obviate but merely externalized and socialized the physicochemical complexity of the miniature chemical plant that was the carbonaceous fuel cell/reformer system.

The nuances of the fuel question would largely be lost in the public discourse that emerged in the wake of the Ford Motor Company's historic decision in late 1997 to invest in fuel cell power. In becoming the first American automaker to make a significant commitment to the technology, Ford made a dramatic statement that stood in stark relief to the company's reputation for conservatism. Like its industry peers, Ford had long engaged in crude dilatory tactics to avoid building cleaner and safer automobiles, combining appeals to individual liberty and fears of big government with disinformation and the threat of job losses. The corporation did its share to roll back existing fuel economy standards and cast doubt on climate science, joining its competitors in forming the Climate Change Coalition in 1989 and the anti-CAFE Coalition for Vehicle Choice in 1991.[98]

By mid-decade, however, Detroit was attracting increasing scrutiny from environmentalists and supporters of electric vehicles, and Ford was not immune. Activists acquired internal memos that seemed to indicate that the company was inflating the price of its electric Ranger pickup truck. This, in turn, suggested that Ford suspected a market for electric vehicles did exist and was trying to suppress it.[99] By late 1997, however, the automaker was showing signs of changing its approach to environmental and pollution politics. In a major speech at an executive retreat in October, William Clay Ford, Jr., the great-grandson of Henry Ford and then chair of the finance committee publicly acknowledged the adverse environmental effects of fossil fuel–based automobility and called on the company to address the issue. Only days before, Alex

Trotman, the company's chair, had joined the other Detroit chiefs in personally lobbying President Clinton to scuttle the upcoming greenhouse gas treaty to be considered at Kyoto.[100]

But in December, the Ford Motor Company dramatically repositioned itself as a leader of sustainable automobility. In a seemingly epochal move, the corporation announced that it was purchasing a 15 percent stake in Ballard Power Systems worth $600 million and entering a joint venture with the Canadian firm and Daimler-Benz to commercially produce fuel cells and fuel cell electric drives for commercial automobiles to be available in 2004. Ford executives lavished praise on the device. Trotman remarked that the fuel cell was "one of the most important technologies for the early twenty-first century." William Clay Ford, Jr., suggested fuel cells had finally made electric drive passenger vehicles practical.[101]

The American auto industry seemed to be undergoing deep structural change. The number two U.S. automaker had literally bought into a technology it had previously shown little interest in and that was being pioneered by foreign companies. In May 1998, Daimler-Benz acquired Chrysler, aligning the smallest American auto manufacturer with the agenda of the German corporate giant. And preoccupied though it was with the EV-1 in this period, GM expanded its own research in fuel cells. The corporation's investment in the technology, held J. Byron McCormick, GM's chief of alternative propulsion and a long-standing proponent of fuel cell power, was at least as large as its commitments to hybrid and battery electric propulsion.[102]

These moves were reflected in the research priorities of the PNGV. By the end of calendar year 1997, representatives of the federal government and USCAR had narrowed the choice of candidate power source technologies to the compression ignition direct injection advanced diesel engine and the fuel cell. True, noted a standing review committee of the National Research Council, cost projections of the gasoline fuel cell system—about $500 per kilowatt—were ten times higher than the PNGV target for 2004. But of the advanced power sources under consideration, the committee reported it could find evidence the consortium had stimulated increased investment only in the fuel cell technology sector.[103]

In fact, the fuel cell was becoming attractive to industry even as the fuel question grew more opaque. Worried that oil companies would not willingly build a methanol infrastructure, Daimler-Benz had begun investigating gasoline reformer technology prior to the merger with Chrysler.[104] Now invested in two kinds of fuel reformer systems, the company said it would wait until late 1999 before deciding which one to pursue. In the meantime, however, it continued to use the hydrogen-powered NECAR II in public demonstrations although its methanol-fueled NECAR III, introduced in late 1997, ostensibly represented the state of the art. This was not a trivial matter, given the different operating characteristics and infrastructure requirements of hydrogen fuel cell and reformer/fuel cell systems and the impressionability of the media and investors to "break-through" fuel cell vehicle demonstrations.[105]

Some observers sensed disquieting trends in the emerging fuel cell electric sector. Ballard had come some way in improving performance and cutting costs, noted *The Economist* in October 1998. But the influential news magazine was doubtful of the parallel Rasul had invoked between his firm and semiconductor giant Intel. The Canadian research laboratory still faced huge obstacles in moving from R&D to manufacturing. As early as 1993 Ballard had developed a proprietary membrane, designed as a cheap low-fluorine alternative to the industrial polymers. But producing this substance in volume was difficult for a small company not optimized for industrial materials production. In 1997, ionomer membrane cost between $400 and $1,000 per kilowatt, or $50–$70 per square foot, a figure that Alfred Steck, Ballard's chemist in charge of the membrane program, believed had to be reduced to between $5 and $15 per square foot for the fuel cell electric automobile to be considered commercially viable.[106] It was a bad sign, noted *The Economist*, that Ballard's stock price was as responsive to green political rhetoric as it was to the latest technological and industrial developments. Still, the company had an impressive array of backers including six carmakers it supplied with prototype cells. And Ballard "would have all the top people on board" once the oil industry moved forcefully into the field.[107]

Indeed, as the oil and car companies cooperated in their anti–battery electric campaign, they found concord with air quality regulators in the principle, if not the reality, of the fuel cell as a fuel-flexible super-engine. In November 1998, industry obtained another concession from the CARB that tacitly recognized fuel cell electric power as a potential solution to the air quality problem. This was the so-called LEV II revision. Only 40 percent of the 2003 quota would now have to consist of true zero-emission vehicles; the other 60 percent could be filled by the partial zero-emission vehicle (PZEV). Based on the new "super ultra low emission vehicle" (SULEV) exhaust standard, the most stringent emission category after the ZEV, the PZEV class was eligible for partial ZEV credit. The CARB had drafted the original mandate, noted ITS researchers, at a time when the battery electric was believed to be the only viable technology that met the definition of a ZEV. The rule change, they held, acknowledged the progress that had been made in the intervening years on a number of alternative automotive propulsion systems including advanced conventional gasoline, advanced natural gas, hybrid electric, and methanol fuel cell electric drive. And it gave automakers the incentive to further develop these technologies.[108]

As air quality officials broadened the technological scope of the mandate, industry and government cooperated in crystallizing fuel cell power as the new exemplar of electric automobility. In March 1999, DaimlerChrysler chose Washington, DC, as the backdrop to showcase the NECAR IV, its latest fuel cell demonstrator. Billed as the first fuel cell vehicle to be driven on public roads in the United States, the automobile was hailed by Environmental Protection Agency Administrator Carol Browner as a step toward sustainable transportation.[109] Actually, this liquid-hydrogen-fueled vehicle was a sudden retrograde move for DaimlerChrysler, whose design arc had been trending toward carbonaceous fuel cell systems. But the car had a range of 450 kilometers, the best performance of any of the company's fuel cell demonstration vehicles to that point.[110]

One month later at the State Capitol in Sacramento, in the presence of air quality officials and California governor Gray Davis and flanked by NECAR IV and Ford's P-2000 hydrogen fuel cell electric automobile, auto and oil companies inaugurated the California Fuel Cell Partnership (CAFCP). Comprising DaimlerChrysler, Ford, Ballard Power Systems, Shell, Texaco, Atlantic Richfield, and the CARB, the CAFCP was a large demonstration program designed to test new technology, develop product standards, and educate consumers in preparation for the commercialization of the fuel cell electric automobile. It was, said Governor Davis, the first part of a long-term plan toward achieving zero emissions in the state.[111] Shortly afterward, GM and Toyota announced a five-year cooperative agreement. The CAFCP, wrote *New York Times* journalist Andrew Pollack, represented a "milestone," a sign that fuel cells were "stepping out of the laboratory and onto the road."[112] After many bitter disappointments, the dream of the universal electrochemical energy converter appeared to be closer to reality than ever before.

Back to the Future

The CAFCP was a landmark in the history of the fuel cell. Never before had mainstream industry mounted such an ambitious, well-resourced, and well-publicized effort to develop the technology. Fuel cell electric programs attracted much attention and commentary. Yet doubts persisted. In a front-page story in the *Wall Street Journal* in March 1999, reporter Jeffrey Ball was struck both by industry's interest in fuel cell power at a time of cheap gasoline and the efforts of automakers to outdo each other in the field. He attributed the events as a product of impending government regulation, technological breakthroughs, and "one-upmanship." Was there an end to pursuing a power source that cost some 8 to 15 times more than the internal combustion engine? Perhaps the prestige that came with being first to market with a potentially revolutionary product, opined Mark J. Amstock, Toyota's national environmental vehicle project manager. DaimlerChrysler's bold 2004 deadline for commercialization suggested that decisions in production capacity were imminent for a host of industry players.[113]

All these factors helped impart momentum to the automotive fuel cell project. In technical terms, the CAFCP represented the beginning of real engineering research on the technology. As a political statement asserting the superiority of fuel cell over battery, the program affirmed the primacy of the technical expertise of the American automobile industry as it related to the issue of sustainability. Ironically, the Clinton administration had helped cue Detroit in taking this technopolitical stance. Through its support of cost-shared research and development, the federal government encouraged manufacturers to take up the innovation of esoteric hardware in lieu of forcing them to use more prosaic systems to ameliorate air pollution. But via the PNGV, it had also prompted automakers to frame the fuel cell as a power panacea, a single solution to the air quality crisis. In so doing, the federal government helped industry undermine the capacity of the public, above all the state of California, to dictate solutions to the problem of sustainable transport.

As events would demonstrate, the federal government was doing Detroit no favors by this strategy. Crucially, the respective states of knowledge of battery and fuel cell power were inversely proportional to the cultural status of each technology. Industry's distaste for batteries as a primary power source was in some cases well founded, particularly where such problematic technologies as the sodium-sulfur combination were concerned. In instances involving more promising designs, this skepticism informed prudent engineering research. John Williams, the GM executive in charge of technical affairs at the USABC, constantly questioned Stanford Ovshinsky's cost, power, and lifetime estimates for his nickel–metal hydride battery, noting that claims made on single cells were not relevant for cells arrayed together in a battery.[114] A slow, methodical approach made good sense in light of such complexities.

Most major automakers gained a wealth of experience building and operating the mandated battery electric fleets in the late 1990s and early 2000s. But although all were circumspect about the potential of this technology, they were not nearly so cautious about fuel cell electric drive, a field of which they knew far less. Only DaimlerChrysler had some familiarity with the system and had yet to subject it to exhaustive operational testing. Nevertheless, the rhetoric of fuel cell power heated up at precisely the moment the electrochemical research and development programs approached a crossroads. Industry was determined to limit production of battery electric automobiles. And real progress in the fuel cell electric projects had occurred mainly with the energy conversion unit itself, conforming to historical trends. A miniaturized electrochemical energy conversion plant on wheels capable of delivering the kind of performance Americans had come to expect was beyond the engineering capabilities of the time. Although engineers had made some advances in reformer technology by the late 1990s, the overall results were disappointing. Gasoline reforming proved impracticable owing to the extreme difficulty of quickly cracking the fuel and extracting the various poisonous by-products, a task further complicated by the determination of oil interests to fight the Environmental Protection Agency's low-sulfur standard for gasoline.[115] Methanol reformers had been improved but fell far short of expectations. In 1999, Arthur D. Little reported that fuel cell vehicles using the technology could not yet be cold-started, requiring up to 30 minutes to warm up before they could be driven.[116]

And researchers also began to realize that the existing fuel infrastructure was not as adaptable as they had first thought. Methanol is more flammable than gasoline when exposed to air in a confined space, and alcohol fuels become highly corrosive when they encounter residual water in gasoline pipelines. For these reasons, Exxon researchers believed extensive modifications to fuel production and distribution systems would be necessary if methanol was to replace gasoline. They estimated this would cost between $15 billion and $23 billion for each million barrels per day and hundreds of billions of dollars for complete duplication of production.[117] Environmental considerations posed yet another obstacle. Some observers wondered whether the massive use of methanol was possible in California, where public opinion was already highly sensitive to groundwater contamination from the gasoline additive methyl tertiary butyl ether (MTBE).[118] This was an ironic twist, given industry's efforts to frame the methanol fuel cell as a green machine.

Detroit's growing affinity for the fuel cell electric automobile as its preferred ZEV, hence, clashed with the realities of the industrial-technological scene in the late 1990s. Producing such a vehicle on a commercial scale promised far greater manufacturing challenges than those automakers were coping with in their battery electric programs. And despite having developed sophisticated hybrid passenger automobile demonstration technology in the PNGV, American companies had chosen not to commercialize this "bridge" to a zero-emission light-duty fleet. Honda and Toyota, on the other hand, marketed such vehicles in the United States in 1999 and 2000 respectively.

American automakers said they wanted a "leapfrog" technology. The carbonaceous fuel cell electric project had foundered. But now hydrogen systems came back into vogue. Thirty years earlier, hydrogen had mainly been the province of futurists, regarded as a dead end by those few entrepreneurs willing to make a go of commercial fuel cell power. Near the turn of the millennium, however, a number of influential actors began to reconsider this fuel. A 1999 report by the Hydrogen Technical Advisory Panel (HTAP) to the DOE concluded that direct hydrogen was feasible for fuel cell automobiles and carried no penalties in comfort or performance.[119] In 2000, Arthur D. Little released a study suggesting that "off-board" fuel reforming might be preferable to miniaturized automobile chemical plants.[120] The National Research Council held similar views, recommending the PNGV consider ways to produce hydrogen from natural gas or gasoline at existing service stations.[121]

To be sure, many companies scaled back their fuel cell work as it became evident that only hydrogen systems could deliver the sort of performance pundits said consumers demanded. These technologies implied costs far greater than most would-be manufacturers were willing to pay. But there was no exodus from the field as in the past, because at this crucial juncture the federal government intervened in support of hydrogen and hydrogen fuel cell technology on an unprecedented scale. Players that hitherto had little exposure to the field such as Honda and above all General Motors would position themselves in the vanguard of what they termed a technological revolution. For a time in the early 2000s, it seemed as if the federal government might guide industry in developing the sort of hydrogen economy the futurists had long dreamed of. The result was the creation of an institutional framework within which fuel cell technopolitics would flourish on a larger scale than ever before.

7

Electrochemical Millennium

The way I think about it, we're close today to switching to a hydrogen economy, so it won't matter what mileage we get on cars now. I think that will happen soon.

—Mark Rutecki, Hummer owner, November 22, 2003

The buildup of fuel cell research formed part of a cresting wave of technoscientific positivism during the 1990s, a widespread belief that advanced science and technology had an almost unlimited potential to reshape society for the better. For the industrially advanced countries, it had been a decade of relative peace, prosperity, and achievement in science and technology, particularly in the fields of biotechnology and, above all, electronics. There was a pervasive air of expectation, a mix of euphoria and apprehension at what the new century held in store, a feeling among observers in political, industry, and environmental circles that a saltation was imminent not only in technology but also in values. An important element of the millennial zeitgeist was rising awareness of the need for social and environmental justice, although the formulas for achieving this remained hotly disputed. This spirit penetrated even the most rarified corporate sanctums. After years as industry's main bulwark against global warming science, the Global Climate Coalition crumbled. Du Pont was the first to defect in 1997, followed by Royal Dutch/Shell in 1998, Ford in 1999, and DaimlerChrysler, General Motors, Texaco, and the Southern Company in 2000.[1]

In no field did high technology and environmental evangelism coalesce more seamlessly than in fuel cell automobility, a system planners in government and industry were increasingly associating with a hydrogen economy by the turn of the millennium. Alan C. Lloyd, appointed chair of the California Air Resources Board in February 1999, was a noted enthusiast of hydrogen and hydrogen fuel cells. So was Amory Lovins, the influential physicist and advocate of sustainable energy technology. Popularizer of the notion of the "soft energy path" and cofounder of the Rocky Mountain Institute, a leading alternative energy think tank, Lovins paid little attention to fuel cells until the late 1990s, when he began promoting the technology as the cleanest and most efficient power source for his conceptual "Hypercar," an electric automobile 60 percent lighter than conventional automobiles.[2] No less a figure than Al Gore spoke glowingly of the fuel cell. In an interview with *Rolling Stone* in November 1999, he remarked that the technology yielded only "clean water" in its waste stream. The hybrid gasoline-electric, said the presidential candidate, was merely an interim step on the road to the fuel cell automobile, some varieties of which would be available for sale "within the next few years."[3] The most important hydrogen advocates, however, were the federal government and the auto industry, above all General Motors. In the 2000s, the manufacturing giant would devote itself to hydrogen fuel cell power with all the zealotry of the recently converted.

As in the carbonaceous fuel cell program, fuel cell–centric hydrogen futurism owed its rise to the ambitions of dissimilar groups with distinct and not always transparent interests in the technology. For the industrial gas industry, the apparent shift to the hydrogen fuel cell automobile was manna from heaven, promising massive growth and huge profits. For idealists, it was the realization of a dream. And automakers and their supporters in industry and government framed their motivations in utilitarian and moralistic terms. To be sure, interest in hydrogen and hydrogen fuel cell technology had arisen mainly because the project to electrochemically convert carbonaceous fuels had stalled. Yet it was convenient for auto manufacturers who had adamantly rejected all near-term alternatives to invest in a technology with utopian overtones but infrastructure requirements that rendered it a deep-future proposition. Although this point did not escape keen observers of energy and transportation technopolitics, it was often lost on futurists themselves, whose perspective of history, after all, was that of the *longue durée*. To them, it was a victory simply that mainstream manufacturers finally appeared to be taking hydrogen seriously.

Indeed, industry's enchantment with hydrogen and the hydrogen fuel cell was not all it appeared to be. One clue was the rapidity with which these technologies assumed the dominant place in green automobile discourse in the early 2000s. The accompanying dramaturgy featured a more calculated stagecraft than in the past, one that explicitly reinforced the federal government's domestic and foreign policy goals. More than any other single device, the fuel cell electric automobile projected the idea of the hydrogen economy to the public, the telegenic waste water issuing from its tailpipe a favored image exploited by boosters in the technopolitical theater of sustainable automobility. In this way, the hydrogen fuel cell acquired a potent symbolism that, for a time, overshadowed all other alternative power sources including its carbonaceous cousin, despite its absence from the market. Relentless promotion by pundits, investors, futurists, environmental advocates, and politicians made the association between hydrogen and fuel cells seem as intuitive as that between gasoline and the internal combustion engine. The place of hydrogen in the energy transformation chain and the merits of the hydrogen economy would be debated as never before. As in the past, however, action would concentrate only on the smallest downstream segment of that chain, the point where hydrogen was converted into electricity in a fuel cell automobile. This reductive vision of hydrogen would serve a score of disparate technopolitical enterprises. Ultimately, it would contribute to the demise of the commercial fuel cell automobile project.

Hydrogen Power Prelude

As the fortunes of the boosters of the membrane fuel cell rose precipitously in the 1990s, bound firmly to the parallel project of developing a miniaturized carbonaceous fuel processor, so, too, did hydrogen futurists prosper, although not quite so dramatically. Ongoing low-level bipartisan congressional interest helped keep hydrogen in the public eye throughout the decade, allowing advocates to consolidate influence and bases of support. The establishment of the Hydrogen Technical Advisory Panel (HTAP) in 1992 as directed by the Matsunaga Act placed hydrogen boosters in a better position than ever to influence energy policy. Responsible to the Department of Energy, this board of experts was charged with guiding the energy secretary's conduct of federal hydrogen and hydrogen-related research and development. The broader goal of the legislation had been to help prepare a five-year plan to domestically produce hydrogen, assess relevant resources from all federal agencies, develop renewable energy sources, and demonstrate relevant technologies. The HTAP's vision statement declared that hydrogen would complement electricity as an energy carrier in the twenty-first century and that both would ultimately be derived from renewable energy sources but also from fossil fuels, which would remain as a long-term transitional resource.[4]

In the mid-1990s, however, the basic dynamics governing federal hydrogen research and development had changed little from those of the 1970s and 1980s. As in the past, conversion technologies were far better developed than hydrogen storage and production systems. The latter would not be ready for more than a generation, noted Neil P. Rossmeissl, manager of the DOE's hydrogen program, in April 1995. But, he emphasized, fuel cells and other end-use technologies existed now.[5] And in fact, the agency's hydrogen effort was devoted mainly to fuel cell technology, although most of the money went to direct and indirect carbonaceous fuel systems. The federal fuel cell effort had a small basic research component but, overall, was aimed squarely at technology development. The largest part lay in the Office of Fossil Energy, devoted to high-temperature stationary molten carbonate and solid oxide fuel cells intended to use coal-based gases. This effort received around half of the $500 million the federal government spent on hydrogen research from 1992 to 1996. Far smaller was the transportation program. Built around the low-temperature membrane fuel cell, it received about $100 million in the same period. Run by the Office of

Energy Efficiency and Renewable Energy (EERE), this was the heart of the PNGV's fuel cell effort. The actual "hydrogen program," the part dedicated to developing pure hydrogen as a fuel and energy carrier, was a fraction of the size of the fuel cell programs, receiving only around $10 million annually in this period. Nevertheless, this was a sharp increase from 1992, when a mere $1.4 million was spent.[6]

The federal government's conflation of what were quite dissimilar enterprises—a not-insubstantial carbonaceous fuel cell program and a small hydrogen program—was already sowing skepticism among some environmentalists. This tendency would greatly complicate the politics of hydrogen futurism, where a patina of unanimity in praise of the simplest element obscured deep divisions among stakeholders, not only along the functional lines of energy carrier and fuel, but also over what types of primary energy should be used to produce hydrogen. Warren E. Leary of the *New York Times* believed hydrogen's political future suddenly brightened when Robert S. Walker, one of the most ardent hydrogen boosters in Congress, became chair of the House Science Committee in early 1995, replacing George E. Brown, Jr., another leading hydrogen luminary on Capitol Hill. An important player in the "Republican Revolution" of 1994, Walker had an expansive political vision of hydrogen remarkably akin to the Clinton administration's technocratic approach to policy. He claimed that it could bring energy independence to the United States and do away with the need for antipollution regulations, thus advancing Republican goals while remaining true to the party's philosophy of smaller government. Walker backed a bill (HR 655), passed by voice vote in May 1995, that would have dramatically boosted expenditures on hydrogen from the then-current annual level of $10 million to $100 million over three years. Perhaps surprisingly, his Democratic colleague Brown opposed the legislation on the grounds that such a drastic increase would deprive other energy programs of resources.[7]

The two hydrogen advocates found compromise in the resulting Hydrogen Future Act of 1996. This law underscored the bipartisan appeal of hydrogen, the growing association between hydrogen and fuel cell as the preferred energy conversion technology, and the contentious issue of government-sponsored technological change. Introduced by Walker and cosponsored by Brown, the three-title act was ostensibly designed to fund exploration of the basic precepts of the hydrogen economy. It allocated $164.5 million over six years to demonstrate the technical feasibility of producing, storing, transporting, and using hydrogen in industrial, residential, transportation, and utility applications. Notably, the only hardware the act specifically referred to was referenced in Title II—"Fuel Cells"—which directed that the technology be integrated with photovoltaic and solid waste gasification systems. The section authorized $50 million for fiscal years 1997 and 1998 to this end. The legislation aimed at enabling the private sector to carry out the demonstration, supporting only those aspects of hydrogen technology the business community was not currently investigating.[8] But the money allocated for mating fuel cells with renewable hydrogen systems was never appropriated, illustrating the difficulties of bridging the long-standing gap between the upstream and downstream portions of the futurist energy conversion chain.[9]

Here lay the chief quandary for hydrogen futurists. They had come a long way in a quarter of a century, gaining some prestige and a degree of influence within the federal apparatus. But the U.S. hydrogen lobby's greatest concern was that it had no real constituency in either government or industry. True, it had people such as Brown, Senator Tom Harkin (D-Iowa), and especially Walker, recognized in 1994 by the National Hydrogen Association for his work as a hydrogen lobbyist. But the NHA itself remained relatively small in this period. It added only a few new members in the 1990s, although these included the prestigious "class 1" companies BMW of North America and Shell Hydrogen.[10] Similarly, hydrogen had been increasingly marginalized in the PNGV in favor of other fuels and energy conversion systems over the course of the 1990s.[11]

The Hydrogen Future Act of 1996 was a step toward reversing this trend. The legislation augmented the power of hydrogen advocacy, freed the HTAP from DOE oversight, and gave the panel the right to make recommendations directly to Congress as well as the energy secretary. The HTAP attempted to coordinate an overall vision for hydrogen energy technology that reconciled the interests of the federal government and the hydrogen industry. In 1998, it claimed credit for helping build consensus within the "hydrogen community" on the need to develop the exotic substance as both an energy carrier and a fuel, bolster the DOE's hydrogen budget, and improve coordination between the various offices engaged in hydrogen research. Despite these perceived successes, the HTAP again stressed that hydrogen demonstrations should be led by industry. Warning of the risk of "white elephants," technological spectaculars that would be technically successful but commercially stillborn, the panel lamented that hydrogen still remained "off the radar" of high-level federal energy planners.[12]

Hydrogen's profile would gradually rise over the next four years thanks to the failure of the carbonaceous fuel cell electric project and the ongoing struggle to shape markets for the green automobile. The turning point was the emergence of the California Fuel Cell Partnership of automakers and oil companies in spring of 1999. For genuine hydrogen idealists, however, this turn proved a mixed blessing. On the one hand, alliance with the automobile industry brought hydrogen to the national stage for the first time. On the other, the unglamorous, tedious question of infrastructure would, as in the past, be set aside and ignored except at the point of use, the hydrogen refueling station. This led to precisely the sort of white elephants the HTAP so feared.

Greater disappointment was in store for the true believers. Among hydrogen's leading proponents were those who saw the panoply of associated

technologies as a means of maintaining the status quo in the energy transportation regime. While his colleagues Matsunaga and Brown supported renewable hydrogen production technology, Walker staunchly backed nuclear power
and had close ties with the military-industrial and aerospace communities. He
consistently voted against research in technologies harnessing wind, solar, and
geothermal energy, repeating the conservative mantra that only basic research
and development not conducted by the private sector was worthy of state aid.
Support for all other activities was "corporate welfare." Critics such as Harold L.
Volkmer (D-Missouri) had responded to the passage of HR 655 by questioning
whether the research it supported could be considered basic, suggesting that it
was incremental.[13] Indeed, like the futurists, Walker framed hydrogen as a
technological solution for social problems. What set him apart was his frank
interest in hydrogen's political potential. Retiring from Congress in 1997, he
remained influential in Washington as chairman of the Wexler-Walker lobbying
firm and joined the presidential campaign of George W. Bush as head of its
Space, Science, and Technology Task Force in 2000. In the fall, Walker delivered
a speech at an international symposium on automotive technology reprising
the precepts of political hydrogen honed in the late 1980s and early 1990s. Used
successfully in spacecraft, its environmental benefits undeniable, hydrogen
was "inexhaustible." Unfortunately, said Walker, the Clinton administration
had done little to advance hydrogen technology. But times were changing. The
hydrogen-powered vehicle, he claimed, was the solution to the problems facing
the automobile industry, particularly environmental activism, for it would ease
political pressure for the expansion of mass transit and put Detroit's "most vir-
ulent critics on the defensive." Thus would the automobile industry's viability
be ensured for years to come.[14]

In early 2000, General Motors entered the arena of political hydrogen.
Almost exactly ten years after the company had rolled the Impact into the spot-
light, it displayed its latest supercar at the North American International Auto
Show in Detroit in January. The fruit of GM's participation in the PNGV, and
reputedly one of the costliest vehicles ever built, the "Precept" was actually a
pair of advanced concept automobiles, one using hybrid drive and the other
employing hydrogen fuel cell electric drive. The occasion coincided with the
company's decision to halt production of the EV-1. The pure battery electric
vehicle program, noted GM vice chair Harry J. Pearce, had demonstrated the
limitations of the technology. But it had also provided valuable experience in
electric drive propulsion that would be useful in developing hybrids and fuel
cell electrics. Indeed, the company ultimately intended, remarked Pearce, to
develop fuel cell electric power with pure hydrogen.[15]

In counterpoising a wildly impractical one-off supercar with the cancellation of the EV-1, GM foreshadowed its strategy for coping with the politics of sustainable automobility in the first decade of the twenty-first century. In a sense,

the Precept was a harbinger of things to come. Over the next several years, General Motors would take the lead in wielding the hydrogen fuel cell as an expressly political instrument designed to quash all attempts at regulating the auto industry. The result was the beginning of a second phase of ZEV politics featuring spiraling competition in the game of green automobile dramaturgy that sowed confusion, division, and eventually disillusion among the ranks of environmentalists.

Hydrogen Fuel Cell Futurism and the Politics of Green Automobility

In the 1990s, the auto industry had helped make the fuel cell a household term. But new players in stationary hydrogen fuel cell power began to appear thanks in part to the frenzy of speculation that seized the market at the end of the decade. In the last run of the post-1987 bull market in the twentieth century, the capitalization of hydrogen fuel cell developers, many founded only a few years before, swelled along with the tech bubble. Plug Power, a General Electric–backed outfit specializing in the membrane fuel cell for domestic use, saw its stock price rocket from $15 to nearly $150 between October 1999 and March 2000. On paper, the company was worth $5.2 billion despite its unprofitability, a phenomenon as prevalent in the world of fuel cell start-ups as in the so-called dot-coms of this period.[16] Some reasons for this were suggested in a perceptive article by the journalist Barnaby J. Feder. Barriers to entry in the hydrogen fuel cell field were not as low as in the Internet world, he noted, but were not nearly as high as in, say, fusion, another miracle energy technology. Fuel cells were "teetering on the edge" of becoming a highly profitable business, bringing these devices to market was another matter entirely.[17] For the rest of the year, stocks would fluctuate wildly until, in early 2001, they began a long decline from the heights of the 1990s.

The market swoon coincided with a marked change in the character of the automobile fuel cell project. Stalled by technical obstacles and the sagging economy, it received a significant boost from the CARB, revealing how the board's technocratic/actuarial approach and deference to auto industry expertise served to bolster futurism in the context of pollution politics. Ever since the agency had begun to compromise with the auto industry in the memoranda of agreement that had paved the way for the first small fleets of battery electric vehicles, it had grown increasingly doubtful about battery technology. Under increasing industry pressure, the CARB had begun to consider flexible formulas for fulfilling the 10 percent ZEV quota for 2003. In November 1998, the board had introduced the partial zero-emission vehicle (PZEV) category, tacitly recognizing methanol fuel cell electric drive. In January 2001, the CARB again overhauled the mandate, devising an even more complicated system that allowed a broader mix of emission-class vehicles. There was the "pure ZEV," the "gold" standard, referring to an electric vehicle using only one primary electrochemical power source, either a battery *or* a hydrogen fuel cell. Only 2 percent of an automaker's 2003 production for California had to be in this category. Another 2 percent had to be "silver" or advanced technology partial zero-emission vehicles (ATPZEV) using electric drive, meaning either hybrids or carbonaceous fuel cell electrics. And the final 6 percent had to be "bronze" partial zero-emission vehicles (PZEV) equipped with advanced internal combustion engines.[18]

In attempting to placate the various interests and their competing technological claims, the CARB further legitimated the fuel cell electric automobile. Oddly enough, it did so at a time when investors were losing confidence in the technology's commercial potential. Throughout 2000, Ballard Power Systems rode the market swells, including a notable trough in December resulting from rumors that the CARB might weaken the 10 percent ZEV quota for 2003. Word that the regulatory agency would maintain this standard led to a brief rally, highlighting the importance of the rule changes to Ballard's potential bottom line. But even this brought no lasting benefits. By April, Ballard shares had lost 68 percent of their value from their 52-week high and analysts were now recommending the stock as a "hold." With the automotive fuel cell project in doubt and investors reluctant to purchase equity, the company suddenly began placing more emphasis on its line of stationary utility fuel cells.[19] By late 2001, noted the *New York Times*, industry analysts were increasingly speaking of the generator set as the first commercial fuel cell application.[20]

But General Motors remained bullish on hydrogen fuel cell power. In 2000, it became the first American automaker to join the National Hydrogen Association. In the summer of 2001, GM purchased stakes in companies that made hydrogen production and storage systems, including 15 percent of Vancouver's General Hydrogen, a company founded by Geoffrey Ballard following his ouster from the company that bore his name. In October, the auto giant struck an alliance with Hydrogenics, a Canadian maker of fuel cells and electrolyzers and one of the better-performing start-up companies. Like Ford and DaimlerChrysler, General Motors mounted demonstrations of hydrogen fuel cell automobiles in this period. To be sure, these were not especially distinguished. A display of technological pageantry held in Washington, DC, to celebrate the deal with General Hydrogen fell flat when the two GM fuel cell electric vehicles on hand broke down.[21]

Indeed, the possibilities of fuel cell power, far more than its actual performance, began to impress a number of opinion shapers. Among these were a panel of academic and industry experts assembled to review the benefits of research and development supported by the DOE's Office of Energy Efficiency and Renewable Energy and Office of Fossil Energy (OFE) since 1978. Formed in 2000 under the aegis of the National Research Council (NRC) at the request of the House Appropriations Subcommittee on the Interior, the experts invoked the familiar principle that government intervention should occur only if it could yield benefits the private sector could not deliver. But they judged the efficacy of collaborative R&D primarily on the basis of industry's willingness to contribute to it. Hence, the NRC panel rated fuel cell electric drive as risky but worthy of the $210 million the federal government had invested in the technology over two decades thanks to "significant" (although unspecified) business contributions to the field. Far less deserving, held the experts, was stationary fuel cell power. Since 1978, the government had lavished around $1.2 billion on technologies that inspired only tepid industry interest. Some of this aid helped bring UTC's 200-kilowatt phosphoric acid power plant to market on a limited basis in the late 1980s and early 1990s before federal managers turned to higher-temperature types believed capable of directly using the dirtiest and cheapest fuels. In fall of 1999, the OFE initiated the Solid State Energy Conversion Alliance (SECA), a collaborative project in solid oxide fuel cell technology. But industry contributed only $292 million to stationary fuel cell power over 20 years, and real commercial success in this field had proved elusive. In this instance, hinted the NRC panel, there was no justification for government involvement.[22]

This analysis raised the question of the roles government and manufacturers respectively had played in developing stationary and automobile fuel cell power, the sorts of results yielded by "balanced" partnership, and historical parallels. Government agencies had generally been the more eager of the parties to initiate most major projects, providing up-front funding and testing and diagnostic support. On the other hand, manufacturers had always been responsible for most of the actual innovations. And the most notable success up to that time—UTC's phosphoric acid power plant—was a relatively unsophisticated stationary generator with a simple load profile that had nevertheless required almost 30 years of development before making limited inroads into the market. The NRC panel chose not to draw conclusions from this history. It accepted at face value the potential of the membrane fuel cell in much more demanding roles, largely on the basis of the auto industry's interest in the technology. And it was uncritical of the DOE's role in promoting collaborative development of the membrane fuel cell in the relatively unexplored stationary application, a role with its own complications and trade-offs.[23] As in the past, fuel cell futurism would emerge from the gap between engineering research and the broader technopolitical considerations of the patrons of the power panacea.

Political Hydrogen Writ Large

The NRC panel's judgment of the automobile fuel cell was shared by the new presidential administration, with a twist. The Bush administration would soon link automotive fuel cell power with hydrogen fuel systems—technologies the NRC review had not dealt with in depth—and make the transportation-based hydrogen economy a key plank of both federal energy and federal energy research and development policy. In so doing, the White House took a cue from General Motors.

Like its predecessor, the Bush administration showed relatively little interest in hydrogen and hydrogen fuel cells, at least at first. The federal government remained preoccupied with high-temperature carbonaceous fuel cells. In August 2001, the DOE announced that it would extend the SECA in a ten-year, $500 million effort.[24] In contrast, spending on hydrogen research was then paltry in comparison. The energy agency's hydrogen budget in fiscal year 2001 was only $27 million, a figure the Bush administration proposed to cut to $13.9 million in fiscal year 2002.[25] So, too, was Bush's overall energy plan vague, although it offered a signpost of what was to come. Ostensibly treading the middle ground between industrialism and environmentalism, it offered a major tax credit to purchasers of hybrid and fuel cell electric automobiles. In reference to the latter, this was, of course, pure rhetoric, as the technology was not yet available on the market. Automakers professed themselves pleased with the plan. But as Jeffrey Ball of the *Wall Street Journal* reported, they were uneasy about how the tax credit would be structured and especially concerned about what the plan did not mention, namely, the federal government's intentions for CAFE. Although the Bush administration did not support scrapping fuel efficiency legislation outright, it accepted industry's argument that such regulation discriminated against Detroit for its greater reliance on trucks and sport utility vehicles than its foreign competitors and, far worse, that it endangered lives by encouraging the development of smaller, less crash-resistant automobiles.[26] Equally worrisome for both the oil and auto industries was the groundswell of support for energy independence in Congress following the terror attacks of September 11, 2001, and the possibility that politicians might draft more technology-forcing legislation.

In response to these fears, noted Ball, one of the more astute observers of the technopolitics of sustainable automobility, the auto industry prompted the federal government to dramatically elevate hydrogen and hydrogen fuel cells for automobiles to new prominence. On January 9, 2002, the administration canceled the PNGV and introduced a new $500 million joint effort between the DOE and auto manufacturers to research hydrogen fuel cell automobiles known as FreedomCAR (Cooperative Automotive Research). The program inherited the assumptions and objectives of its predecessor. It aimed, said Energy Secretary Spencer Abraham at the North American International Auto Show, to eliminate automobile pollution and reduce America's dependence on oil without building a vehicle that "succeeds in the laboratory and fails in the marketplace," a clear allusion to the battery electric programs. There was no mention of infrastructure.[27]

Once again, politicians and industrialists renewed their vow of faith in cooperative research and development as a substitute for regulation. The logic of FreedomCAR, explained Senator Carl Levin (D-Michigan), was that the resources automakers would otherwise be forced to devote to meeting higher CAFE standards could be put to better use, and yield faster results, if they were voluntarily invested in fuel cell technology. One General Motors official echoed Walker's stated strategy, remarking to Ball that the program would help industry in its fight against environmentalists and their campaign for tougher CAFE standards.[28] An anonymous Bush administration official claimed that any effort on the part of the Department of Transportation to increase fuel efficiency would take so long that the fuel cell project would have by then borne fruit, rendering legislation unnecessary.[29] It was far better to "drive toward the hydrogen economy and try to really solve the problem" rather than "just playing the numbers" with CAFE, held GM product development official Tom Davis in a February briefing of the Alliance of Automobile Manufacturers.[30]

But now such claims evoked widespread skepticism. The timing of the Bush administration's announcement of FreedomCAR was notable, for it came in the same month that GM and DaimlerChrysler sued the CARB to repeal its requirement for advanced technology partial zero-emission vehicles (ATPZEV, or hybrid electrics) on the grounds that matters pertaining to legislation of fuel economy (as opposed to air quality) were a federal responsibility.[31] The new program drew skepticism from activists and the press as a political vehicle designed to forestall immediate meaningful action on sustainable automobility at a time when Detroit was profiting handsomely from sport utility vehicles and light trucks. The *Wall Street Journal's* Ball rooted the program in Detroit's "futuristic technological crusade" against higher CAFE standards.[32] Describing FreedomCAR as "almost revolutionary," the *New York Times* editorial page questioned Abraham's assumptions, noting that without a hydrogen infrastructure, the plan would have no effect on energy consumption or climate change because it relieved automakers of the obligation to pursue technological breakthroughs in the short term. At the same time, said the editorial, the technology "had immense promise," a point on which many environmentalists agreed.[33]

The problem, noted observers such as *Fortune*'s Stuart F. Brown and Danny Hakim of the *New York Times*, was that GM's commitment to fuel cell automobility appeared less than sincere.[34] The company's latest hydrogen fuel cell concept-car, the AUTOnomy, unveiled at the Detroit Auto Show two days before Abraham introduced FreedomCAR, was splashy but nonfunctioning. Its centerpiece was the "skateboard," a 203-millimeter-thick slab chassis containing the electric motor, power source, and control equipment, upon which consumers could mount their choice of a wide variety of body shells capable of satisfying every mood, taste, and identity whim, from boy racer to frontier wrangler. Repeating the now-familiar formula, GM chief executive officer Rick Wagoner remarked that the fuel cell electric automobile would remove the automobile from the "environmental equation," rendering pollution politics obsolete.[35]

The vehicle might even be used as a mobile power plant, speculated Christopher Borroni-Bird. Formerly a fuel cell guru at DaimlerChrysler, he had jumped ship in 2000 and landed a job as the chief of GM's fuel cell concept automobile program. Now serving as the director of "design and technology fusion," he mused that the future fuel cell automobile might produce electricity and feed it back to the grid when not in use.[36] In April, Wagoner remarked that fuel cells were very important to the future of General Motors. The automaker planned to be the first to sell a million fuel cell cars, although not before the end of the decade.[37]

But GM left little doubt that the AUTOnomy would be no reprise of the Impact/EV-1. There would be no pilot production of the baroque techno-phantasm. Indeed, the conduct of automobile manufacturers in hydrogen fuel cell R&D in the early 2000s was a study in distinct dramaturgical styles. In an age when the information and advanced materials sectors had supplanted heavy manufacturing in the vanguard of American industrial progress, GM had rather abruptly attempted to make over its conservative image by adopting the fuel cell as one of its chief claims to technological leadership, hoping to attract the best and brightest minds in the field to its research headquarters in Warren, Michigan. Borroni-Bird had been enticed in this way.[38]

And General Motors was firmly wedded to the paradigm of the solitary concept car. The next challenge was to turn the simulacrum of the AUTOnomy into a functional hydrogen fuel cell supercar. Dubbed the "Hy-Wire," this was a veritable rolling laboratory of high-tech gadgetry. Based on the skateboard chassis, the vehicle featured "drive-by-wire," an electronic control system that replaced all mechanical linkages and was normally used only in advanced commercial and military aircraft. By September 2002, the Hy-Wire was ready to do battle in the "futuristic technological crusade," debuting at the Paris Motor Show. But General Motors had made plain it did not plan to commercially produce the automobile. Environmental groups such as the Sierra Club were quick to notice the disparity between the company's current goals and its vision of the future.[39] As greenwash, the Hy-Wire was ineffectual.

Japanese automakers, meanwhile, were taking a much more pragmatic approach to fuel cell power. Already by mid-2002 Honda had outstripped the competition. The company's philosophy was to develop a relatively simple standardized platform, incrementally improve it, and build a small demonstration fleet as quickly as possible. In June, the same month a federal district judge sided with GM and DaimlerChrysler and prevented the CARB from enforcing the 2001 ZEV amendments for the 2003 and 2004 model years, Honda's humble FCX subcompact, equipped with a Ballard-built power pack, became the first fuel cell electric automobile to received certification as a market-ready zero-emission vehicle from U.S. federal and California state regulators. A relative latecomer to the fuel cell game, rolling out its first FCX demonstrators in 1999, Honda took demonstration technology seriously. By the end of 2002, both

Honda and Toyota were ready to place small fleets of fuel cell electrics in the hands of Californians, making a joint announcement on December 2. Although DaimlerChrysler and the American automakers also had similar fleets, these were not nearly as well rationalized or promoted.[40]

Japanese manufacturers had long since topped their American rivals in product dynamism and diversity. Now they were proving that they were as adept at developing working concept automobiles as they were at producing commercial automobiles. Given the opposition of Honda and Toyota to battery electric drive, the lethargy of their rivals in marketing alternative automobile products, and the lack of progress in building hydrogen infrastructure, their fuel cell electric programs probably represented no greater ambition than to excel in all forms of competition including sustainable automobile dramaturgy. But the Japanese companies may also have intended these programs as hedges in case the winds of pollution politics suddenly shifted in America and planners in Detroit and Washington decided—or were forced—to match words with deeds and build a hydrogen economy.

Hydrogen Fuel Cell: Technology Policy, Policy for Technology

For a time in 2003, it seemed as if such a development was in the offing. On January 28, only months after the Justice Department filed an amicus curiae brief against CARB in support of GM and DaimlerChrysler's suit, George W. Bush used the occasion of his second State of the Union address to announce the Hydrogen Fuel Initiative (HFI). This was the infrastructure portion of the fuel cell automobile program. The federal government, said the president, would devote $1.2 billion for research over five years, making the hydrogen fuel cell automobile the basis of energy, transportation, infrastructure, and foreign policy. In language reminiscent of the first generation of hydrogen visionaries, Bush dwelt on the miracle properties of hydrogen, citing the "simple chemical reaction" with oxygen that produced energy. Bush made clear he viewed the project as a national effort. In powering the fuel cell automobile of the future, hydrogen would help solve all the problems that came with the country's dependence on fossil fuel.[41]

Predictably enough, given the widespread skepticism engendered by the PNGV and FreedomCAR, this ambitious scheme was immediately attacked from both the left and the right. Environmentalists noted that the HFI did not expressly call for the construction of specialized new hydrogen infrastructure using renewable energy, pointing out that the Bush administration planned to produce hydrogen from existing energy resources, initially natural gas, with a larger role for coal and nuclear systems in the future.[42] A *Wall Street Journal* editorial slammed the plan as dishonest and ineffective, a combination of Karl Rove–style trickery and welfare-state porkbarrelry. Not only did the HFI have only a slim chance of winning over environmental groups, it subsidized work on a technology that the private sector was developing very well on its own, scolded the newspaper, citing Honda's program as an example.[43] As in the past, few condemned hydrogen and hydrogen fuel cell technology outright; rather, the critics heaped scorn on the administration's approach to their development. The commercial fuel cell electric vehicle would remain mere speculation, said John M. DeCicco, a senior fellow at Environmental Defense, a New York–based advocacy group, simply because the Bush plan didn't provide nearly enough money, either for fuel cell technology or hydrogen infrastructure, to achieve its stated goals. A far more worthy project in the interim, he remarked, was the advanced diesel engine.[44]

Hearings held by the House Science Committee in March raised these and other issues. Critics noted that only $700 million of the HFI consisted of new money, the rest having been reallocated from existing programs including renewable energy. Authorities agreed that $700 million, although an unprecedented investment in hydrogen technology, was still far too small. Princeton energy analyst Joan Ogden estimated that a system supporting 100 million vehicles could cost between $50 billion and $100 billion. Donald Huberts, Shell's chief of hydrogen operations, held that $20 billion would be needed to supply 2 percent of the U.S. automobile fleet with hydrogen by 2020, and hundreds of billions thereafter. No transportation revolution had ever occurred in the United States, added committee chair Sherwood Boehlert (R-New York) without "massive government involvement."[45]

Government was indeed central to the vision of CARB chair Alan C. Lloyd. Identified by filmmaker Chris Paine as the proximate cause of the "death" of the ZEV mandate, Lloyd was not exactly the stereotype of the corporate hatchet man. He was a species of hydrogen futurist, a true believer in hydrogen fuel cell power. Lloyd's role in automobile pollution politics instead revealed the limitations of the CARB's technocratic approach to emissions control. Holding a doctorate in gas kinetics, Lloyd had had a long career as an air pollution scientist and policy-maker in California. He had served as the assistant director of the Statewide Air Pollution Research Center at the University of California from 1972 to 1975, as senior staff scientist and general manager of Environmental Services of Camarillo, California, from 1976 to 1988, and as chief scientist of the South Coast Air Quality Management District from 1988 to 1995. In 1990, he joined the National Hydrogen Association and became a member of the Hydrogen Technical Advisory Panel, chairing its subcommittee on demonstrations and then the panel itself between 1996 and 1998. Appointed chair of the CARB in February 1999 by Gray Davis, Lloyd disagreed with the carbonaceous fuel approach in the automobile fuel cell program, arguing that auxiliary onboard fuel systems spoiled the "elegant" pure hydrogen fuel cell by adding weight and complexity.[46] Here he put himself at odds with the oil industry's vision of fuel cell power.

In December 2002, while still chair of the CARB, Lloyd was appointed the chair of the California Fuel Cell Partnership in a move some critics believed was a conflict of interest. Lloyd saw this demonstration program, then comprising 26 hydrogen fuel cell vehicles and seven refueling stations, as a means of introducing as many people as possible to the physical reality of a fuel cell automobile. He reasoned that allowing consumers to literally "kick the tires" of fuel cell electrics would help "cultivate a mindset" that zero-emission vehicles were on the verge of being pressed into service. In 2002 alone, he claimed, nearly 200,000 people had been "reached" by the CAFCP through its demonstration vehicles, refueling installations, and events, especially vehicle rallies. This, said Lloyd, showed that California was committed to its policy of using technology to solve the problem of air quality. All that was needed to launch this revolution was to give the public a vision that they would find irresistible. Ultimately, he noted, industry had to build something the public would be willing to buy. David Garman, head of the EERE, agreed. Automotive fuel cell technology first had to be made affordable and reliable before questions of infrastructure could even be discussed. To this end, said Lloyd, the federal government needed to stimulate the market for fuel cell vehicles by purchasing them for use in its own fleets.[47]

In March and April of 2003, mandate pollution politics and hydrogen fuel cell boosterism were even more tightly coupled. Under pressure from automakers and the federal government, and feeling that no zero-emission vehicle technology of any kind, either battery or fuel cell electric, was ready for commercialization, the CARB again drastically modified the mandate. Resetting the ZEV mandate deadline to 2005, it introduced a new credit-based formula. In a response to industry and federal pressure, all references to fuel efficiency were removed in calculating ZEV credits for hybrids. Large automakers were given two choices: they could opt for the revised 2001 formula, the so-called "base compliance path," allowing them to produce either battery or hydrogen fuel cell vehicles for their 2 percent "gold" pure ZEV requirement; or they could choose the "alternative compliance path," allowing them to build "silver" vehicles to gain "gold" credits if the vehicle in question used carbonaceous fuel cell electric drive. The requirement for the model years 2001–2008 was 250 such automobiles, but the quota increased by an order of magnitude each for the 2009–2011 and 2012–2014 time periods respectively, topping out at 50,000 units in 2015–2017.[48]

Strictly speaking, this plan did not kill the electric car (as Paine suggests), because the fuel cell automobile *was* an electric car. What the new CARB rules did do was undermine *battery* electric drive. The regulator rolled back the ZEV deployment deadline by a few more years. And in acknowledging the hydrogen fuel cell electric as a full-fledged ZEV, the CARB provided industry the option of pursuing a technology with a range of unsolved cost, reliability, and infrastructure issues that justified the indefinite postponement of deployment. Between

2003 and 2005, the major automakers began winding down their battery electric programs, efforts that had placed over 4,000 ZEVs with customers with very little advertising. Because most of these vehicles had been available only for lease, not for sale, companies were able to recall and destroy most of them.[49]

Collantes holds that the mandate failed in part because CARB had shifted from its role as a setter of standards to an assessor of technology, work for which it was not well equipped. The result, he says, was to impart a bias for certain lines of technological development.[50] The board had played a largely reactive role, refining its definition of a zero-emission vehicle as a host of new and largely untried automotive technologies, some inspired by the PNGV, appeared in precommercial trials and on the market. When Roger Smith declared the battery electric automobile commercially feasible in 1990, a time when the Impact was a rough-edged concept vehicle, the CARB developed the ZEV standard in response. It acted similarly when automakers declared fuel cell electric drive commercially feasible.

The reality, however, was that in 2003 automakers were still struggling to make the hydrogen fuel cell electric drive as cheap and dependable as the battery electric drive, let alone the gasoline engine. Despite Ballard's success in improving power output in the 1990s, fuel cell automobiles faced a host of problems relating to cost, durability, and range. For decades, researchers had been surprised and dismayed by unanticipated phenomena manifesting in fuel cell power plants undergoing long-term tests, and the automobile program was no exception. As Ballard researchers discovered, the automotive duty cycle exerted punishing effects on a membrane fuel cell. When the power unit was not drawing hydrogen, a condition known as fuel starvation, it would emit carbon dioxide. Workers subsequently learned that in the absence of hydrogen, the platinum catalyst was oxidizing the carbon support structure on which it was mounted, resulting in the slow degradation of the electrode, loss of active catalytic surface area, and plummeting performance. Fuel starvation was a common occurrence in the stop-and-go operation of fuel cell automobiles, since fuel cells did not "idle," drawing no fuel at rest, unlike internal combustion engines. On paper, automotive fuel cells of the early 2000s had a lifetime of around 5,000 hours, although manufacturers typically published much less information on durability than on power output.[51]

There were other difficulties with integrated fuel cell automotive systems. Onboard hydrogen storage was still not capable of providing the minimum 480-kilometer (300 mile) range automakers insisted consumers demanded in an automobile. Compression was the most mature storage technology but offered limited range owing to hydrogen's low energy density in this form. For example, 3,600 pound per square inch (245 atmosphere) storage, the standard in the late 1990s, gave ranges of 160 kilometers (Ford Focus FCV) to 250 kilometers (NECAR II). Liquid hydrogen had more energy capacity and, on paper, longer range (NECAR IV, 450 kilometers), but it cost energy to liquefy hydrogen. Liquid hydrogen also rapidly evaporated through containment materials, resulting in considerable wastage. The advent of 5,000 pound per square inch (340 atmosphere) compression storage in the mid-2000s provided ranges of anywhere between 145 kilometers (DaimlerChrysler F-Cell) and 355 kilometers (Honda FCX), still well short of the 480-kilometer benchmark.[52]

Still, real advances had been made in the power, durability, and range of fuel cell electric automobiles over a decade. The parameter that remained frustratingly resistant to progress was cost. In the early to mid-2000s, fully integrated automotive fuel cell systems were worth around $4,000–$5,000 per kilowatt, far from the Department of Energy target of $35–$45 per kilowatt. In 2004, ionomer membrane cost about $200 per square meter, a figure Ballard believed would have to be cut to $5–$7 per square meter to make commercialization feasible. But opportunities for reducing costs were circumscribed by physical-economic barriers. A chief means of improving the performance of the membrane fuel cell was to make the membrane thinner, which improved conductivity and performance. Yet this also helped ensure the material could never be produced in quantities large enough to foster economies of scale, even if millions of fuel cell electric automobiles were built, because the total volume of material was too small to tempt manufacturers into developing it.[53]

By the early 2000s, a number of experts were attacking the basic premises of a fuel cell–based hydrogen economy. Even if the most optimistic forecasts of commercial fuel cell electric drive proved correct, wrote Vaclav Smil in 2003, a million such vehicles constituted less than one-half of one percent of the U.S. light-duty fleet. Despite the hyperbole, characteristic mainly of popular accounts that were nevertheless sometimes traceable to science authorities, Smil wrote, society did not stand on the verge of a paradigm shift in energy conversion technology.[54] Former DOE undersecretary Joseph J. Romm had similar views. He believed that hydrogen fuel cell electric automobiles would have few environmental benefits and criticized plans to use natural gas as a feedstock for hydrogen in the near term. In 2004, Romm published a book devoted to refuting the economic and environmental premises of the hydrogen economy and fuel cell power.[55]

A fundamental question was what the transition to hydrogen energy would look like, a matter on which the HFI, ostensibly devoted to infrastructure, was notably vague. Echoing the sponsors of the Bacon cell a half century before, automakers and government believed the question of primary energy resources could be set aside for the time being because hydrogen could be produced from such a wide variety of substances using a variety of means that were being constantly improved. As critics had long pointed out, however, these had very different industrial and environmental implications.[56] Observers took note of dissimilar approaches in the wake of a joint European-U.S. hydrogen fuel cell research program announced in June 2003. Committed to obtaining 12 percent of its energy and 22 percent of its electricity from renewable sources by 2010 under their Kyoto obligations, European planners had a vision similar to that of the original hydrogen futurists, where hydrogen functioned as a fuel, storage medium, and energy carrier.[57] On the other hand, the Americans saw hydrogen primarily as an automotive fuel and planned to rely much more heavily on fossil and nuclear energy to produce it. Prominent hydrogen booster Jeremy Rifkin, then advising European Commission president Romano Prodi on energy matters, feared that the European program might be "hijacked" by its American partner, observing that the joint research deal compelled Europe to abandon the goal of using pipeline hydrogen as an energy carrier.[58]

In early 2004, the National Academy of Sciences (NAS) published a comprehensive study addressing these questions. Authored by a committee of experts from academe, environmental groups, and industry struck by the NAS's National Research Council in fall 2002, it concluded that there were major technological and economic barriers to a hydrogen economy, defined as a fuel cell–based transportation system that would draw from a variety of primary energy resources. Hardly any of the crucial components of this system were without need of further research and development before they could be produced. Breakthroughs, the committee indicated, were still needed to lower costs and improve the durability of fuel cells and electrolyzers, renewable primary energy systems, and carbon sequestration in concert with coal-based hydrogen production. As for managing the crucial shift from old to new energy regimes, the committee recommended developing distributed hydrogen production based on electrolysis and natural gas reforming systems, thus avoiding the high cost of constructing centralized infrastructure. But this path depended on preexisting centralized gas and electric systems; subsequent transition phases were opaque and "could not be identified and defined." Echoing the findings of other experts, the committee concluded that such a transition would have little impact on emissions and energy use for several decades unless R&D proved successful and major investments were made in hydrogen systems of all sorts. That, in turn, suggested the committee, required significant regulatory efforts by governments to create the proper climate for business investment.[59]

The report could be read, as it was by the *New York Times*'s Matthew Wald, as a veiled critique of the Bush administration's hydrogen program.[60] But its findings were also indicative of trends in U.S. energy and energy research and development policy. Unsurprisingly, the NAS-NRC committee did not rule out the hydrogen economy as a bad idea that should be abandoned outright. Indeed, it held that feasibility studies were a worthy national goal in itself. Perhaps most importantly, in light of the White House's stated energy strategy, the committee identified the fuel cell as the key energy conversion system of a hydrogen economy.[61] Only a few years earlier, Peter Hoffmann, the hydrogen historian and cofounder and editor of the *Hydrogen & Fuel Cell Letter*, had made a similar connection in a revised and updated version of his pioneering 1981 book, *The Forever Fuel*. He had been motivated to write his new book, he suggested, mainly by progress in fuel cell technology during the preceding 20 years.[62] All this showed how important hydrogen and fuel cell futurism had become in energy discourse. For specialized would-be manufacturers of the power source that had mobilized utopian imagery in building their brand, such symbolism existed in uneasy tension with the realpolitik of the established industrial order.

What Does It Take to Change the World?

The experience of Ballard Power Systems perfectly illustrated this tension between the real and the ideal, revealing the company's vulnerability to the changing technopolitical imperatives of its partners. In most of their demonstrations of fuel cell automobility, Ballard and Daimler-Benz/Daimler/Chrysler used pure hydrogen, a fuel desirable not only for its operational characteristics but also for its centrality to the clean energy imagery the companies self-consciously constructed. Work on the direct methanol fuel cell and reformer technology constituted a distinctly secondary aspect of their partnership, with efforts in these fields being led by other companies. For its part, Ballard specialized in developing the fuel cell energy converter, not the auxiliary technologies needed to store, produce, or process fuel for electric automobiles. And while Ballard periodically released data on improvements in the power output of its membrane fuel cells, it was much less forthcoming with information on durability.[63] This did not prevent company executives from framing the fuel cell as a kind of universal chemical energy converter capable of operating on a variety of fuels—hydrogen, methane, methanol, and gasoline—while saying little about the respective technical and infrastructure requirements of stationary and vehicular applications.[64] Such claims resonated among American energy and transportation entrepreneurs, investors, and policy planners but also produced misleading claims about what was possible given the state of the art.

But Ballard's efforts to shape perceptions of what was possible in the field of energy conversion ran against the agenda of its industry partners. Their relationship was asymmetrically symbiotic. On the one hand, Ballard fuel cells allowed automakers to mount a credible defense—at least for a time—that they were actively engaged in developing a commercial zero-emission vehicle after they ended their battery electric programs. In turn, for a decade the automakers had been Ballard's largest investors and customers for its precommercial fuel cell power packs, vaulting the company to prominence in the process. On the other, Daimler and Ford had joined other automakers in spending millions lobbying to roll back and loosen the mandate quotas, contributing to Ballard's crisis of expectations.

Heavily invested in a technology that required a radical shift in the ways automobiles were industrially produced, deriving most of its revenue selling the promise of revolution in the form of demonstration power units and equity stakes, Ballard was beholden to external events over which it had little control. The company could not afford to develop its own specialized materials for membrane fuel cells.[65] And the collapse of the automotive fuel cell bubble also forced Ballard to adapt its technology to the niche applications it had previously neglected. Here, it faced a difficult challenge. The likeliest near-term automotive application was the industrial electric truck, a situation recalling the fate of battery electric drive after the Second World War.[66] The other role for membrane fuel cells was providing power and heat for buildings. But Ballard had staked its future on a technology better suited to the demands of the automotive duty cycle. In that application, the membrane fuel cell had to function reliably for around 4,000 hours, assuming that most commuters used their vehicles for about an hour each day over the average ten-year life span of an automobile. In contrast, utility fuel cells were in almost constant use, requiring a lifetime of at least 40,000 hours. Given the state of the art in the early 2000s, the only way utility units could operate more efficiently than existing conventional generators, at least in principle, was to utilize their by-product heat for water and space heating. Yet the low-temperature membrane fuel cell produced low-quality waste heat. To compete as a stationary generator, its operating temperature and durability had to be increased. This posed "an extreme challenge" for the technology, observed Ballard's vice president for research and development Charles Stone in 2004.[67]

The long-term ideological consequences of the shift in applications were perhaps even more serious for Ballard. Like many fuel cell research and development communities, the company molded its message for the target audience of the moment. To investors, Ballard CEO Firoz Rasul had long spoken of the potential of the fuel cell electric automobile to compete with the internal combustion engine automobile in all areas of performance, highlighting its presumed ability to consume hydrocarbons.[68] Like the civil servants and auto executives who helped build the fuel cell electric program and elevate Ballard from obscurity, company executives framed the membrane fuel cell as a market-driven solution to the problem of sustainable transport. But the essence of the company's mystique, and the source of much of its early strength, was that it was selling not simply a better power source but a whole new lifestyle. In appealing to popular dissatisfaction with fossil fuel civilization, the company pitched its products to a younger generation, invoking hydrogen and the hydrogen fuel cell as a decisive break with an unsustainable past.

The realities of ZEV politics eventually forced Ballard out of the field of automotive fuel cell power. As years passed and Ford and DaimlerChrysler continued to delay production, Ballard invested capital in research and development that it could not recoup. In late 2007, the company sold most its automotive fuel cell business back to its partners, bringing the company full circle.[69] By linking environmental awareness with individual consumer choice, the Canadian firm was part of a larger corporate movement that, as Matthew Paterson notes, sought to green the automotive subject as much as automotive technology, appealing to consumers to act as socially conscious agents of change.[70] When Ballard was forced to revert to low-profile niche applications, its credo of "Power to Change the World" rang hollow. Neither range nor climate change were factors in the decisions of prospective purchasers of industrial electric trucks, who operated purely on the cost-effectiveness criterion, one that fuel cells historically had been unable to meet.

Hydrogen Fuel Cell Power in an Age of Crises

Hydrogen and fuel cell futurism and green automobile dramaturgy may be seen as related technopolitical practices rooted in the material and political interests of a cluster of diverse groups. Sometimes collaborating, sometimes striking out on their own, these organizations pursued their objectives with varying degrees of success until, at a crucial juncture in the history of the automobile industry, they coalesced and firmly reinforced one another. Futurism gained momentum in a complex, dynamic society where consumerism, as the sociologist David Gartman has noted, was a surrogate for "identity, autonomy, and individuality." In this market, American automakers had long since lost the ability to dominate style but remained committed to the production of the large gasoline-engine passenger automobile.[71] They were acutely aware of the relationship between technology and individual identity and of the role of the consumer as both a shaper and an object of style. A number of scholars have pointed out that the designers of sport utility vehicles and their accompanying advertising campaigns appealed to powerful, deep-seated, and sometimes contradictory drives in the American social psyche such as freedom, personal security, and environmentalism. But this does not mean, as Paterson implies, that many or even most American consumers necessarily interpreted products and promotional efforts in the ways their creators intended.[72] Fickle motorists imbued with a green sensibility or simply angered by rising fuel costs helped foster new market realities by purchasing small foreign imports and supporting the efforts of legislators to force technological change. Detroit and the North American divisions of the German automakers responded to the nascent ethos of virtuous consumption with technological symbolism. Their Japanese counterparts realized it represented a demand they set out to meet, and that proved highly profitable.

For all their novelty, then, the fuel cell electric auto programs (with the exception of the Japanese ones) were enabled by relatively conservative corporate cultures that paid homage to the *idea* of high technology but were reluctant to make radical changes to the status quo. Such programs were intimately bound up in battery technopolitics and followed an even more convoluted route from laboratory bench to test track than their electrochemical counterparts. As with battery technology, dedicated workers from outside the industrial mainstream advanced fuel cell technology in good faith until they caught the attention of sponsors. Daimler played a major role in legitimizing the idea of fuel cell automobility. Over time, the company's expectations of the technology changed considerably. Begun during Edzard Reuter's term as chief executive officer, fuel cell R&D seems to have been originally intended for experimental purposes. In contrast, Jürgen Schrempp, Reuter's successor, appears to have genuinely hoped that a carbonaceous fuel cell power system could be commercially developed. The company's efforts in this realm may be interpreted as one of a number of money-losing missteps committed under Schrempp's tenure including the purchase of the Dutch aircraft company Fokker, the Chrysler merger, and investment in Mitsubishi Motors. Of course, had researchers been able to develop a viable gasoline or even methanol automobile fuel cell system, perhaps even American manufacturers may have produced this in response to the Japanese hybrid programs.

But when this project encountered irreconcilable physical and political obstacles in the early 2000s, automakers crafted a calculated dramaturgy around hydrogen fuel cell electric drive, a technology they well knew could not be marketed without first constructing new infrastructure. Here, General Motors took the lead, repackaging fuel cell rhetoric in an anticampaign for the hydrogen fuel cell automobile.[73] The effects of green automobile dramaturgy and hydrogen fuel cell futurism on consumer attitudes are difficult to know for certain in the absence of a thoroughgoing sociopsychological study. The automobile historian Tom McCarthy is skeptical of the power of conventional advertising, opining that trends emerged simply because consumers emulated one another.[74] Analyzing the marketing of sport utility vehicles, William Rollins claims that greenwashing did resonate among consumers possessing an inchoate environmental consciousness. Battery and fuel cell–based green automobile theater can be seen as springing from the same impulse, the flip side of industry's Janus-faced campaign to produce large, powerful, and dirty vehicles, a grudging acknowledgment of the diverse nature of demand and the complex and sometimes inimical value systems at play in American society. If SUVs conjured powerful, contradictory human desires regarding wilderness—the desire to commune with it but also master it—the battery and fuel cell electrics embodied the techno-progressive facet of the American character. They were a far more credible "machine in the garden" than the hulking fleet pouring forth from Detroit.

As with industry's marketing campaign for the SUV, however, demonstrations of electrochemical automobility valorized nature while falsely representing technology's ecological effects.[75] Detroit treated such displays as a means of showing Americans what they should and shouldn't value in a commercial vehicle. If automakers hoped to convince consumers to adhere to traditional fare, however, they were disappointed. As forms of deception, these programs were risky, even counterproductive, for, as in the case of battery electric drive, putting a popular vehicle into the hands of consumers and then whisking it away generated resentment and bad press. Confined to well-publicized and interminable though still quite necessary engineering research, the fuel cell auto programs were no substitute for battery electric drive, alienating and embittering supporters of electric drive and engendering considerable cynicism, if Paine's 2006 documentary can be taken as a barometer of broader attitudes. If the demonstration programs—both battery and fuel cell electric drive—inadvertently nurtured an environmental ethos among consumers, European and especially Japanese manufacturers were in a much better position to satisfy it than Detroit. By 2005, expensive gasoline had busted the truck boom, driving consumers to smaller cars supplied largely by the foreign companies. The economic crisis that began in the fall of 2008 dealt a blow to all automakers, especially American ones. As Chrysler and General Motors faced bankruptcy, the *New York Times* pronounced the end of the era of the SUV.[76]

Hydrogen fuel cell electric drive remained central to California pollution politics in this period. Among the state's air quality bureaucrats, the technology retained its status as the exemplar of zero-emission automobility, helping ensure that major automakers would continue to keep a hand in this field. General Motors, Toyota, Daimler, and Honda seemed intent on maintaining demonstration fleets into the 2010s. By 2008, the technology had been greatly improved in terms of performance and convenience, if not yet in cost and durability. Honda's 2008 FCX Clarity, for example, had a maximum range of 448 kilometers and, reported Martin Fackler of the *New York Times*, a degree of refinement comparable to that of the latest gasoline automobiles. That year, Honda offered the first of 200 to lessees in Japan and California.[77] For a time, Arnold Schwarzenegger's state government made hydrogen and hydrogen fuel cells key pillars of its energy policy, promoting the idea that Californians could have their sustainable energy cake and eat it too. The governor famously drove this idea home with his hydrogen Hummer. By the end of 2008, around 250 fuel cell electrics of various makes cruised the state's public roads. But there were only 26 hydrogen refueling stations, mostly in the San Francisco and Los Angeles metropolitan regions, and none on major intra- and interstate highways.[78] This compared well with Germany, which boasted 30 refueling stations in by 2009. For its part, Daimler then had a fleet of around 100 fuel cell vehicles including 60 A-Class F-cell subcompacts. In late 2010, the company introduced its B-Class-based successor, planning to produce 70 available for lease for select customers in California.[79] But prospects for rapid expansion of these demonstration projects in the 2010s were clouded by the global recession.

Costly infrastructure requirements that ruled out hydrogen fuel cell electric drive as a near-future solution to air quality problems had suited American automakers during the halcyon years of the truck and SUV market, particularly GM. But the dramaturgical function of hydrogen fuel cell R&D as a dilatory political tool in ZEV mandate politics became less important to them as market conditions shifted by the late 2000s. It slowly dawned on Detroit that the commercial hybrid automobile had become an even more potent symbol of alternative automobility thanks to its popularity and profitability. True, American manufacturers began to produce commercial hybrid vehicles around the mid-2000s, although in small numbers and all on truck or SUV chassis. But they eschewed the market for hybrid passenger automobiles, one pioneered in the late 1990s by Japanese manufacturers. Although Toyota had sold the first Prius models at a loss, it claimed to be making money from the line as early as 2001.[80] By the end of 2006, the company had sold nearly 342,000 units in the United States and 640,000 globally.[81] Stung by this success and by the rise of California-based battery electric start-up Tesla Motors, recounted GM vice chair Bob Lutz in a *Newsweek* interview, he initiated the Chevrolet Volt hybrid program that year. Displayed as a concept unit at the 2007 Detroit Auto Show, the Volt drew predictable criticism that it was yet another GM exercise in public relations.[82]

If that was the case, the company was prepared to take dramaturgy to considerable new heights. Beyond Lutz's professed intentions to commercialize the vehicle, there were reasons to suspect the Volt did represent a shift in GM's philosophy of electric automobility. Lutz claimed he had overridden internal resistance from a variety of quarters including the company's fuel cell research unit. Moreover, the Volt combined elements of conservative and radical engineering that suggested a genuine, if halting, effort to compete in a market segment that Toyota had created and dominated. The production Prius was a "full" parallel hybrid system utilizing two motors, one battery electric and one gasoline internal combustion. The vehicle could be powered by one or the other, or both, depending on conditions. But it could not be operated as a pure electric for very long and did not run at all without gasoline. In contrast, the Volt was planned as a plug-in hybrid electric vehicle (PHEV), a type closer to a pure ZEV than conventional hybrid technology. It was designed to operate as a pure electric, using a large 16-kilowatt-hour capacity lithium ion battery with a range of about 56 kilometers, and then as a series hybrid thereafter, using gasoline to drive the electric motor for another 550 or so kilometers. But the car could also operate in series-parallel mode, with the gasoline engine assisting the electric motor in certain circumstances. Home owners could recharge the battery using standard 110-volt sockets. Days after the appearance of the Volt concept car in Detroit, Toyota announced that it would develop a plug-in variant of the Prius, with some Japanese media hinting that the auto giant had already secretly been working on the technology.[83] In July, the company said that it would deploy a prototype Prius PHEV demonstration fleet for testing and evaluation by the University of California at its Berkeley and Irvine campuses.[84]

The Volt program bore all the hallmarks of an attempt to steal a march on an archrival. In concert with the latest demonstrations of fuel cell electric drive, it may have helped influence the CARB in ways that recalled the technopolitics of the Impact. In March 2008, the board devised yet another, even more tortuously complex ZEV formula, this time using a range-based credit system. It created two new ZEV categories, types IV and V, defined as fast-refueling vehicles with a range of more than 320 and 480 kilometers respectively. By then, the CARB had abandoned the pretense of not identifying technologies per se but only range and fuel parameters, so it alternatively referred to types IV and V as fuel cell electrics. And although the CARB classified both battery and hydrogen fuel cell electric drive as "gold" ZEVs, the latter had a higher credit rating owing to recent test improvements in range. Hence, the ZEV gold quota for 2012–2014 based on annual sales of 1.4 million passenger cars was 25,000 type IV ZEVs. But automakers had the option of achieving the quota with 7,500 type IV fuel cell electrics, or 5,375 type V fuel cell electrics, or 12,100 short-range battery electrics, or a mix therein that added up to the baseline range credit represented by the 7,500 type IV vehicles, *if* the balance of the quota was made up by 58,000 "enhanced ATPZEV." This was a new category of vehicle using a "ZEV fuel," either hydrogen in an internal combustion vehicle or electricity in a plug-in hybrid. Such vehicles could also be used to obtain up to 50 percent of the gold requirement for the 2015–2017 time frame.[85]

This latest attempt by the CARB to reconcile conflicting technopolitical interests posited the plug-in hybrid as a sort of interim or bridge technology on the road to the ideal ZEV, the as yet unperfected fuel cell electric automobile. The shift accommodated automakers that had invested in both technologies, notably Toyota and GM. For the American firm, however, the obstacles, and stakes, were higher. If the Volt indeed represented its future, as executives frequently claimed, GM faced immense challenges. The company had premised the program on an advanced rechargeable lithium ion battery that did not exist in 2006 and that it did not plan to manufacture itself. Indeed, Asian companies dominated the production of this class of power source technologies. After launching a competition between LG Chem and U.S. start-up A123, GM selected the South Korean battery giant to supply the Volt. In February 2008, Nissan and Renault upped the ante in the business of sustainable automobility, launching their Leaf project, a commercial pure battery electric passenger auto.

A national crisis once more prompted the federal government to intervene in the affairs of auto and power source makers, this time on a massive scale. As the global financial system approached meltdown in the fall of 2008, Congress launched the Advanced Technology Vehicles Manufacturing Loan Program (ATVMP). Authorized under the Energy Independence and Security Act of 2007, the program provided $25 billion in loans managed by the Department of Energy, to be used by American automakers and their suppliers to develop the capability to manufacture "advanced technology vehicles" that met the federal Bin 5 Tier II emission standard (roughly equivalent to a low-emission vehicle under the LEV II system for California and the northeastern states) and that were 25 percent more fuel efficient than otherwise similar vehicles from the 2005 model year.[86] In effect, the ATVMP was a new PNGV, a government instrument intended to guide heavy industry into the twenty-first century, only now bolstered with significantly more resources, at least on paper. It starkly illuminated the paradoxes of American-style "quasi planning." In the 1990s, an outwardly healthy Detroit spurned the mild medicine offered by the paternal hand of the state. A decade of wasting competition in the 2000s brought the Motor City to its knees. Government was all that stood in the way of total collapse, yet confusion reigned as it maneuvered ponderously to break the fall. The assumption underpinning the ATVMP was that an emergency injection of new production technology could cure the prostrate corporate technosaur. But the legislation stipulated that money could only be loaned to "financially viable" companies. And Chrysler and General Motors had immediate solvency problems that only a government bailout—de facto partial nationalization—could solve in the short term. By mid-2011, most ATVMP loans had gone to Ford ($6 billion) and Nissan ($1.5 billion).[87]

To be sure, the Obama administration was not strictly relying on the market to take up those advanced technologies needed to achieve policy ends. In May 2009, the president announced a new national average light-duty fleet (passenger automobiles and light trucks) mileage rule, increasing the standard from around 25 to 35.5 miles per gallon by 2016.[88] Nevertheless, politicians, bureaucrats, and entrepreneurs remained powerfully attracted to the idea of a technical fix. Once again, they favored a particular approach, viewing the lithium ion rechargeable battery as the basis of alternative automobility. In 2009, the White House committed $2.4 billion to develop an advanced battery manufacturing complex virtually from scratch, triggering a familiar pattern of expectation generation. A new electrochemical power panacea supplanted the hydrogen fuel cell as the lynchpin of national security and the avatar of American industrial rebirth. As the idealized hydrogen economy faded from the discourse of sustainable energy and automobility, in its place shimmered the prospect of what the science journalist Seth Fletcher called the "lithium economy."[89]

Whither the fuel cell in these tumultuous developments? Its latest revival had been the most dramatic and most disappointing of all. More engineering progress had been made on the technology since the early 1990s than at any point in its long history. But two decades of research suggested that the membrane fuel cell was not the most practical power source for the commercial electric passenger automobile. Parallel developments in advanced battery chemistry in the 1990s swung the pendulum in favor of electrochemical storage in this role. This is not to subscribe to the fallacy that commercial technologies are "naturally selected." Historically, the question of energy efficiency was never the sole or even the most important consideration for developers of the automobile system after the First World War, as the history of the gasoline car attests. Engineers, bureaucrats, and politicians collaborated to build markets and shape demand, and the choices they made were conditioned as much by social factors as by the physical limits of materials. Fuel cell electric power was constrained by the availability of certain strategic materials, but so was the permanent-magnet electric motor, which required rare earth metals. In 2011, users of fuel cell electrics in the demonstration programs had to moderate expectations of durability, convenience, and performance conditioned by decades of experience with the fossil fuel system, especially in cold climates. But this was no argument that some commercial, nondemonstration form of this mode of transportation could not be built and deployed. At the beginning of the second decade of the twenty-first century, U.S. political and industrial elites simply did not believe it was in their interests to do so.

Nevertheless, the historical legacy of the fuel cell in the history of American automobile technopolitics seems assured. Thanks to a variety of factors, academics, entrepreneurs, and bureaucrats made the fuel cell the model of the sustainable supercar of the new millennium. The numerous laboratory successes recorded by seekers of the power panacea over the years validated the notion of technological progress as social progress. And fuel cell futurism played a central role in industry's experiment with commercial electric automobile technology. It undermined efforts by California to force the creation of a market for zero-emission vehicles. When manufacturers decided to again experiment with commercial electric automobility near the end of the 2000s, the onset of recession severely restricted the scope of this enterprise. On an ideological level, the dream of fuel cell power reflected trends in the U.S. and global economies in the 1990s and 2000s, especially the bubble cycle that characterized this period and the federal government's penchant for framing cooperative R&D and innovation policy as a substitute for industrial policy. In 2011, few pundits remained under any illusions that the fuel cell would have an impact on sustainable energy in the immediate future. But although history seemed once more to pass the technology by, at least for the time being, the forces that inspired the pursuit of the power panacea remained very much in play.

Conclusion

The question of why certain technologies failed to meet commercial expectations is voiced frequently in media, business, political, and science and engineering circles, although the onus to offer a convincing explanation is often shirked. That question is one of the organizing principles of this book. In this and other examples, however, it may not be entirely appropriate, for it can reinforce the assumption that because material ubiquity is the criterion of commercial success, then that, hence, is what makes artifacts historically significant. Success is a relative term, and as history demonstrates, fuel cells have given some practical service over the years, although never on the scale their leading protagonists have longed for.

But the fact that fuel cells did not become widely commercialized despite widespread expectations is itself interesting and worthy of investigation. It formed the point of departure for a more thoroughgoing study of the persistence of the idea of fuel cell power and the phenomenon of fuel cell futurism. Over the years, the idea of the fuel cell as a power panacea was fashioned from the real and imagined qualities of not a single device but a range of technologies that used a variety of materials at different temperatures and pressures. This mythical fuel cell helped justify or otherwise advance a host of other technoscientific enterprises large and small, directly and indirectly relating to energy, power, and transportation. Despite its relative rarity in the marketplace, then, the fuel cell influenced many aspects of modern life *as an idea*. Deconstructing this idea unpacks and assesses in a new light a familiar set of questions relating to the history of science, technology, and society in twentieth-century American society including the relationship between the state and industry in national growth and the use of technoscience to harness energy and shape social landscapes.

Invented in Europe and investigated and adopted in a score of countries, fuel cell technology was most fully developed—materially and rhetorically—in the United States. It was pursued initially to satisfy national security imperatives and sustained thanks to problems arising from American energy, transportation, and environmental policies. But there were also powerful cultural antecedents that made American society especially receptive to the idea of a power panacea. The record of fuel cell research and development may be seen as part of national narratives that elevate technology as a key determining factor, stories David Nye sees rooted in the contradictory tension between the ideas of individual liberty and collective well-being that constitute the *volk* philosophy. These narratives are iterations of foundation or second-creation stories explaining the genesis and regeneration of the imagined American community in an age when it was increasingly common to apply metaphors drawn from science and engineering to other sociocultural endeavors. Authors of these stories were guided by four constants: geometrical space, natural abundance, the free market, and increasing access to force and the efficiency of its application, what in the twentieth century would be defined as energy and power respectively. Harnessed successively by ax, mill, canal, and steam-driven railroad in the classic narratives, energy and power in ever-larger quantities fueled the nation-building project. The result, suggests Nye, was that energy and power technology became transcendent symbols of progress indicative of manifest destiny, the teleological philosophy of history of the national elite. Doubters of the narrative of perpetual progress such as the writer Henry Adams and the inventor Nikola Tesla held that evolutionary law and social progress were incompatible with the laws of thermodynamics predicting the eventual heat death of the universe, conflating rules governing matter in the physical, biological, and social realms.[1] For them, progress was transitory and fleeting.

The specter of entropy was no barrier, however, to the imagination of innovators such as Edison, Ford, Sloan, and even Tesla, who believed in and embodied technological progress. It receded even further as competing interest groups used the latest technologies to discover and exploit America's vast reserves of energy, leading to the simultaneous utilization of old and new resources—the traditional water and coal but also, increasingly, oil and natural gas. The result was a period of cheap, plentiful energy that lasted for a half century, from the end of the First World War until the early 1970s.

It was not until after the Second World War, however, that high-energy civilization appeared. And it was in this vast military-industrial, technoscientific, and consumer complex, where the wildest fancies of techno-futurists of all stripes finally seemed possible, that partisans of fuel cell power would find receptive audiences and generous patrons.[2] In this period of energy surplus, there was no consumer demand for fuel cell power. Nevertheless, a few engineers and scientists were drawn to the technology, which seemed to them the first truly original idea for a practical power source since the nuclear reactor. Unlike that prodigiously costly and complex enterprise, fuel cells could be explored relatively cheaply, yielding remarkable results in modestly equipped laboratories. Yet planners and pundits considered these devices suitably advanced in an age of burgeoning popular technophilia. Even individual scaled-up stacks of fuel cells could be built and tested comparatively inexpensively. Serial production, of course, was another matter entirely. In an age of corporate and government-backed big technoscience, the fuel cell field was among a handful offering space for what the historian Eric S. Hintz, focusing on alkaline battery development, calls the "post-heroic" individual inventor.[3]

But fuel cell researchers faced a dilemma. They were haunted by the vision of Mond and Langer, Jacques, and Baur of a device that could compress the complex material requirements and social relations of a variety of energy regimes into an autonomous black box, a "closed system" universal chemical energy converter. The problem was that fuel cells of all classes worked very well only when using pure hydrogen. And this "open system" required major changes in existing energy systems and, hence, human power relations, in order to find broad use. Fuel cell researchers in a host of milieus never fully compartmentalized the dream technology from the electrochemical reality, blurring the crucial distinction between hydrogen and hydrocarbon fuel cells.

In the 1950s, work was conducted on a limited basis by a few American companies, but it was the Bacon cell that excited researchers the world over. Efforts to develop this system did much to inspire durable false analogies and assumptions about the nature of fuel cell technology, forging the behavioral template of fuel cell research and development for decades to come. As in many other technology projects, promoters tailored their justifications to accord with changing economic, political, and cultural circumstances. Faced with the Electrical Research Association's resistance to the "reversible cell," Bacon agreed to reconfigure his technology as a "fuel cell." This semantic shift reframed the device as an electrochemical engine that used hydrogen not as an energy carrier, a medium for the storage of electricity, but as a fuel, a chemical oxidized in an irreversible reaction, with the waste product discarded rather than dissociated back into its constituent elements through electrolysis. In Britain, and in the smattering of similar programs then under way in the United States, hydrogen fuel cells ruled the laboratory demonstration rack while workers and their supporters speculated about the possibilities of electro-oxidizing carbonaceous fuels. At an early date, they deduced that the fuel cell could be operated as long as fuel was supplied, the presumed quality above all others that gave rise to the notion of the fuel cell as a hybrid of battery and heat engine.

The dramaturgy of early-stage demonstrations became an important way fuel cell research communities supported themselves. Expectations for long-lived and cheap full-size devices helped win patronage. This modest form of fuel cell futurism in turn merged with bureaucratic politics in fledgling government science and technology institutions in the 1950s and 1960s. Planners in the

NRDC, ARPA, and NASA used the presumed revolutionary potential of the fuel cell to stake out turf and identify themselves and the institutions they worked for as standard-bearers of technological and social progress. Almost invariably, researchers and patrons were unprepared for the arduous journey of engineering research. It was then that fuel cell research and development became a capital-intensive activity.

At a time of cheap energy, the only initial market for fuel cell power was in the military-industrial complex. Harvey Brooks has observed that linearly organized, state-backed research and development proved fruitful in circumstances where a product was completely novel, where product differentiation was unnecessary, and when a developer had much larger resources than its competitors and was producing low-volume goods in a protected market where performance was valued over cost.[4] Such conditions were present in the weapons, telecommunications, nuclear, computer, and electronics sectors, at least in the early stages of the Cold War. Out of these interlinked socio-technoscientific and industrial spheres grew requirements for advanced military power sources. Aerospace and portable electronics represented two promising applications.

Yet there was relatively little interest in the hydrogen fuel cell in the defense research establishment outside the federal agencies involved in space technology. Hence, it was the developers of the aerospace fuel cell that helped give impetus to the dream of the terrestrial carbonaceous fuel cell. Stripped of its space responsibilities in 1958, ARPA viewed the fuel cell as one of its few remaining assets and set about developing it as an autonomous black box. Guided by the Institute for Defense Analyses, ARPA found willing collaborators in the Army's Signal Corps/Electronics Command and Mobility Command. Unable to develop the carbonaceous fuel cell in any role, stationary or mobile, Army researchers instead emphasized the quality of silent operation in soldier-portable radios and experimented with an increasingly exotic series of alternative fuels.

For their part, NASA managers believed fuel cells offered both practical and political advantages. By promoting the terrestrial carbonaceous fuel cell as a space spin-off, they hoped to leverage the success of their aerospace fuel cells into badly needed political capital that would demonstrate the agency's social relevance and deflect criticism that space adventures were an unaffordable luxury in the 1960s. Engineers and planners held the successful aerospace fuel cell as analogous to terrestrial fuel cell technology, a comparison that eventually proved false. But the space fuel cell program did provide the material basis for the development of a fledgling terrestrial fuel cell industry built around the phosphoric acid fuel cell and dominated by Pratt & Whitney/United Aircraft/United Technologies Corporation.

The abrupt end of the era of energy abundance and growing popular concern with environmental degradation beginning in the late 1960s supplied new rationales for fuel cell power. Genitor of the first large-scale terrestrial fuel cell systems, the TARGET program originated in the respective interests of the gas utilities, which posited the fuel cell, distributed throughout the gas grid, as a means of competing with the electric utilities, and Pratt & Whitney's fuel cell division, which was desperate for any kind of support in a period when NASA had drastically scaled back its fuel cell orders. These players, later joined by the electric utilities, emphasized efficiency, clean emissions, and scalability in an attempt to exploit dissatisfaction with large conventional power stations. Such justifications helped convince the federal government to reprise its role as the chief patron of fuel cell research and development in the late 1970s. Such programs represented not insignificant political capital for the fledgling Department of Energy after 1977.

These actors, in turn, helped inform the vision of hydrogen futurists in the 1970s. For advocates of fuel cell power and alternative energy conversion technology of all sorts, however, the 1980s were years spent largely in the wilderness. It was only in a period of deadlocked pollution and transportation politics around the early 1990s that academic researchers and bureaucrats imported the idealized fuel cell into an ongoing discourse of innovation that posited research and development and advanced technology as instruments of environmentally sustainable industrial revival. In this milieu, fuel cell futurism was stimulated by the conflict between the state of California and its traditional technology-forcing interventionism in the auto sector and the Clinton administration's experiment with the laissez-faire approach of voluntary cooperative research and development in this sector. Through the Partnership for a New Generation of Vehicles, the federal government indirectly promoted the fuel cell as an alternative to the battery in the electric automobile, validating the notion of the former as a power panacea in the process. Industry enlisted this idea in its dilatory campaign against the Zero-Emission Vehicle mandate, compelling the CARB to accept its technological estimates and forcing the air quality regulator to essentially abandon its project.

By the early 2000s, the nascent fuel cell industry was characterized by contradiction and confusion. On the one hand, its rapid growth through the 1990s seemed to validate the principle of collaborative R&D. Government programs had stimulated industry investment that outstripped state expenditures from at least 1997. According to one industry survey taken before the Bush administration announced its commitment to a hydrogen economy in January 2003, the federal government had spent some $847 million on fuel cell R&D compared to over $4 billion by the private sector between 1996 and 2002. In the United States some 4,500 to 5,500 people were employed in fuel cell work of some sort in 2002. Oddly, however, there was no consumer demand for fuel cells and, hence, no profits to be had. Venture capitalists, accordingly, played a small role in financing fuel cell work. Public equity was the main source of investment, but the vast majority had been raised in the 20-month period from November 1999 that represented the terminal phase of the Internet bubble. Corporate venture investment was much more consistent, amounting to at least $100 million per year from 1997 until dropping precipitously in 2002. Internal corporate R&D spending was probably much higher, ranging from $300 to $400 million per annum. In the wake of the collapse of the high-tech bubble, respondents in industry and financial circles hoped government would play a much larger role in stimulating the fuel cell market, especially in the automobile sector.[5]

With the failure of the project to develop a hydrocarbon automotive fuel cell, hydrogen fuel cell futurism flourished and reached its zenith. The science writer Scott L. Montgomery probed the essence of the appeal of the hydrogen fuel cell when he wrote that it had "no social enemies." Hydrogen's centrality to modern industry, he believed, meant that the device would have a place in any future energy regime as a kind of technological passe-partout, a universal chemical energy converter that would not displace any existing interest groups.[6] Indeed, this was a frequently voiced justification, yet the converse proved true. A product of industry/government collaboration, the hydrogen fuel cell never really had a firm constituency either in industry or government, ironically enough. Although some entrepreneurs and politicians found it politically expedient to talk up the fuel cell–based hydrogen economy, none seriously considered investing in it.

Nevertheless, some visionaries took the rhetoric at face value. Occupying the lowest niche downstream in the energy conversion chain—the point of use— the hydrogen fuel cell captured the foreground of imagination in their minds. They viewed the technology not simply as a better energy converter but as a means of finally reconciling social progress with the laws of thermodynamics. Influential pundits such as Jeremy Rifkin interpreted the hydrogen economy as social justice in the form of thermodynamic equilibrium, the hydrogen fuel cell as a means of literally and figuratively placing power in the hands of the people. Distributed generation was recast as social leveling. Electrochemical energy converters sprinkled throughout the grid and in the automobile fleet would allow citizens to produce their own electricity in their own homes and offices and sell the surplus, breaking utility monopolies once and for all. Power, held Rifkin, would be exchanged over an energy web much like information over the Internet. He saw a fuel cell–centered hydrogen economy as the culmination of the historical progression from heavy fossil fuels to the lightest element. True democracy would be attained by staving off entropy.[7]

Such analogies were utterly misleading. Energy exchange via net metering and distributed generation was akin to information exchange over phone, electrical, and fiber-optic lines only on the most superficial level, entailing considerably more complex engineering and economic issues. And distributed fuel cell generation could not be considered truly decentralized unless the power plants were directly linked to renewable energy sources, systems that were impractical in large cities, particularly at northern latitudes in winter, where they were capable of meeting only a tiny fraction of demand. In order for fuel cells to displace incumbent power sources, they required large quantities of primary energy, presumably drawn from the gas grid, itself a highly centralized system. Indeed, Rifkin assumed that the electrical grid would have to remain in place in order to allow for power trading.

Inappropriate, too, were parallels between fuel cells and semiconductors. Especially specious were suggestions that the developmental arc of the power source was determined by a version of "Moore's law," a comparison favored by some Ballard executives during the heyday of the fuel cell boom. This expression refers to a long-term trend in industrial microelectronics engineering said to have emerged in the late 1950s, whereby workers were able to double the number of transistors placed on a microchip or, in an alternate version, the overall performance of integrated circuits, about every two years. As Vaclav Smil notes, the term has been widely misinterpreted as representing the norm in modern inventiveness.[8] Solid-state computer memory is not at all the same thing as electrochemical current density. Although researchers spectacularly improved the power of fuel cells in the 1990s and early 2000s, the barriers to exponential growth of this parameter were encountered far sooner than with semiconductor technology.[9]

It is perhaps not a coincidence that other forms of techno-futurism flourished in the United States in this period.[10] The nanotechnology movement is an important example. Like hydrogen and fuel cell futurists, nanotechnologists had central defining machines—the scanning tunneling microscope and atomic force microscopy—from which issued a whole range of possibilities in manipulating matter and the accompanying analogies abstracted to the social sphere. No less than with fuel cells, government played a vanguard role in adopting nanotechnology as a subject of research and development policy.[11] The historian W. Patrick McCray has noted how articulate and charismatic nanotechnology advocates acquired champions inside the government science and technology apparatus, making often-breathtaking claims to help win them.[12]

In these technopolities, a certain interpretive flexibility was at play. Futurist apostles operated within dense "exchange networks" where hydrogen, fuel cell, and nano technologies served as symbolic and social capital with the potential of being converted into real capital, benefiting a range of interests.[13] These networks were facilitated by the physical and social multivalence of utopian techno-futurism. As in the hydrogen and fuel cell fields, nanotechnology had its own evangelists who sought alliances with a variety of interest groups, selecting from among a variety of arguments to suit the particular audience at hand. True believers and, later, planners, politicians, and industrialists recognized that hydrogen's ubiquity made it central to many aspects of academic and industrial chemistry, a fact they tried to leverage for their respective purposes at every opportunity. Like nanotech advocates, fuel cell promoters frequently made the technology all things to all people, appealing to various theoretical qualities attractive to particular interest groups with contrasting, sometimes conflicting visions of society. Each saw technology as an instrumental means of creating an ideal sociotechnical landscape. The fuel cell could be a superior kind of battery, one that would enable infantry to survive on the nuclear battlefield or make portable electronics even more convenient for consumers. It could power aircraft, spacecraft, submarines, or green automobiles, and facilitate distributed generation in places where the risk of large generating stations was unacceptable. Some saw the hydrocarbon fuel cell as a way to make the fossil fuel system more sustainable. Others perceived a hydrogen fuel cell as a means of sweeping away the old fossil fuel order. For hydrogen futurists, the basis for this grand vision was either nuclear or "clean coal" power. For some environmentalists, it was hydrogen produced by renewable energy. No less than the hydrogen and fuel cell futurists, nanotech boosters drove home the idea that nanotechnology was applicable in practically every facet of daily life and manufacturing.[14]

Moreover, some researchers saw technical complementarity between these technologies. They promoted novel nanomaterials such as carbon nanotubes as a solution to the reverse salients plaguing fuel cell and hydrogen systems, as well as other alternative energy conversion devices.[15] In these examples, techno-futurists of all stripes claimed their work would utterly reshape society, not immediately, but in the near future. A crucial shared bond was a philosophy of history that saw technological hardware (often elided with evolutionary biology by the nanotechnology movement) as the engine of progress.[16] Nowhere were these shared assumptions more boldly illustrated than in the 2010 coffee-table book *Reinventing the Automobile*. Coauthored by GM electrochemical visionaries Christopher Borroni-Bird and Lawrence Burns, along with William J. Mitchell, a professor of architecture and media arts and sciences at MIT's Media Lab, one of the intellectual birthplaces of the nanotechnology movement, the book synthesized the most popular techno-tropes of the time in a breathless vision of sustainable personal automobility. Artificial intelligence, biotechnology, and technological convergence, key themes of the nanotechnology and singularity movements, were applied to urban transportation. Change would come through altering automobility's "DNA," shifting it from internal combustion to electric power. In the future, no single technology would dominate; fuel cell and electric power would coexist in their respective socio-ecological niches, contributing and drawing hydrogen and electricity as interchangeable energy carriers to and from a distributed-energy smart grid that would also support a networked fleet of autonomous robot vehicles.[17]

Studiously avoiding the play of interest groups involved in the history of electric automobility, the book was written with an exuberance that contrasted starkly with the political and economic realities of the world of 2011. Such thinking reflected what had by then become conventional wisdom within science and technology policy circles. For the planners, innovation policy, not social policy, was the key to the future sustainable energy conversion chain. But in the United States, decades of intermittent support by the federal government and a more recent and equally irregular investment pattern by the private sector had yielded only a small, unprofitable fuel cell industry that seemed destined to serve niche markets for the foreseeable future. A product of the crises of the American socioeconomic system—concurrent deindustrialization, environmental devastation, and financial speculation—the latest generation of fuel cell futurists promised solutions they could not deliver. A telling tableau unfolding in Santa Barbara, a city with a storied history as a high-profile theater of environmental politics following the notorious 1969 oil spill, highlighted the temptations and perhaps insurmountable weaknesses of low-temperature fuel cells. Five years after installing a fuel cell in its wastewater treatment plant in a cogeneration role consuming methane, the city removed the unit in 2010 after it became fouled by sulfur. In an ironic object lesson of the merits of incremental versus disjunctive technological development, councilors chose to replace the fuel cell with an advanced internal combustion engine.[18]

Gijs Mom has observed that one effect of the long project to commercialize the battery electric vehicle is that its promoters focused scrutiny on existing technological systems.[19] This was also the case with the fuel cell project. Attempts to apply the various types in practically every known application cast into relief the physical limitations of the entire energy transformation chain and transportation system, as well as the human political relations that shaped these technosocial networks. That so many researchers, pundits, and politicians at the end of the twentieth century subscribed to the same assumptions harbored by an earlier generation of fuel cell researchers and sponsors is a testament not only to the seductive power of the idea of the power panacea but also to the pervasiveness of the ahistorical philosophy of progress animating science, technology, and policy elites.

The history of fuel cell research and development shows how futurism was generated in the dialogue between science and technology communities and the political representatives of a social order built around the fossil fuel–based energy conversion regime. The precedent of technological progress was an important driver of these relations. It underpinned the political economy of fuel cell communities in a period when electrochemical engineering was a marginal technoscience in the United States. And it lent credence to more hopeful auguries of a society enabled by fuel cell power, occluding the limitations of the technology. Fuel cell futurism served a variety of interests even as it helped generate unrealistic expectations. In many cases, such promises were co-opted by tribunes of the socioeconomic establishment. Yet the denouement of the latest episode of fuel cell futurism demonstrated the poverty of techno-determinist notions of progress and history to a broader audience than ever before. In the history of the fuel cell, we may read not simply a sociology of innovation—the social overpotential that mobilizes an array of actors from many backgrounds—but also one of the concluding chapters in the chronicle of the American Century. As an unfolding narrative, it has the potential, at least, to inform new understandings of the role and limitations of technology in liberal democracy.

NOTES

INTRODUCTION: FUEL CELL FUTURISM

1. A. J. Appleby, a leading fuel cell researcher, used the term "electrochemical engine" in the title of an article for popular consumption; see "The Electrochemical Engine for Vehicles," *Scientific American* 281, no. 1 (July 1999): 74–80. Ballard Power Systems, a leading developer of automotive fuel cell demonstration power plants in the 1990s and 2000s, long promoted this understanding of the fuel cell; Ballard Power Systems, annual reports, 1996–2002.

2. Some of the more evocative fuel cell headlines in this period included "Breathtaking . . . the Vehicle Powered by Air" (Nick Nuttall, *Times* [London], May 15, 1996, Home News, 7); "The Great Green Hope: Are Fuel Cells the Key to Cleaner Energy?" (Anthony DePalma, *New York Times*, October 8, 1997, D1); "Intel on Wheels: Can a Small Canadian Company Overthrow the Internal-Combustion Engine?" (*Economist*, October 31, 1998).

3. George W. Bush, "President Delivers 'State of the Union,'" January 28, 2003, http://www.whitehouse.gov/news/releases/2003/01/20030128-19.html; George W. Bush, "Hydrogen Fuel Initiative Can Make 'Fundamental Difference': Remarks by the President on Energy Independence," February 6, 2003, http://www.whitehouse.gov/news/releases/2003/02/20030206-12.html; David K. Garman, House Committee on Science, "The Path to a Hydrogen Economy: Hearing before the Committee on Science," 108th Cong. 1st sess., March 5, 2003, serial no. 108-4, 34–53.

4. Wolf Vielstich, Arnold Lamm, and Hubert A. Gasteiger, eds., "Part 4: Fuel Cell Principles, Systems and Applications," in *Handbook of Fuel Cells: Fundamentals, Technology and Applications; Volume I: Fundamentals and Survey of Systems* (Chichester, UK: John Wiley and Sons, 2003), 161–162.

5. Emil Baur, *Bulletin Schweiz ETV* 30, no. 17 (1939): 478.

6. See, for example, Lawrence R. Samuel, *Future: A Recent History* (Austin: University of Texas Press, 2009), 4; and Matthew Connelly, "Future Shock: The End of the World as They Knew It," in *The Shock of the Global: The 1970s in Perspective*, ed. Niall Ferguson, Charles S. Maier, Erez Manela, and Daniel J. Sargent (Cambridge, MA: The Belknap Press of Harvard University Press, 2010), 340.

7. E. H. Carr, *What Is History? The George Macaulay Trevelyan Lectures Delivered in the University of Cambridge, January–March 1961* (New York: Alfred A. Knopf, 1963), 86–91.

8. I draw this typology of futurism from Samuel; see *Future*. The historian Matthew Connelly observes that in addition to rational forms of prediction and forecasting, socioeconomic upheaval around the world from the late 1960s stimulated an interest in occult "prevision" (astrology and extrasensory perception) in both elite and popular circles; see Connelly, "Future Shock," 343.

9. For exemplary explorations of this phenomenon, see, for example, Hyungsub Choi and Cyrus C. M. Mody, "The Long History of Molecular Electronics: Microelectronics Origins of Nanotechnology," *Social Studies of Science* 39, no. 1 (February 2009): 11–50; Richard Barbrook, "New York Prophecies: The Imaginary Future of Artificial Intelligence," *Science as Culture* 16, no. 2 (June 2007): 151–167; W. Patrick McCray, "Will Small Be Beautiful? Making Policies for our Nanotech Future," *History and Technology* 21, no. 2 (June 2005): 177–203; Cyrus C. M. Mody, "Small, but Determined: Technological Determinism in Nanoscience," *HYLE* 10, no. 2 (2004): 99–128; Michael Fortun, "Mediated Speculations in the Genomics Futures Markets," *New Genetics and Society* 20, no. 2 (2001): 139–156; Joseph J. Corn, "Introduction," in *Imagining Tomorrow: History, Technology, and the American Future*, ed. Joseph J. Corn (Cambridge, MA: MIT Press, 1986), 2.

10. Marita Sturken and Douglas Thomas, "Introduction: Technological Visions and the Rhetoric of the New," in *Technological Visions: The Hopes and Fears That Shape New Technologies*, ed. Marita Sturken, Douglas Thomas, and Sandra J. Ball-Rokeach (Philadelphia: Temple University Press, 2004), 6.

11. Langdon Winner, "Sow's Ears from Silk Purses: The Strange Alchemy of Technological Visionaries," in *Technological Visions*, 34–47.

12. In some ways, the fuel cell futurists of the 1990s resembled a group of visionary technologists devoted to space colonization, nanotechnology, and other speculative projects that began to emerge in the 1970s and 1980s, actors that, as the historian W. Patrick McCray notes, combined the roles of futurist, researcher, and promoter. I thank McCray for fruitful discussions on this subject at the Center for Nanotechnology in Society at the University of California at Santa Barbara.

13. David E. Nye, "Technological Prediction: A Promethean Problem," in *Technological Visions*, 159–161.

14. I draw this definition of utopia from Michael D. Gordin, Helen Tilley, and Gyan Prakash, "Utopia and Dystopia beyond Space and Time," in *Utopia/Dystopia: Conditions of Historical Possibility*, ed. Michael D. Gordin, Helen Tilley, and Gyan Prakash (Princeton, NJ: Princeton University Press, 2010), 1–4.

15. Thomas More, *Utopia*, trans. Clarence H. Miller (New Haven, CT: Yale University Press, 2001).

16. Karl Marx, "Theses on Feuerbach" in *Ludwig Feuerbach and the Outcome of Classical German Philosophy by Frederick Engels* (New York: International Publishers, 1941), 84. In its combination of philosophy and logical method (dialectical materialism), sociological theory (historical materialism), and political economy and revolutionary practice (scientific socialism), Marxism is distinguished from the vastly less ambitious social practices of futurism and utopianism.

17. See Howard P. Segal, *Technological Utopianism in American Culture* (Chicago: University of Chicago Press, 1985), 21–22.

18. Folke T. Kihlstedt, "Utopia Realized: The World's Fairs of the 1930s," in *Imagining Tomorrow*, 97–118.

19. Scholars have long criticized the tendency of social scientists and humanists to privilege the study of inventions over the ways inventions come to be manufactured and used on a large scale, or fall out of use, as the case may be. In a 1999 paper, for example, Edgerton noted that when scholars did focus on technology-in-use, they tended to concentrate on a handful of iconic successes (automobiles, antibiotics, electric production and distribution, and electronics and information technology),

truncating the tortuous, nonlinear processes by which an idea conceived by mechanics or scientists or engineers is made into an artifact in use. Yet most innovations are never used, he suggests, partly because systems of technology-in-use and their socioeconomic relations are so resilient and persistent; see David Edgerton, "From Innovation to Use: Ten Eclectic Theses on the Historiography of Technology," *History and Technology* 16, no. 2 (1999): 117–136; see also Edgerton, "Innovation, Technology, or History: What Is the Historiography of Technology About?" *Technology and Culture* 51, no. 3 (July 2010): 681; David C. Mowery, "Collaborative R&D: How Effective Is It?" *Issues in Science and Technology* (Fall 1998): 37–44; David C. Mowery and Nathan Rosenberg, *Paths of Innovation: Technological Change in Twentieth-Century America* (Cambridge: Cambridge University Press, 1998).

20. Langdon Winner, *Autonomous Technology: Technics Out-of-Control as a Theme in Political Thought* (Cambridge, MA: MIT Press, 1977), 237.

21. David E. Nye, *Consuming Power: A Social History of American Energies* (Cambridge, MA: MIT Press, 1998), 188–216.

22. Timothy Mitchell, "Hydrocarbon Utopia," in *Utopia/Dystopia*, 131–138.

23. Jan Nederveen Pieterse, "Innovate, Innovate! Here comes American Rebirth," in *Education in the Creative Economy*, ed. Daniel Araya and Michael A. Peters (New York: Peter Lang, 2010), 401–419.

24. See, for example, National Academy of Sciences, the National Academy of Engineering, and the Institute of Medicine of the National Academies, *Rising Above the Gathering Storm: Energizing and Employing America for a Brighter Economic Future* (Washington, DC: National Academies Press, 2007), 1–8. Christophe Lécuyer holds that universities were "more beneficiaries than causes of" regional growth of semiconductor and computer industries in the United States after the Second World War; see "What Do Universities Really Owe Industry? The Case of Solid State Electronics at Stanford," *Minerva* 43 (2005): 69–70.

25. Koonin repeatedly made this point in a series of presentations in 2010; "Sustainability and Maintaining US Competitiveness: A Presentation by Under Secretary Steven Koonin, June 25, 2010." http://science.energy.gov/s-4/speeches-and-presentations/?p=1.

26. David A. Kirsch, *The Electric Vehicle and the Burden of History* (New Brunswick, NJ: Rutgers University Press, 2000), 203–208.

27. See, for example, Karin Knorr Cetina, *Epistemic Cultures: How the Sciences Make Knowledge* (Cambridge, MA: Harvard University Press, 1999), Peter Galison and David Stump, eds., *The Disunity of Science: Boundaries, Contexts, and Power* (Stanford, CA: Stanford University Press, 1996); Walter Vincenti, *What Engineers Know and How They Know It: Analytical Studies from Aeronautical History* (Baltimore: Johns Hopkins University Press, 1990); Edward Constant, *The Origins of the Turbojet Revolution* (Baltimore: Johns Hopkins University Press, 1980); Edwin Layton, "Mirror-Image Twins: The Communities of Science and Technology in 19th-Century America," *Technology and Culture* 12, no. 4 (1971): 562–580.

28. Christophe Lécuyer and David C. Brock, "The Materiality of Microelectronics," *History and Technology* 22, no. 3 (September 2006): 307.

29. Karl V. Kordesch, "25 Years of Fuel Cell Development (1951–1976)," *Journal of the Electrochemical Society* 125, no. 3 (March 1978): 78C–79C.

30. Richard H. Schallenberg, *Bottled Energy: Electrical Engineering and the Evolution of Chemical Energy Storage* (Philadelphia: American Philosophical Society, 1982), 391–392.

31. Ibid., 286–287.

32. I thank University of California Los Angeles professor of materials science and engineering Bruce S. Dunn for this insight into the American battery industry; interview with the author, September 7, 2010, UCLA.

33. Seth Fletcher, *Bottled Lightning: Superbatteries, Electric Cars, and the New Lithium Economy* (New York: Hill and Wang, 2011), 45–59; Michael M. Thackeray, "20 Golden Years of Battery R&D at CSIR, 1974–1994: A Contribution to the History of Science in South Africa," unpublished manuscript.

34. Schallenberg, *Bottled Energy*, 1.

35. For notable work on this subject, see Hyungsub Choi, "The Boundaries of Industrial Research: Making Transistors at RCA, 1948–1960," *Technology and Culture* 48, no. 4 (October 2007): 758–782; Lécuyer and Brock, "The Materiality of Microelectronics"; Leslie Berlin, *The Man behind the Microchip: Robert Noyce and the Invention of Silicon Valley* (New York: Oxford University Press USA, 2005); and Stuart W. Leslie, "Blue Collar Science: Bringing the Transistor to Life in the Lehigh Valley," *Historical Studies in the Physical and Biological Sciences* 32, no. 1 (2001): 71–113.

36. Benoît Godin, "The Linear Model of Innovation: The Historical Construction of an Analytical Framework," *Science, Technology, and Human Values* 31, no. 6 (November 2006): 646; Leonard S. Reich, "Irving Langmuir and the Pursuit of Science and Technology in the Corporate Environment," *Technology and Culture* 24, no. 2 (April 1983): 199–221; George Wise, "A New Role for Professional Scientists in Industry: Industrial Research at General Electric, 1900–1916," *Technology and Culture* 21, no. 3 (July 1980): 408–429.

37. Ronald Kline, "Construing 'Technology' as 'Applied Science': Public Rhetoric of Scientists and Engineers in the United States, 1880–1945," *Isis* 86, no. 2 (June 1995): 194–221, 218.

38. See Choi, "The Boundaries of Industrial Research," 775–777.

39. See Choi, "The Boundaries of Industrial Research," 775–777; and Glen R. Asner, "The Linear Model, the U.S. Department of Defense, and the Golden Age of Industrial Research," in *The Science-Industry Nexus: History, Policy, Implications*, ed. Karl Grandin, Nina Wormbs, and Sven Widmalm (Sagamore Beach, MA: Science History Publications/USA 2004), 1–12.

40. See the review essay of Robert W. Smith, "Introduction," in *Historical Studies in the Physical and Biological Sciences* 32, no. 1 (Second Laboratory History Conference, 2001): 3–9.

41. For classic sociological and/or historical analyses of tests and experiments, see Steve Woolgar and Bruno Latour's *Laboratory Life* (1979), Edward Constant's *The Origins of the Turbojet Revolution* (1980), Trevor Pinch's *Confronting Nature: The Sociology of Solar Neutrino Detection* (1986), Steven Shapin and Simon Schaffer's *Leviathan and the Air-Pump: Hobbes, Boyle and the Experimental Life* (1985), and Donald MacKenzie's *Inventing Accuracy: A Historical Sociology of Nuclear Missile Guidance* (1990).

42. Trevor Pinch, "'Testing—One, Two, Three . . . Testing!' Toward a Sociology of Testing," *Science, Technology and Human Values* 18, no. 1 (Winter 1993): 27–31.

43. For studies of dramaturgy in science and engineering communities, see Steven Shapin and Simon Schaffer, *Leviathan and the Air-Pump: Hobbes, Boyle and the Experimental Life* (Princeton, NJ: Princeton University Press, 1985); and Stephen Hilgartner, *Science on Stage: Expert Advice as Public Drama* (Stanford, CA: Stanford University Press, 2000).

44. George E. Evans and Karl V. Kordesch, "Hydrazine-Air Fuel Cells: Hydrazine-Air Fuel Cells Emerge from the Laboratory," *Science*, new series, 158, no. 3805 (December 1, 1967): 1148.

45. Thomas P. Hughes, *Networks of Power: Electrification in Western Society, 1880–1930* (Baltimore: Johns Hopkins University Press, 1983), 14, 80.

46. Langdon Winner, *The Whale and the Reactor: A Search for Limits in an Age of High Technology* (Chicago: University of Chicago Press, 1986), 12, 17.

47. Edmund Russell, James Allison, Thomas Finger, John K. Brown, Brian Balogh, and W. Bernard Carlson, "The Nature of Power: Synthesizing the History of Technology and Environmental History," *Technology and Culture* 52, no. 2 (April 2011): 246–259.

48. Bruce Podobnik, *Global Energy Shifts: Fostering Sustainability in a Turbulent Age* (Philadelphia: Temple University Press, 2006), 3–4.

49. Ronald Kline and Trevor Pinch developed this concept in reference to actors who used technologies to serve purposes other than what their designers intended; see "Users as Agents of Technological Change: The Social Construction of the Automobile in the Rural United States," *Technology and Culture* 37, no. 4 (October 1996): 767–777.

50. Jeremy Rifkin, *The Hydrogen Economy: The Creation of the World-Wide Energy Web and the Redistribution of Power on Earth* (New York: Jeremy P. Tarcher/Penguin, 2002), 9.

CHAPTER 1   DEVICE IN SEARCH OF A ROLE

A version of this chapter was published as "A Modern Philosopher's Stone: Techno-Analogy and the Bacon Cell," *Technology and Culture* 50, no. 2 (April 2009): 345–365.

1. It is also possible that Bacon was related to the family of Roger Bacon, the thirteenth-century natural philosopher; see K. R. Williams, "Francis Thomas Bacon, December 21, 1904–May 24, 1992," *Biographical Memoirs of Fellows of the Royal Society* 39 (February 1994): 3, 11–12.

2. Practically all accounts of Bacon's work take an internalist approach, examining the record of technology development largely at the laboratory level without considering the broader sociocultural and economic context. As might be expected, Bacon himself adopted this view in a career retrospective delivered to the Vittorio de Nora-Diamond Shamrock society in Seattle, Washington, in May 1978, the most detailed account of his career; see Francis Thomas Bacon, "The Fuel Cell: Some Thoughts and Recollections," *Journal of the Electrochemical Society* 126, no. 1 (January 1979): 7C–17C.

3. David E. Nye, *American Technological Sublime* (Cambridge, MA: MIT Press, 1994).

4. D.E.H. Edgerton, "Science and Technology in British Business History," *Business History* 29, no. 4 (1987): 84.

5. Robert Bud, "Penicillin and the New Elizabethans," *British Journal for the History of Science* 31, no. 3 (September 1998): 305–306.

6. Ibid, 316. In his study of the French electric automobile program of the 1960s, Michel Callon noted similar dynamics. Impressed by British and American fuel cell projects in the late 1950s and convinced that a technological revolution was in the offing, French science and technology elites forged a broad fuel cell coalition in which "everyone's interests were taken care of," a process that involved simplifying the technical issues at stake; see "The State and Technical Innovation: A Case Study of the Electrical Vehicle in France," *Research Policy* 9 (1980): 362–365; and "The Sociology of an Actor-Network: The Case of the Electric Vehicle," in *Mapping the Dynamics of Science and*

Technology: Sociology of Science in the Real World, ed. Michel Callon, John Law, and Arie Rip (London: Macmillan Press, 1986), 28–31.

7. S. Waqar H. Zaidi notes similar discursive uses of technology in postwar Britain in his study of the engineer Barnes Wallis and his unsuccessful schemes for a supersonic jet and submarine transport; see "The Janus-Face of Techno-nationalism: Barnes Wallis and the 'Strength of England,'" *Technology and Culture* 49, no. 1 (January 2008): 62–88.

8. Sally M. Horrocks, "Enthusiasm Constrained? British Industrial R&D and the Transition from War to Peace, 1942–1951," *Business History* 41, no. 3 (1999): 42–63. Relatively little attention has been paid to the activities of the research associations after the Second World War, and the extant literature raises intriguing questions regarding continuity and change in their policies and practices. The most notable works dedicated specifically to research associations include Ronald S. Edwards's *Co-operative Industrial Research: A Study of the Economic Aspects of the Research Associations, Grant-Aided by the Department of Scientific and Industrial Research* (London: Sir Isaac Pitman & Sons, 1950); P. S. Johnson's *Co-operative Research in Industry: An Economic Study* (London: Martin Robertson, 1973); and Ian Varcoe's "Co-operative Research Associations in British Industry, 1918–34," *Minerva: Review of Science, Learning and Policy* 19, no. 3 (Autumn 1981): 433–463.

9. William Robert Grove, "On a Gaseous Voltaic Battery," *Philosophical Magazine and Journal of Science* 21, S.3 (December 1842): 417.

10. H. A. Liebhafsky and E. J. Cairns, *Fuel Cells and Fuel Batteries: A Guide to Their Research and Development* (New York: John Wiley & Sons, 1968), 23–24, 29–42; John O'M. Bockris and S. Srinivasan, *Fuel Cells: Their Electrochemistry* (New York: McGraw-Hill, 1969), 25–26.

11. Bacon, "The Fuel Cell: Some Thoughts and Recollections," 7C.

12. Williams, "Francis Thomas Bacon," 6.

13. Francis Thomas Bacon, "Research into the Properties of the Hydrogen-Oxygen Fuel Cell," *British Electrical and Allied Manufacturers Association Journal* 61, no. 6 (Summer 1954): 6–8.

14. Bacon, "The Fuel Cell: Some Thoughts and Recollections," 7C.

15. Horrocks, "Enthusiasm Constrained?" 58–59; David Edgerton, "Whatever Happened to the British Warfare State? The Ministry of Supply, 1945–1951," in *Labour Governments and Private Industry: The Experience of 1945–1951*, ed. Helen Mercer, Neil Rollings, and Jim Tomlinson (Edinburgh: Edinburgh University Press, 1992), 91–93; Edgerton, *Warfare State: Britain, 1920–1970* (Cambridge: Cambridge University Press, 2006), 95–107.

16. Varcoe, "Co-operative Research Associations in British Industry," 433–463; Edgerton, "British Industrial R&D, 1900–1970," *Journal of European Economic History* 23, no. 1 (Spring 1994): 51–52; Edgerton, *Science, Technology and the British Industrial "Decline," 1870–1970* (Cambridge: Cambridge University Press, 1996), 42–43.

17. Varcoe, "Co-operative Research Associations in British Industry," 463; Horrocks, "Enthusiasm Constrained?" 44.

18. Edwards, *Co-operative Industrial Research*, 1–5.

19. Johnson, *Co-operative Research in Industry*, 116–117.

20. Leslie Hannah notes that in the early 1920s the ERA invested £4,000 in investigating the loading requirements of underground cables, saving the electrical supply industry some £4 million. Hannah describes this as some of the most cost-effective research in

the industry's history; see Leslie Hannah, *Electricity before Nationalisation: A Study of the Development of the Electricity Supply Industry in Britain to 1948* (London: Macmillan Press, 1979), 235.

21. C. W. Marshall, "ERA and the Electricity Supply Industry," *Cooperative Electrical Research: The Journal of The British Electrical & Allied Industries Research Association* 1 (July 1956): 7–9.

22. Varcoe, "Co-operative Research Associations in British Industry," 445, 458; Edgerton, *Science, Technology and the British Industrial "Decline,"* 42–43.

23. Bacon, "Research into the Properties of the Hydrogen-Oxygen Fuel Cell," 8.

24. Unconfirmed Minutes of the Second Meeting, Section Z, Sub-Committee F, Electrical Research Association, November 28, 1947, Papers and Correspondence of Francis Thomas Bacon, Reference Code NCUACS 68.6.97, Churchill College Archives Centre, University of Cambridge, Cambridge, England (hereafter cited as Bacon typescript [TS] or manuscript [MS]), Section B, Research and Development, Electrical Research Association/University of Cambridge, Agendas and Minutes of Meetings, B.127, 4–5.

25. Hannah, *Electricity before Nationalisation*, 311–323.

26. Memorandum from A. J. Allmand, November 19, 1947, Bacon TS, Electrical Research Association/University of Cambridge, Agendas and Minutes of Meetings, B.127, 1. With assistance from the physical chemist and Imperial College lecturer H.J.T. Ellingham, Allmand had authored the classic *Principles of Applied Electrochemistry* in 1924, a work that had a major influence on Bacon. The development of a practical fuel cell—a device consuming carbonaceous or hydrocarbon fuel—would, they had written, trigger a new industrial revolution comparable to that wrought by the steam engine; see Allmand and Ellingham, *The Principles of Applied Electrochemistry* (London: Edward Arnold & Co., 1924), 217–218.

27. Unconfirmed Minutes of the Second Meeting, Section Z, Sub-Committee F, 2.

28. Bacon, "The Fuel Cell: Some Thoughts and Recollections," 8C.

29. Bacon gave relatively little thought to applications in the early days. Shortly after the war, he approached a chemical company for support and was at a loss when its executives asked how he intended to use the reversible cell. The "best he could say" was that, supplied with electrolytic hydrogen, it might power an electric automobile, a proposal the executives scornfully rejected; see Bacon, "The Fuel Cell: Some Thoughts and Recollections," 8C.

30. Unconfirmed Minutes of the Second Meeting, Section Z, Sub-Committee F, 4–5.

31. The consultant C. W. Marshall ventured that there might be no need for a research program at all, since British lead-acid battery makers had achieved a world monopoly on their four-year lifetime guarantee almost exclusively by "commercial means," presumably meaning simple empirical methods; Unconfirmed Minutes of the Fourth Meeting, Section Z, Sub-Committee F, Electrical Research Association, June 8, 1949, Bacon TS, Electrical Research Association/University of Cambridge, Agendas and Minutes of Meetings, B.128, 6–7.

32. In his copy of the minutes of the seventh meeting of Subcommittee F, Bacon scribbled a correction, noting that the cell was comparable to the lead-acid battery only for discharge periods of between one and five hours; Unconfirmed Minutes of the Seventh Meeting, Section Z, Sub-Committee F, Electrical Research Association, November 17, 1950, Bacon TS, Electrical Research Association/University of Cambridge, Agendas and Minutes of Meetings, B.129, 5.

33. Unconfirmed Minutes of the Fifth Meeting, Section Z, Sub-Committee F, Electrical Research Association, November 29, 1949, Bacon TS, Section B, Research and Development, Electrical Research Association/University of Cambridge, Agendas and Minutes of Meetings, B.128, 6.

34. Unconfirmed Minutes of the Seventh Meeting, Section Z, Sub-Committee F, 5–8.

35. Ibid., 3, 7.

36. Aide Memoire by A. P. Paton, July 19, 1955, Bacon TS, Electrical Research Association/University of Cambridge, Correspondence, B.143, 1. By 1953, these agencies had together committed a total of £6,560, equivalent to £100,000 in 2007.

37. Memorandum, "Admiralty Interest in British Provisional Patents 25801/56 and 25802/56," August 17, 1957, Bacon TS, Research Centres, Laboratories and Sponsors, "Admiralty and C.J.B.," B.1080, 3.

38. George Szasz to Stanley Whitehead, April 15, 1953, Electrical Research Association/University of Cambridge, Correspondence, B.136.

39. Whitehead to Szasz, April 20, 1953, Bacon TS, Electrical Research Association/University of Cambridge, Correspondence, B.136.

40. R.G.H. Watson, "Electrochemical Generation of Electricity," *Research* 7, no. 1 (January 1954): 34, 39–40.

41. Bacon, "Research into the Properties of the Hydrogen-Oxygen Fuel Cell," 12.

42. Bacon, "The Fuel Cell: Some Thoughts and Recollections," 9C.

43. Whitehead to Bacon, May 20, 1954, Bacon TS, Electrical Research Association/University of Cambridge, Correspondence, B.139.

44. Whitehead to Bacon, September 10, 1954, Bacon TS, Electrical Research Association/University of Cambridge, Correspondence, B.140.

45. In October, a retired National Carbon Company researcher observed in the *Journal of the Electrochemical Society* that the economics of the fuel cell were "confused" and that the technology might never become competitive with the heat engine; see George W. Heise, "Research in Industry," *Journal of the Electrochemical Society* 101, no. 12 (December 1954): 293C.

46. Whitehead to T. E. Allibone, November 10, 1955, Bacon TS, Electrical Research Association/University of Cambridge, Correspondence, B.144, 2; Unconfirmed Minutes of the Twelfth Meeting, Section Z, Sub-Committee F, Electrical Research Association, January 24, 1956, Bacon TS, Electrical Research Association/University of Cambridge, Agendas and Minutes of Meetings, B.132, 6–8.

47. Whitehead to Bacon, September 13, 1955, Bacon TS, Electrical Research Association/University of Cambridge, Correspondence, B.144. Trained as a physicist at Oxford before taking his doctorate in electrical engineering at London University, Whitehead admitted he had only a "layman's knowledge" of electrochemical engineering; "Obituary: Dr. Whitehead," *Cooperative Electrical Research: The Journal of the British Electrical & Allied Industries Research Association* 1 (July 1956): 2–3; Whitehead to T. E. Allibone, November 10, 1955, Bacon TS, Electrical Research Association/University of Cambridge, Correspondence, B.144, 2.

48. Whitehead to Bacon, November 11, 1955, Bacon TS, Electrical Research Association/University of Cambridge, Correspondence, B.144.

49. Bacon to Earl of Halsbury, February 23, 1956, Bacon TS, National Research Development Corporation/Marshall's of Cambridge, "NRDC I," Correspondence, B.273.

50. Whitehead to Allibone, November 10, 1955, Bacon TS, Electrical Research Association/ University of Cambridge, Correspondence, B.144, 2.

51. Sir David Brunt to Bacon, May 15, 1956, Bacon TS, Research Centres, Laboratories and Sponsors, Ministry of Fuel and Power/Ministry of Power, British Electricity Authority/Central Electricity Authority, B.1164.

52. Bacon to Earl of Halsbury, February 23, 1956, Bacon TS, National Research Development Corporation/Marshall's of Cambridge, Correspondence, B.273.

53. Bud, "Penicillin and the New Elizabethans," 320–321.

54. S. T. Keith, "Inventions, Patents and Commercial Development from Governmentally Financed Research in Great Britain: The Origins of the National Research Development Corporation," *Minerva: Review of Science, Learning and Policy* 19, no. 1 (Spring 1981): 113, 117.

55. *An Introduction to the National Research Development Corporation* (London: Waterlow & Sons, 1956), 1.

56. Earl of Halsbury, "Strategy of Research," *Research* 7, no. 1 (December 1954): 480.

57. Earl of Halsbury to Bacon, February 28, 1956, Bacon TS, National Research Development Corporation/Marshall's of Cambridge, "NRDC I," Correspondence, B.273.

58. *Introduction to the National Research Development Corporation*, 4–9.

59. Earl of Halsbury, "Strategy of Research," 480.

60. Keith, "Inventions, Patents and Commercial Development," 120–121.

61. In a follow-up letter to Halsbury, Bacon played the nationalist card, advising the NRDC chief of Patterson Moos's interest and remarking it would be "most disappointing" if the device was developed in the United States rather than Britain; Bacon to Earl of Halsbury, June 12, 1956; Earl of Halsbury to Bacon, June 14, 1956, Bacon TS, National Research Development Corporation/Marshall's of Cambridge, "NRDC I," Correspondence, B.273.

62. Francis Thomas Bacon, "Details of Research and Development Work for the Hydrogen-Oxygen Cell, Leading up to the Construction of a Unit of About 10 kW," January 16, 1957, Bacon TS, National Research Development Corporation/Marshall's of Cambridge, "NRDC I," Correspondence, B.275, 5.

63. Bacon linked the British electrical industry's lack of interest in the hydrogen fuel cell to the expansion of the civilian nuclear energy program and its consequent monopolization of qualified technical staff; Bacon to Halsbury, June 12, 1956, Bacon TS, National Research Development Corporation/Marshall's of Cambridge, "NRDC I," Correspondence, B.273.

64. "Meeting at Patterson-Moos Division, Universal Winding Company, N.Y., March 15, 1957," March 18, 1957, Bacon TS, National Research Development Corporation/ Marshall's of Cambridge, "NRDC I," Correspondence, B.275, 1.

65. Patterson-Moos wanted the NRDC's electrode recipes indicating the pressures necessary for compacting nickel powder into disks and the temperatures and times necessary to bake them in such a way that only the jagged outer edges of the tiny metal crystals fused together, leaving sufficient space between them to create uniform pore distribution without melting part or all of the disks into a solid mass. The British team was determined not to release information on this delicate process—known as "sintering"—in the absence of a formal agreement; D. Hennessey to Anthony Moos, November 6, 1957; H. J. Crawley to Bacon, November 21, 1957, Bacon TS, Section B,

Research and Development, National Research Development Corporation/Marshall's of Cambridge, Correspondence, "NRDC I," B.279.

66. Blackwell C. Dunnam, "Air Force Experience in the Use of Liquid Hydrogen as an Aircraft Fuel," in *Hydrogen Energy Part B: Proceedings of the Hydrogen Economy Miami Energy (THEME) Conference, March 18–20, 1974*, ed. T. Nejat Veziroglu (New York: Plenum Press, 1975), 993–1001; J. C. Frost, "Report on Visit to the United States from 10th to 24th September," October 2, 1959, Bacon TS, "Letters to and from NRDC, 2nd File,'" B.318, 1; Bacon, "The Fuel Cell: Some Thoughts and Recollections," 10C.

67. "U.S. Interest in Fuel Cell," *Manchester Guardian Weekly*, August 27, 1959, 4. The NRDC reported that a total of £100,000 was spent in developing the Bacon cell, equivalent to about £1.53 million in 2007.

68. *Fuel Cells: A Symposium Held by the Gas and Fuel Division of the American Chemical Society at the 136th National Meeting in Atlantic City*, ed. G. J. Young (New York: Reinhold Publishing Corporation, 1960), 150–152. In the introduction to these conference proceedings, General Electric's H. A. Liebhafsky and D. L. Douglas authoritatively defined a fuel cell as a device that converted the chemical energy of a conventional or hydrocarbon fuel directly into DC energy (p. 1).

69. E. Gorin and H. L. Recht, "Fuel Cells," in *Proceedings: Tenth Annual Battery Research and Development Conference, May 23, 1956* (Fort Monmouth, NJ: Battery Conference Committee, 1956), 54.

70. In his study of U.S. government-backed research and development programs, Richard R. Nelson noted that the most productive fields—aviation and electronics—were those in which the government had a "well defined procurement interest" in the mature technologies; see Richard R. Nelson, *Government and Technical Progress* (New York: Pergamon Press, 1982), 453, 456.

71. Bacon to John Duckworth, April 26, 1960; Duckworth to Bacon, May 11, 1960, Bacon TS, National Research Development Corporation/Marshall's of Cambridge, "NRDC II," Correspondence, B.289.

72. This was a completely different design, thought by some capable of employing unreformed hydrocarbon fuels but plagued by corrosion and severe thermal expansion.

73. For example, see "U.K. Consortium to Develop Fuel Cells," *Financial Times*, October 30, 1961; John Maddox, "British Firm to Develop New Sources of Power: A Future for the Fuel Cell," *Guardian*, October 30, 1961.

74. Edgerton, *Warfare State*, 218.

75. Bacon to Karl V. Kordesch, February 23, 1957, Bacon TS, Section G, Fuel Cell Correspondence, G.120; see also Bacon, "The Fuel Cell: Power Source of the Future," *New Scientist* 6, no. 145 (August 1959): 274.

76. See Bacon, "The Fuel Cell: Some Thoughts and Recollections," 7C–17C; Bacon and T. M. Fry, "Review Lecture: The Development and Practical Application of Fuel Cells," *Proceedings of the Royal Society of London: Series A, Mathematical and Physical Sciences* 334, no. 1599 (September 1973): 427–452.

77. Edgerton, *Warfare State*, 233.

78. A 1961 article in the *London Times* referred to fuel cells as one of the NRDC's "projects of importance"; see "Finding Parents for Inventors' Brain-Children: Many Ideas, but Few Are Financed," *Times* (London), June 21, 1961. In 1962, the Royal Society sponsored a technology conference in which the Bacon cell figured prominently; see V. T. Saunders, "Some Exhibits at the Royal Society Conversazione, May 1962," *Contemporary*

Physics 4, no. 3 (February 1963): 228; Harold Wilson cited fuel cells, along with the hovercraft and the Atlas computer, as evidence of the vitality of British innovation in 1963 and 1964; Edgerton, *Warfare State*, 218.

79. Anthony M. Moos, "Introduction," in *Fuel Cells Volume II: A Symposium Held by the Divisions of Fuel Chemistry and Petroleum Chemistry of the American Chemical Society at the 140th National Meeting in Chicago*, ed. G. J. Young (New York: Reinhold Publishing Corporation, 1963), 2.

80. John H. Huth, "Program Plan for Electrochemistry: Program Plan No. 4," February 1, 1963, 1958–1966, Official Correspondence Files—Materials Sciences Office, Advanced Research Projects Agency, accession number (AN) 68-A-2658, record group (RG) 330, Box 2, AO 247—Monsanto Research Corporation (DA 36-039-SC-88945), 1, National Archives and Records Administration, College Park, MD.

CHAPTER 2   MILITARY MIRACLE BATTERY

1. David R. Adams et al., *Fuel Cells: Power for the Future* (Cambridge, MA: Fuel Cell Research Associates, 1960), 9.

2. Gijs Mom, *The Electric Vehicle: Technology and Expectations in the Automobile Age* (Baltimore: Johns Hopkins University Press, 2004), 238–241.

3. Earle F. Cook, "Introductory Remarks," in *Proceedings: 11th Annual Battery Research and Development Conference, May 22–23, 1957* (Fort Monmouth, NJ: Battery Conference Committee, 1957), 1–2.

4. See, for example, Fred Block, "Innovation and the Invisible Hand of Government," in *State of Innovation: The U.S. Government's Role in Technology Development*, ed. Fred Block and Matthew R. Keller (Boulder, CO: Paradigm Publishers, 2011), 8–14; and Michael Belfiore, *The Department of Mad Scientists: How DARPA Is Remaking Our World, from the Internet to Artificial Limbs* (New York: Smithsonian Books/HarperCollins, 2009), xiii–xxiii. Alex Roland and Philip Shiman point out that DARPA simply contracted out its bureaucratization to other institutions; see *Strategic Computing: DARPA and the Quest for Machine Intelligence, 1983–1993* (Cambridge, MA: MIT Press, 2002), xiii.

5. ARPA-E, http://arpa-e.energy.gov/About/About.aspx.

6. Herbert F. York, *Making Weapons, Talking Peace: A Physicist's Odyssey from Hiroshima to Geneva* (New York: Basic Books, 1987), 136–138.

7. Urner Liddel, October 24, 1960, Box 4, Project Lorraine—Energy Conversion, 1958–1966 Official Correspondence Files—Materials Sciences Office, Advanced Research Projects Agency (OCF-MSO, ARPA), accession number (AN) 68-A-2658, record group (RG) 330, National Archives and Records Administration, College Park, MD (hereafter cited as 1958–1966 OCF-MSO, ARPA).

8. For a classic study of this phenomenon, see Brian Balogh, *Chain Reaction: Expert Debate and Public Participation in American Commercial Nuclear Power, 1945–1975* (Cambridge: Cambridge University Press, 1991).

9. Dwayne A. Day, "Invitation to Struggle: The History of Civilian-Military Relations in Space," in *Exploring the Unknown: Selected Documents in the History of the U.S. Civilian Space Program, Volume II: External Relationships*, ed. John M. Logsdon (Washington, DC: NASA History Office, 1996), 244–249, 255. After ARPA was stripped of its space programs in September 1959, the armed services were allowed to develop their own satellites.

10. Cook, "Introductory Remarks," 1–2.

11. David Linden, "The Next Ten Years," in *Proceedings: 10th Annual Battery Research and Development Conference, May 23, 1956* (Fort Monmouth, NJ: Battery Conference Committee, 1956), 63.

12. *Proceedings: 10th* Annual Battery Research and Development Conference, 53–56; *Proceedings: 11th Annual Battery Research and Development Conference*, 5–7.

13. Patterson-Moos's Kenneth Rapp claimed that the hydrogen-oxygen cell would compete favorably with battery systems in "mobile military applications" requiring power at the minimum weight, although he admitted the economics were then unclear; see Kenneth Rapp, "Hydrogen-Oxygen Fuel Cells," in *Proceedings: 12th Annual Battery Research and Development Conference, May 21–22, 1958* (Fort Monmouth, NJ: Battery Conference Committee, 1958), 11. The National Carbon representative noted that testing at that time suggested that a single power source could be used to "operate a variety of different loads," lending credence to the idea of a general-purpose fuel cell. Interestingly, the National Carbon Company's program had proceeded under completely different assumptions than Cook indicated. George E. Evans claimed that the fuel cell the company had unveiled in 1957 had been produced through a "laborious collection of information" and long-term testing; it was not "some sudden scientific breakthrough." G. E. Evans, "An Experimental Hydrogen-Oxygen Fuel Cell System," in *Proceedings: 12th Annual Battery Research and Development Conference*, 5–7.

14. In the late 1960s, the assistant director of Pratt & Whitney's research and development laboratory asked Bacon to help them recruit electrochemical talent in Britain on their behalf; H. M. Hershenson to Bacon, July 22, 1968, Papers and Correspondence of Francis Thomas Bacon, Reference Code NCUACS 68.6.97, Section B, Research and Development, Research Centres, Laboratories and Sponsors, "Leesona-Moos and P&W," B.1147, Churchill College Archives Centre, University of Cambridge, Cambridge, England (hereafter cited as Bacon typescript [TS] or manuscript [MS]).

15. Stuart W. Leslie and Robert H. Kargon observe this phenomenon in Frederick Terman's efforts to enlist the participation of eastern U.S. industrial firms as part of an attempt to replicate Stanford University's successful academic-industrial complex in New Jersey in the mid-1960s; see "Selling Silicon Valley: Frederick Terman's Model for Regional Advantage," *Business History Review* 70, no. 4 (Winter 1996): 447.

16. Francis Thomas Bacon, "Visit by Mr. F. T. Bacon and Mr. K.E.V. Willis of NRDC to Fuel Cell Activities in the USA—July 1961," undated, Bacon TS, Section B, Research and Development, Energy Conversion Limited, Reports, B.672, 1. Bacon did not name the participating companies.

17. Gerrit Jan Schaeffer credits these demonstrations with triggering the fuel cell boom of the 1960s; see his "Fuel Cells for the Future: A Contribution to Technology Forecasting from a Technology Dynamics Perspective" (PhD diss., University of Twente, 1998), 171, http://doc.utwente.nl/fid/1515#search.

18. "Summary of Proposal of Research on Energy Conversion," February 6, 1961, Box 4, Project Lorraine—Energy Conversion, 1958–1966 OCF-MSO, ARPA.

19. W. T. Grubb, "Ion-Exchange Batteries," in *Proceedings: 11th Annual Battery Research and Development Conference*, 5; M. L. Perry and T. F. Fuller, "A Historical Perspective of Fuel Cell Technology in the 20th Century," *Journal of the Electrochemical Society* 149, no. 7 (2002): S60.

20. U.S. Army Signal Research and Development Laboratory, "Fuel Cell for Radar Set AN/TPS-26 (XN-1)," *Research and Development Summary* 8, no. 1 (January 1, 1961): 2.

Labeling the 84-pound, 200-watt hydrogen-fueled unit a backpack system, the Signal Laboratory considered the device a breakthrough in power sources for forward-area radars. The Signal Laboratory also sponsored work on the membrane fuel cell in the regenerative role for possible use in space. The Army abandoned work on membrane fuel cells around 1963, and thereafter General Electric focused on developing them for space and commercial terrestrial applications. In the late 1980s, Ballard Power Systems developed an advanced variant, sparking a resurgence of interest in the technology.

21. Herbert F. Hunger, "Ion Exchange Liquid Fuel Cells," in *Proceedings: 14th Annual Power Sources Conference, May 17–18–19, 1960* (Red Bank, NJ: PSC Publications Committee, 1960), 55–56.

22. James E. Wynn, "Liquid Fuel Cells," in *Proceedings: 14th Annual Power Sources Conference*, 52–55.

23. For example, Hunger included partially oxygenated hydrocarbons and hydrocarbons as one class of fuels distinct from "exotic" liquid fuels; see Hunger, "Ion Exchange Liquid Fuel Cells," 56.

24. In the spring of 1960, Hans K. Ziegler, the laboratory's chief scientist, considered recent work the most spectacular advance of a battery technology "previously considered hardly worthwhile"; see Ziegler, "Keynote Address," in *Proceedings: 14th Annual Power Sources Conference*, 2.

25. Nathan W. Snyder, "IDA IM 155: Research in Advanced Energy Conversion," January 4, 1960, Box 4, Project Lorraine—Energy Conversion, 24–26, OCF-MSO, ARPA, NARA College Park. In addition to fuel cells, ARPA funded work on thermionic, thermo-electric, and photo-electric devices. The Army's Signal Corps managed the photo-electric and electrochemical projects while the Office of Naval Research (ONR) managed thermionic contracts and shared responsibility for thermo-electric projects with the Bureau of Ships (BuS). Of the four technologies included in ARPA's fiscal year 1959 operations (fall 1958–spring 1959), fuel cells placed third in funding under the rubrics of electrochemistry: photoelectricity ($566,450), thermo-electricity ($285,000), electrochemistry ($117,610), and thermionics ($100,000).

26. Silvan S. Schweber, "The Mutual Embrace of Science and the Military: ONR and the Growth of Physics in the United States after World War II," in *Science, Technology and the Military*, vol. 1, ed. Everett Mendelsohn, Merritt Roe Smith, and Peter Weingart (Dordrecht, Netherlands: Kluwer Academic Publishers, 1988), 34; Alexander Kossiakoff, "Conception of New Defense Systems and the Role of Government R&D Centers," in *The Genesis of New Weapons: Decision-Making for Military R&D*, ed. Franklin A. Long and Judith Reppy (New York: Pergamon Press, 1980), 65.

27. "About IDA," www.ida.org/IDAnew/Welcome/history.html.

28. York, *Making Weapons, Talking Peace*, 141.

29. Richard J. Barber Associates, *The Advanced Research Projects Agency, 1958–1974* (Washington, DC: ARPA, 1975), IV-13–IV-15; James R. Killian, Jr., *Sputnik, Scientists and Eisenhower: A Memoir of the First Special Assistant to the President for Science and Technology* (Cambridge, MA: MIT Press, 1977), 129; York, *Making Weapons, Talking Peace*, 141. Although York had been appointed as ARPA's chief scientist on the recommendation of James Killian, head of the President's Scientific Advisory Committee, he had actually been hired through the IDA. Other major figures in defense policy such as George W. Rathjens and Jack Ruina also worked at both ARPA and the IDA.

30. Richard J. Barber Associates, *The Advanced Research Projects Agency*, IV-14.

31. Ibid., III-2, III-70–III-76. ARPA's budget peaked at $520 million in fiscal year 1959 before declining to around $250 million per annum during the 1960s.

32. Snyder, "IDA IM 155: Research in Advanced Energy Conversion," January 4, 1960, Box 4, Project Lorraine—Energy Conversion, 1–4, 1958–1966 OCF-MSO, ARPA. In fiscal year 1959, ARPA spent $2.34 million on space power research.

33. This was a common refrain among scientists engaged in fundamental research; see Daniel S. Greenberg, *The Politics of Pure Science*, new ed. (Chicago: University of Chicago Press, 1999), 215–218.

34. "Project Lorraine Summary," undated, Box 4, Project Lorraine—Energy Conversion, 1958–1966 OCF-MSO, ARPA.

35. Memorandum by Urner Liddel, October 24, 1960, Box 4, Project Lorraine—Energy Conversion, 1958–1966 OCF-MSO, ARPA. Nearly half of its resources were to be devoted to solid-state physics. The remainder was to support research in solar energy and heat transfer (5 percent apiece), electrochemical conversion and energy storage (5 to 10 and 15 percent respectively), and magnetohydrodynamics (20 percent).

36. Founded in 1951, the U.S. Army Office of Ordnance Research (AOOR) became Army Research Office (ARO) Durham in February 1961. Located at Duke University and supervised by the ARO Physical Sciences Division, this office was responsible for coordinating and contracting for basic research in the physical sciences; *Proceedings: 15th Annual Power Sources Conference, May 9–11, 1961* (Red Bank, NJ: PSC Publications Committee, 1961), 2–3. The role of Ernst M. Cohn is especially noteworthy, both in this episode and in the broader history of U.S. fuel cell research and development in the 1960s. Cohn was one of the two Ordnance Corps personnel who had helped ARPA's Liddel formulate the Army's hydrocarbon fuel cell policy and very likely had coauthored the proposal on energy conversion. As a member of the ARO, Cohn subsequently played an important role in administering this program before moving to NASA in 1962–63, where he became head of the agency's in-house fuel cell research and development program. Cohn's involvement in both Army and NASA fuel cell research over a decade and a half provides a valuable window of insight into the disparate technical challenges facing these agencies and the changing economic and technological fortunes of the federal government's fuel cell programs well into the 1970s.

37. "Summary of Proposal of Research on Energy Conversion," 1–6.

38. Glen R. Asner, "The Linear Model, the U.S. Department of Defense, and the Golden Age of Industrial Research," in *The Science-Industry Nexus: History, Policy, Implications*, ed. Karl Grandin, Nina Wormbs, and Sven Widmalm (Sagamore Beach, MA: Science History Publications/USA 2004), 3–12.

39. William O. Baker, "Advances in Materials Research and Development," in *Advancing Materials Research*, eds. Peter A. Psaras and H. Dale Langford (Washington, DC: National Academy Press, 1987), 5–7.

40. Stuart W. Leslie, *The Cold War and American Science: The Military-Industrial-Academic Complex at MIT and Stanford* (New York: Columbia University Press, 1993), 213.

41. Lyle H. Schwartz, "Materials Research Laboratories: Reviewing the First Twenty-Five Years," in *Advancing Materials Research*, 35–48.

42. Richard J. Barber Associates, *The Advanced Research Projects Agency*, V-46.

43. Robert L. Sproull, "Materials Research Laboratories," in *Advancing Materials Research*, ed. Psaras and Langford, 32.

44. Raymond H. Bates, "Welcome," in *Proceedings: 15th Annual Power Sources Conference*, 1.

45. "Project Lorraine: Project Objectives," June 9, 1961, Box 4, Project Lorraine—Energy Conversion, 1958–1966 OCF-MSO, ARPA.

46. R. P. Buck, L. R. Griffith, R. T. MacDonald, and M. J. Schlatter, "Mechanisms and Kinetics of Reactive Groups in Organic Fuels," in *Proceedings: 15th Annual Power Sources Conference*, 16–20.

47. George W. Rathjens to Chief, Research and Development, Department of the Army, July 26, 1961, Box 2, AO 247—Esso Research and Engineering, 1958–1966 OCF-MSO, ARPA.

48. Ziegler, "Keynote Address," in *Proceedings: 14th Annual Power Sources Conference*, 2.

49. Memorandum by Charles F. Yost, July 13, 1962, Box 2, AO 247—Esso Research & Engineering, 1958–1966 OCF-MSO, ARPA.

50. One noted chemist observed that American companies involved in fuel cell research in the early 1960s generally knew very little of electrochemistry and its related fields; Ernest Yeager to F. T. Bacon, February 2, 1961, Bacon TS, Section G, Correspondence, Fuel Cell Correspondence, G.125.

51. J. C. Frost, "Report on Visit to the United States from 10–24 September," October 2, 1959, Bacon TS, Section B, Research and Development, National Research Development Corporation/Marshall's of Cambridge, Correspondence, B.318, 2.

52. Bacon, "Visit by Mr. F. T. Bacon and Mr. K.E.V. Willis," 5.

53. Charles F. Yost, "Renewal of Program in Electrochemistry, AO 247—General Electric Company, Contract DA-44-009, ENG-4853," August 13, 1962, Box 2, AO 247-General Electric, 1958–1966 OCF-MSO, ARPA.

54. General Electric, "Saturated Hydrocarbon Fuel Cell Program (ERDL), Quarterly Letter Report Number 1, December 1, 1961–March 31, 1962," undated, Box 2, AO 247—General Electric, 2, 1958–1966 OCF-MSO, ARPA; W. T. Grubb, "Hydrocarbon Fuel Cells," in *Proceedings: 16th Annual Power Sources Conference, May 22–24, 1962* (Red Bank, NJ: PSC Publications Committee, 1962), 31–32.

55. Yost, "Renewal of Program in Electrochemistry."

56. J. P. Ruina, September 25, 1962, Box 2, AO 247—General Electric, 1958–1966 OCF-MSO, ARPA.

57. Hamilton wanted to reduce funding from $1.7 million to $1 million for fiscal year 1963; G. C. Szego to S. S. Penner, September 20, 1962, Box 4, Project Lorraine—Energy Conversion, 1–7, 1958–1966 OCF-MSO, ARPA.

58. Ibid.

59. John C. Orth to George Szego, September 17, 1962, Box 4, Project Lorraine—Energy Conversion, 1958–1966 OCF-MSO, ARPA.

60. J. W. Babcock to George Szego, September 17, 1962, Box 4, Project Lorraine—Energy Conversion, 1, 1958–1966 OCF-MSO, ARPA.

61. Szego to Penner, September 20, 1962.

62. "Project Lorraine," December 1962, Box 4, Project Lorraine—Energy Conversion, 1958–1966 OCF-MSO, ARPA.

63. Yost to R. L. Sproull, "Memorandum for Dr. Sproull: Subject: Project Lorraine," September 12, 1963, Box 4, Project Lorraine—Energy Conversion, 1958–1966 OCF-MSO, ARPA.

64. Yost to Ruina, "Subject: Expansion of Areas in the Energy Conversion Program (Lorraine): Hydrogen-Reformer Fuel Cell Unit," October 16, 1962, Box 4, Project Lorraine—Energy Conversion, 1958–1966 OCF-MSO, ARPA.

65. ARPA's gradual emphasis on hardware is reflected in its summary lists of funding disbursements to contractors. In 1960 and 1961, it simply grouped all fuel cell contractors together under the rubric of "electrochemistry." By 1963, it was classifying fuel cell contractors under those working on hardware or "complete systems" (General Electric, Esso, and Allis-Chalmers/Engelhard), components or "catalysts" (Monsanto and California Research Corporation), and "basic work" (Stanford University, General Atomic, AMOCO, Tyco, Westinghouse, and the University of California); "General Comments," undated, 3–4, in John H. Huth and Charles F. Yost, "Status Report—Project Lorraine," October 31, 1963, Box 4, Project Lorraine—Energy Conversion, 7, 1958–1966 OCF-MSO, ARPA.

66. Yost, "Increased Funding for Applied Fuel Cell Research at General Electric—ARPA Order 247, Amendment 5," April 8, 1963, Box 2, AO 247—General Electric, 1958–1966 OCF-MSO, ARPA.

67. Bacon, "Visit by Mr. F. T. Bacon and Mr. K.E.V. Willis," 5.

68. Oscar P. Cleaver to Director, Advanced Research Projects Agency, March 25, 1963, Box 2, AO 247—General Electric; Yost, "Increased Funding for Applied Fuel Cell Research at General Electric—AO 247, Amendment 5," 1958–1966 OCF-MSO, ARPA.

69. General Electric, "Quarterly Letter Report Number 3, January 1, 1963–March 31, 1963," undated, Box 2, AO 247—General Electric, 8–9, 1958–1966 OCF-MSO, ARPA.

70. Ibid., 3.

71. John H. Huth, "Program Plan No. 4: Program Plan for Electrochemistry," February 1, 1963, Box 2, AO 247—Monsanto Research Corporation (DA 36-039-SC-88945), 1, 1958–1966 OCF-MSO, ARPA.

72. "GE Fuel Cell Advances," *Wall Street Journal*, April 24, 1963. C. G. Suits, a General Electric vice president, went so far as to invite Harold Brown, the director of defense research and engineering. Brown politely declined on the grounds that he was scheduled to appear before the House Appropriations Subcommittee on that date; C. G. Suits to Harold Brown, March 27, 1963; Brown to Suits, April 12, 1963, Box 2, AO 247—General Electric, 1958–1966 OCF-MSO, ARPA.

73. Bacon to H. A. Liebhafsky, May 29, 1963, Bacon TS, Section G, Correspondence, Fuel Cell Correspondence, G.130.

74. Memorandum by Yost, July 13, 1962.

75. Ernst M. Cohn, "Space Applications," in *Proceedings: 17th Annual Power Sources Conference, May 21–23, 1963* (Red Bank, NJ: PSC Publications Committee, 1963), 86–87.

76. Barry L. Tarmy, "Methanol Fuel Cells," in *Proceedings: 16th Annual Power Sources Conference, May 22–24, 1962* (Red Bank, NJ: PSC Publications Committee, 1962), 30–31.

77. C. E. Heath, "The Methanol-Air Fuel Cell," in *Proceedings: 17th Annual Power Sources Conference*, 96–97.

78. G. C. Szego, "Economic and Logistic Considerations," *Proceedings: 17th Annual Power Sources Conference*, 88. As one of his sources for this presentation, Szego cited *Power for the Future*, the 1960 analysis authored by the Harvard Business School graduate students.

79. The U.S. Army Signal Corps Research and Development Agency was redesignated the U.S. Army Electronics Research and Development Agency and subsumed under the U.S. Army Electronics Command on August 1, 1962.

80. B. C. Almaula, "High Power Ground Applications," in *Proceedings: 17th Annual Power Sources Conference*, 83. Almaula also mentioned the "energy depot" concept. Among

the Army's more speculative futurist power source projects, this was the idea of a small nuclear reactor intended to serve as a mobile chemical factory in a front-line capacity. Parachuted into a combat zone, the plant would either provide electricity directly or electrolyze hydrogen and oxygen from local water sources, which could then be used to power a fuel cell.

81. David Linden, "Introductory Remarks," in *Proceedings: 17th Annual Power Sources Conference*, 80.

82. M. J. Schlatter et al., "Special Report No. 1: Appraisal of State of Development of Low Temperature Hydrocarbon-Oxygen (or Air) Fuel Cells," September 6, 1963, Box 2, AO 247–California Research Corporation (DA-49-186-ORD-929), 1–2, 1958–1966 OCF-MSO, ARPA.

83. M. X. Polk, "Program Plan No. 148," August 14, 1963, Box 2, AO 247—Esso Research & Engineering, 1, 1958–1966 OCF-MSO, ARPA.

84. M. J. Schlatter et al., "Special Report No. 1," 1–2.

85. In 1962, the United States consumed 866,459 ounces, the lion's share of global production of 1.314 million ounces, but produced a mere 28,742 ounces. The U.S. Bureau of Mines indicated that the entire U.S. supply of palladium, the next most common platinoid metal after platinum itself, was sufficient to build 30,000 automotive fuel cells annually; Raymond H. Comyn, November 8, 1963, Box 2, AO 247—California Research Corporation (DA-49-186-ORD-929), 1–4, 1958–1966 OCF-MSO, ARPA.

86. John O. Smith et al., "First Quarterly Progress Report: Research to Improve Electrochemical Catalysts, May 15, 1963 to August 15, 1963," September 10, 1963, Box 4, AO 437—Monsanto Research Corporation (DA-44-009-AMC-202 (T)), 1–2, 1958–1966 OCF-MSO, ARPA.

87. General Electric, "Saturated Hydrocarbon Fuel Cell Program, Quarterly Letter Report Number 4, July 1, 1963–September 31, 1963," undated, Box 2, AO 247—General Electric (DA-44-009-ENG-4909), 1–9, 1958–1966 OCF-MSO, ARPA.

88. U.S. Energy Study Group, *Energy Research and Development and National Progress: Prepared for the Interdepartmental Energy Study by the Energy Study Group under the Direction of Ali Bulent Cambel* (Washington, DC: U.S. Government Printing Office, 1965), v, xviii. The study involved the Office of Science and Technology; the Bureau of the Budget; the Office of Emergency Planning; the departments of Defense, Interior, Commerce, Labor, State, and Health, Education and Welfare; the Atomic Energy Commission; the Federal Power Commission; the National Science Foundation; and the National Aeronautics and Space Administration.

89. Ibid., 418–436. The committee's key findings appear to have been drawn largely from a report authored by Ernst M. Cohn; see Cohn, "Interdepartmental Energy Study: Chemical and Biochemical Fuel Cells," August 19, 1963, Record Number 13761: Propulsion: Auxiliary Power: Fuel Cells, 1961–1999, NASA Headquarters Archive, Washington, DC, 13.

90. U.S. Energy Study Group, *Energy Research and Development and National Progress*, 302–308.

91. Huth and Yost, "Status Report—Project Lorraine," October 31, 1963, 3.

92. Yost to Sproull, "Memorandum for Dr. Sproull," September 12, 1963, 2.

93. Huth, "Comments on the Interdepartmental Energy Study," October 8, 1963, Box 4, Project Lorraine—Energy Conversion, 2–4, 1958–1966 OCF-MSO, ARPA. Huth held that as of 1963 there were only four academicians in the United States specializing in electrode

phenomena specifically relating to fuel cells, with a combined output of well under 40 students per year: Yeager (Western Reserve), Bockris (University of Pennsylvania), Tobias (University of California), and Delahay (Louisiana State University). The conclusions in the published version of the Interdepartmental Energy Study (IES) differed significantly from those in the draft report reviewed by ARPA managers in the fall and appeared to incorporate the agency's views on the hydrocarbon project. Although the published study noted that the shortage of electrochemical workers threatened the long-term viability of the program, it did not contain any of the recommendations cited in the draft, including the funding increases for electrochemical training or for fuel cell research in general. Moreover, the published report seemed to reverse the draft's assessment of the methanol cell, referring to it instead as the most advanced project involving oxygenated hydrocarbons. These revisions reflected the views of Huth and NASA's Cohn respectively, both of whom had sat on the IES fuel cell panel.

94. Yost to Sproull, "Phaseout of the Energy Conversion Program (Project Lorraine)," November 1, 1963, Box 4, Project Lorraine-Energy Conversion, 1958–1966 OCF-MSO, ARPA.

95. Huth and Yost, "Status Report—Project Lorraine," October 31, 1963.

96. Yost to Sproull, "Memorandum for Dr. Sproull," September 12, 1963.

97. As evidence of progress, the IDA analysts had cited extremely high power densities obtained in hydrocarbon and methanol fuel cells but did not account for "parasitic" demand, the power that various accessory systems including pumps, fans, and reformers drew from overall output; see Huth, "Comments on IDA Report R-103 by R. Hamilton and G. Szego," March 2, 1964, Box 4, Project Lorraine—Energy Conversion, I, 1958–1966 OCF-MSO, ARPA.

98. Sproull to Secretary of the Army, March 9, 1964, Box 4, Project Lorraine—Energy Conversion, I–2, 1958–1966 OCF-MSO, ARPA.

99. Huth, "Memo for the Record: Subject: Renewal of Fuel-Cell Contracts with GE & Esso During FY65 (AO 247, Re: Letter to Army), May 6, 1964, Box 2, AO 247—General Electric (DA-44-009-ENG-4909), 1958–1966 OCF-MSO, ARPA.

100. Bacon to Liebhafsky, June 10, 1964, Bacon TS, Section G, Correspondence, Fuel Cell Correspondence, G.133.

101. Notes on meeting with Liebhafsky by Bacon, June 13, 1965, Bacon MS, Section G, Correspondence, Fuel Cell Correspondence, G.135.

102. H. A. Liebhafsky, "Fuel Cells and Fuel Batteries: An Engineering View," *IEEE Spectrum* (December 1966): 53.

103. Cohn, "Interdepartmental Energy Study," August 19, 1963.

104. C. E. Heath, "Methanol Fuel Cells," in *Proceedings: 18th Annual Power Sources Conference, May 19–21, 1964* (Red Bank, NJ: PSC Publications Committee, 1964), 34.

105. Barry L. Tarmy, "Methanol Fuel Cell Battery," in *Proceedings: 19th Annual Power Sources Conference, May 18–20, 1965* (Red Bank, NJ: PSC Publications Committee, 1965), 41–42.

106. Liebhafsky, "Fuel Cells and Fuel Batteries," 54.

107. Esso Research and Engineering Company, Process Research Division, "Proposal for the Continuation of Government Contract Research on Fuel Cells; Program Period—Calendar Year 1965," July 24, 1964, Box 2, AO 247—Esso Research and Engineering Company, 1958–1966 OCF-MSO, ARPA.

108. Expenditures on fuel cells constituted the lion's share of spending on energy conversion technology in Project Lorraine. Of the $14.6 million ARPA invested in power technology research through fiscal year 1964, about $8 million was spent on fuel cells, of which General Electric received the vast majority; Huth, "Comments on IDA Report R-103 by R. Hamilton and G. Szego," March 2, 1964. In comparison, spending on materials sciences in this period averaged around $20–$25 million per year. ARPA's total annual budget in the first half of the 1960s fluctuated between a low of $186 million in fiscal year 1962 to a high of $278 million in fiscal year 1965, remaining in the $270 million range during fiscal years 1964–1966; Richard J. Barber Associates, *The Advanced Research Projects Agency*, V-1, VI-1.

109. Tarmy, "Methanol Fuel Cell Battery," 41–43.

110. Historical Branch Headquarters, United States Army Electronics Command, *Annual Historical Summary of U.S. Army Electronics Command, July 1, 1964–June 30, 1965, March 1, 1967*, 1012, Communications and Electronics Command Historical Research Collection, Communications-Electronics Lifecycle Management Command Historical Office, Fort Monmouth, New Jersey (hereafter cited as CECOM Historical Research Collection).

111. G. Ciprios, "Methanol Fuel Cell Battery," in *Proceedings: 20th Annual Power Sources Conference, May 24–26, 1966* (Red Bank, NJ: PSC Publications Committee, 1966), 48.

112. Stephen J. Bartosh, "500-Watt Indirect Hydrocarbon System," in *Proceedings: 20th Annual Power Sources Conference*, 31; T. G. Kirkland, "5 KW Hydrocarbon-Air Fuel Cell Power Plant," in *Proceedings: 20th Annual Power Sources Conference*, 35.

113. Yost, "Memorandum for the Director, Program Management; Subject: Electrochemistry (5 KW Reformer Fuel-cell Unit)," March 4, 1963, Box 4, AO 430—Allis-Chalmers, 1–2, 1958–1966 OCF-MSO, ARPA.

114. James R. Huff and John C. Orth, "The USAMECOM-MERDC Fuel Cell Electric Power Generation Program," in *Fuel Cell Systems II: 5th Biennial Fuel Cell Symposium Sponsored by the Division of Fuel Chemistry at the 154th Meeting of the American Chemical Society, Chicago, Illinois, September 12–14, 1967* (Washington, DC: American Chemical Society, 1969), 318.

115. T. G. Kirkland and W. G. Smoke, Jr., "5 KVA Hydrocarbon-Air Fuel Cell System Test," in *Proceedings: 19th Annual Power Sources Conference*, 26.

116. Bacon, "Energy Conversion Limited Visit Report VR 136: Visit to America," March 13–24, 1967, Bacon TS, Section B, Research and Development, Energy Conversion Limited, Reports, B.760, 7–8.

117. The Pratt & Whitney reformer fuel cell system was more successful, operating as a unit for about 100 hours at its rated capacity; Stephen J. Bartosh, "500-watt Indirect Hydrocarbon System," in *Proceedings: 20th Annual Power Sources Conference*, 33.

118. T. G. Kirkland, "5 KW Hydrocarbon-Air Fuel Cell Power Plant," 36, 39.

119. Agency for Toxic Substances and Disease Registry, "Toxicological Profile for Jet Fuels JP-4 and JP-8, Chemical and Physical Information," 64, 69, http://www.atsdr.cdc.gov/toxprofiles/tp76.html.

120. News Release, U.S. Army Electronics Command, March 2, 1967, CECOM Historical Research Collection TS, 2.

121. Kirkland and Smoke "5 KVA Hydrocarbon-Air Fuel Cell System Tests," 28.

122. H. A. Liebhafsky, "The Electrocatalyst Problem in the Direct Hydrocarbon System," in *Proceedings: 20th Annual Power Sources Conference*, 1.

123. Huff and Orth, "The USAMECOM-MERDC Fuel Cell Electric Power Generation Program," 321.

124. Edward A. Gillis, "Hydrazine-Air Fuel Cell Power Sources," in *Proceedings: 20th Annual Power Sources Conference*, 41. One 20-kilowatt hydrazine fuel cell was developed for use with an electric truck at Fort Belvoir. In this hybrid system, the fuel cell was combined with a 20-kilowatt battery. When maximum power was called for, the vehicle could draw on both power sources, while during periods of low demand, the fuel cell would recharge the battery. Bacon, "Energy Conversion Limited Visit Report VR 136," 7.

125. Gillis, "Hydrazine-Air Fuel Cell Power Sources," 44.

126. Huff and Orth, "The USAMECOM-MERDC Fuel Cell Electric Power Generation Program," 321; Historical Branch Headquarters, Annual Historical Summary of United States Army Electronics Command, July 1, 1964–June 30, 1965, March 1, 1967, CECOM Historical Research Collection TS, Annual Historical Data, 1012.

127. "DMR-C4 (e) 7451: General Comparison Between Hydrazine Hydrate and Liquid Hydrocarbon Fuels," March 6, 1968, Bacon TS, Section, B, Research and Development, Research and Development Topics, "Hydrazine and Ammonia," B.1050, 7. Materials that react catalytically with hydrazine include Monel (nickel-copper alloy with smaller quantities of iron, manganese, and aluminum), zinc, copper alloys, iron, magnesium, stainless steels with more than 0.5 percent molybdenum, and some aluminum alloys.

128. Gillis, "Hydrazine-Air Fuel Cell Power Sources," 44.

129. Dennis Starks, "So Now You Have a PRC-47," http://www.pacificsites.com/~brooke/NYH_PRC47.html.

130. Historical Branch Headquarters, "Power Sources," *Annual Historical Summary of United States Army Electronics Command, July 1, 1965–June 30, 1966, April 20, 1969*, CECOM Historical Research Collection TS, Annual Historical Data, 1007; S. M. Chodosh, M. G. Rosansky, and B. E. Jagid, "Metal-Air Primary Batteries: Replaceable Zinc Anode Radio Battery," in *Proceedings: 21st Annual Power Sources Conference, May 16–17–18, 1967* (Red Bank, NJ: PSC Publications Committee, 1967), 103.

131. F. G. Perkins, "Experience with Hydrazine Fuel Cells in SEA," in *Proceedings: 24th Annual Power Sources Conference*, May 19–20–21, 1970 (Red Bank, NJ: PSC Publications Committee, 1970), 202.

132. Gillis, "Hydrazine-Air Fuel Cell Power Sources," 44–45.

133. D. P. Gregory, "A Review of Factors Influencing Policy Towards Hydrazine Fuel Cells: Presentation to the ECL Technical Committee," June 8, 1966, Bacon TS, Section B, Research and Development, Research and Development Topics, "Hydrazine and Ammonia," B.1049–1050, 2–3.

134. George E. Evans, "Hydrazine-Air Fuel Cells," in *Proceedings: 21st Annual Power Sources Conference*, 34.

135. Ibid., 36. In 1961, Union Carbide fuel cell researchers had believed that given the current state of the art, general-purpose fuel cells were impossible. Fuel cells then had to be specially built to suit each new application; see H. W. Holland, "Carbon Electrode Fuel Cell Battery," in *Proceedings: 15th Annual Power Sources Conference*, 28.

136. Evans, "Hydrazine-Air Fuel Cells," 35.

137. Preliminary work had not prepared researchers for the problems of scaling up fuel control equipment. One Union Carbide engineer observed that it was one thing to develop a fuel feeder that would operate well "on a relay rack in an air conditioned laboratory," something else again to build one that could withstand "salt-spray test,

humidity-cycling test and . . . environmental extremes." Evans, "Hydrazine-Air Fuel Cell Controls," in *Proceedings: 22nd Annual Power Sources Conference, May 14–15–16, 1968* (Red Bank NJ: PSC Publications Committee, 1968), 1.

138. Perkins, "Experience with Hydrazine Fuel Cells in SEA," 202–203.

139. Evans and Kordesch, "Hydrazine-Air Fuel Cells," 1151–1152.

140. Galen R. Frysinger, "Integrated Cell Stacks," in *Proceedings: 22nd Annual Power Sources Conference*, 26.

141. R. E. Salathe, "Evolution of the Replaceable Hydrazine Module as a Basic Building Block," in *Proceedings: 24th Annual Power Sources Symposium*, 204.

142. Bacon reported that the Army simply threw away the alkaline electrolyte of hydrazine fuel cells when it became contaminated with carbon dioxide in ambient air; Galen R. Frysinger, interview by Francis Thomas Bacon, March 17, 1967, Bacon TS, Section F, Visits and Conferences, "Visit to USA," F.24.

143. Historical Branch Headquarters, "Electronics Components Laboratory," in *Annual Historical Summary of United States Army Electronics Command, July 1, 1969–June 30, 1970*, CECOM Historical Research Collection TS, Annual Historical Data, 24.

144. The Electronic Components Laboratory (ECL) did not have its own chemists and electrochemists and had to recruit entirely from outside Electronics Command, reflecting the larger trend in the United States. As early as 1964, the ECL's request to Army Materiel Command for 28 additional personnel and $2.6 million to support the "intensified research effort" was denied. Repeated requests for more funds and personnel were ignored. Well before the Johnson administration escalated the war in Vietnam in 1964, the ECL laboratory was losing personnel, 117 in all since 1960, representing 21 percent of its overall staff. Although the Power Sources Division avoided the worst of the cuts, it could not avoid making reductions as well. U.S. Army Electronics Command, "Commander's Operations Report," February 1966, CECOM Historical Research Collection TS, Annual Historical Data, 1-15–1-16.

145. Frysinger, interview by Bacon, March 17, 1967; B. S. Baker, "Grove Medal Acceptance Address," *Journal of Power Sources* 86 (2000): 9.

146. Gregory, "A Review of Factors Influencing Policy Towards Hydrazine Fuel Cells," 3.

147. John B. O'Sullivan, "Historical Review of Fuel Cell Technology," in *Proceedings: 25th Annual Power Sources Symposium, May 23–24–25, 1972* (Red Bank, NJ: PSC Publications Committee, 1972), 149–150.

148. David F. Noble, *The Religion of Technology: The Divinity of Man and the Spirit of Invention* (New York: Alfred A. Knopf, 1997), 5.

CHAPTER 3   FUEL CELLS AND THE FINAL FRONTIER

1. John M. Logsdon, *The Decision to Go to the Moon: Project Apollo and the National Interest* (Cambridge, MA: MIT Press, 1970), ix.

2. Joan Lisa Bromberg, *NASA and the Space Industry* (Baltimore: Johns Hopkins University Press, 1999).

3. A. J. Appleby and F. R. Foulkes, *Fuel Cell Handbook* (New York: Van Nostrand Reinhold, 1989), 163.

4. Edward Gottlieb & Associates, "Civilian Dividends from Space Research: A Review of Some of the Peaceful Benefits from Our Satellite Program, Now and in the Future,"

undated, Record Number 18530 IX: Technology Utilization, Addresses, Speeches, 28–29, NASA Headquarters Archive, Washington, DC.

5. Howard E. McCurdy, *Inside NASA: High Technology and Organizational Change in the U.S. Space Program* (Baltimore: Johns Hopkins University Press, 1993), 71–73.

6. Logsdon, *The Decision to Go to the Moon*, 98–99, 174.

7. Courtney G. Brooks, James M. Grimwood, and Loyd S. Swensen, Jr., *Chariots for Apollo: A History of Manned Lunar Spacecraft* (Washington, DC: National Aeronautics and Space Administration, 1979), 9.

8. The STG was established by NASA administrator T. Keith Glennan in 1958 at the Langley Research Center to manage Project Mercury, the first U.S. human-carrying space mission. Responsible for designing the Mercury capsule and supplying the technical criteria for the Gemini and Apollo spacecraft, leaving the actual design to contractors, the STG was renamed the Manned Spacecraft Center and relocated to Houston under the aegis of the Office of Manned Spaceflight in late 1961. This quickly became the most important single bureau within NASA, responsible for spacecraft and rocket boosters and commanding most of the agency's budget and personnel.

9. David Bell III and Fulton M. Plauché, *Apollo Experience Report: Power Generation System* (Washington, DC: National Aeronautics and Space Administration, March 1973), 2.

10. Ernst M. Cohn, "Primary Hydrogen-Oxygen Fuel Cells for Space," text of presentation, June 1967, RN 13761: Propulsion, Auxiliary Power, Fuel Cells, 1961–1999, 2, NASA Headquarters Archive.

11. Thomas F. Dixon to D. Brainerd Holmes, December 12, 1961, RN 13761: Propulsion, Auxiliary Power, Fuel Cells, 1961–1999, NASA Headquarters Archive.

12. This practice emerged during the Second World War as an emergency measure enabling the federal government to develop a direct relationship with a selected contractor by fiat; see W. Henry Lambright, "The NASA-Industry-University Nexus: A Critical Alliance in the Development of Space Exploration," in *Exploring the Unknown: Selected Documents in the History of the U.S. Civilian Space Program, Volume II: External Relationships*, ed. John M. Logsdon (Washington, DC: NASA History Office, 1996), 413.

13. Preston T. Maxwell, "Conference with Lewis Research Center Personnel to Discuss R&D Contract for Hydrogen-Oxygen Fuel Cell; Date of Conference, April 25, 1961," April 27, 1961, Record Number 16307, Program: Apollo, Location: Box 062-35, University of Houston-Clear Lake, Neumann Library.

14. At the same time, Allis-Chalmers, supported by NASA's Office of Advanced Research and Technology, was developing a rival version of the Bacon cell as a backup power source, one specialists would come to regard as superior to Pratt & Whitney's model.

15. Bernard J. Crowe, *Fuel Cells: A Survey* (Washington, DC: National Aeronautics and Space Administration, 1973), 8; Francis Thomas Bacon and T. M. Fry, "Review Lecture: The Development and Practical Application of Fuel Cells," *Proceedings of the Royal Society of London: Series A, Mathematical and Physical Sciences* 334, no. 1599 (September 25, 1973): 431–433; Francis Thomas Bacon, "The Fuel Cell: Some Thoughts and Recollections," *Journal of the Electrochemical Society* 126, no. 1 (January 1979): 8C–9C. Efforts to operate at higher temperatures, pressures, and electrolyte concentrations created severe corrosion and leakage problems.

16. F. T. Bacon, "Report on Visit to USA, July 14–31, 1961 to see the Present State of Development of Fuel Cells," Papers and Correspondence of Francis Thomas Bacon, Section B, Research and Development, Energy Conversion Limited, Reports, B.672, 4,

Churchill College Archives Centre, Cambridge University, England (hereafter cited as Bacon typescript [TS] or manuscript [MS]).

17. R. W. Fahle, "Short Memorandum Report No. 2940: An Investigation of the Performance of the British Bacon Fuel Cell," August 31, 1960, Bacon TS, Section B, Research and Development, Background Material, Fuel Cell Reports, "Leesona-Moos and P&W," B.1321, 3, 15.

18. Bacon and Fry, "Review Lecture," 440.

19. Crowe, *Fuel Cells*, 28. Systems were activated from a minimum of 35 hours (*Apollo 7*) to a maximum of 206 hours (*Apollo 9*) before launch; *Apollo 4 Mission Report*, MSC-PA-R-68-1, Houston, Texas, January 1968; *Apollo 7 Mission Report*, MSC-PA-R-68-15, Houston, Texas, December 1968; *Apollo 8 Mission Report*, MSC-PA-R-69-1, Houston, Texas, February 1969; *Apollo 9 Mission Report*, MSC-PA-R-69-2, Houston, Texas, May 1969; *Apollo 10 Mission Report*, MSC-00126, Houston, Texas, August 1969; *Apollo 11 Mission Report*, MSC-00171, Houston, Texas, November 1969.

20. Bell and Plauché, *Apollo Experience Report*, 10.

21. Bacon grew increasingly exasperated as the program stalled under the range of problems introduced by the decision to lower pressure. In an April 1964 letter to William Podolny, the head of Pratt & Whitney's fuel cell division, Bacon noted that all the various combinations of temperature, pressure, and electrolyte concentration with which Pratt & Whitney was experimenting had been already been tried by his team at the Cambridge laboratory; Bacon to W. M. Podolny, April 28, 1964, Bacon TS, Section B, Research and Development, Research Centres, Laboratories and Sponsors, "Leesona-Moos and P&W," B.1144.

22. McCurdy, *Inside NASA*, 29.

23. Robert B. Hotz, "Apollo and Its Critics," *Aviation Week & Space Technology*, April 29, 1963, 17.

24. "First Fuel Cell Delivery Called Apollo Milestone," *Roundup: NASA Manned Spacecraft Center* 3, no. 6 (January 8, 1964): 1.

25. Harvey J. Schwartz, "Batteries and Fuel Cells," in *Space Power Program Review for Office of Advanced Research and Technology, Lewis Research Center July 13–14, 1964, Volume II: Presentations*, RN 13717: Space Power Review, 7, NASA Headquarters Archive.

26. William E. Rice, interview by Rebecca Wright, NASA Johnson Space Center Oral History Project, March 18, 2004, 7–8.

27. "Apollo Power System in White Sands Test," *Roundup: NASA Manned Spacecraft Center* 5, no. 7 (January 21, 1966): 8.

28. McCurdy, *Inside NASA*, 28–29.

29. William Simon, interview by John Mauer, "Oral History with Bill Simon about Early Space Shuttle Development," March 22, 1984, Record Number 15557, Program: Shuttle, Location: SHU-INT2, 4–5, University of Houston-Clear Lake, Neumann Library; Rice interview by Wright, 11.

30. J. R. Foley to F. T. Bacon, January 3, 1969, Bacon TS, Section B, Research and Development, Laboratories and Sponsors, "Leesona-Moos and P&W," B.1148.

31. Barton C. Hacker and James M. Grimwood, *On the Shoulders of Titans: A History of Project Gemini* (Washington, DC: National Aeronautics and Space Administration, 1977), 103, 148. Hacker and Grimwood claim the actual decision was made by a midranking official named Robert Cohen, who faced criticism when the program became mired in difficulties. However, in a 1967 interview, Cohen claimed the final

decision to use the membrane fuel cell in Gemini spacecraft was made by James Chamberlin, the director of the Gemini program. McDonnell Aircraft's endorsement of the fuel cell must also be considered as an additional factor in the Office of Manned Space Flight's decision-making chain of command; Robert Cohen, interview by Peter J. Voorzimmer, February 21, 1967, Record Number 030076: Oral History Interviews, NASA Headquarters Archive.

32. General Electric, "Saturated Hydrocarbon Fuel Cell Program, Quarterly Letter Report Number 5, January 1, 1964–March 31, 1964," Box 2, AO 247—General Electric, DA-44-009-ENG-4909, 1958–1966 Official Correspondence Files—Materials Sciences Office, Advanced Research Projects Agency (hereafter cited as OCF-MSO, ARPA), accession number (hereafter AN) 68-A-2658, Record Group 330, National Archives and Records Administration, College Park, MD, 1.

33. E. A. Oster, "Ion Exchange Membrane Fuel Cells," in *Proceedings: 16th Annual Power Sources Conference, May 22–24, 1962* (Red Bank, NJ: PSC Publications Committee, 1962), 23.

34. Cohen interview by Voorzimmer, February 21, 1967.

35. Ernst M. Cohn, "The Growth of Fuel Cell Systems," August 1965, Record Number 13761: Propulsion, Auxiliary Power: Fuel Cells, 1961–1999, 5–7, NASA Headquarters Archive; William Simon, interview by John Mauer, "Oral History with Bill Simon about Early Space Shuttle Development," 7; Hacker and Grimwood, *On the Shoulders of Titans*, 149.

36. James M. Grimwood and Barton C. Hacker, *Project Gemini Technology and Operations: A Chronology* (Washington, DC: National Aeronautics and Space Administration, 1969), 104.

37. Hacker and Grimwood, *On the Shoulders of Titans*, 178.

38. Grimwood and Hacker, *Project Gemini Technology and Operations*, 116–117.

39. Hacker and Grimwood, *On the Shoulders of Titans*, 178.

40. Direct Energy Conversion Operation, General Electric, "Fuel Cells for Power in Space," NAM/NASM 1673, Cat. No. 1966-0646 and 0647, National Air and Space Museum Archive, Washington, DC.

41. Chester J. Civin to Edward M. Kennedy, December 30, 1964, NAM/NASM 1673, Cat. Nos. 1966-0646 and 0647, p. 0002, National Air and Space Museum Archive.

42. Frederick C. Durant III to S. Paul Johnston, undated, NAM/NASM 1673, Cat. Nos. 1966–0646 and 0647, pp. 0021–0023, National Air and Space Museum Archive.

43. Grimwood and Hacker, *Project Gemini Technology and Operations*, 179.

44. R. S. Mushrush to F. C. Durant, May 10, 1965, NAM/NASM 1673, Cat. Nos. 1966–0646 and 0647, p. 0020, National Air and Space Museum Archive.

45. "Commercial Uses of Fuel Cell Seen: GE Will Market Device It Built for Gemini 5," *New York Times*, September 2, 1965, 38.

46. P. W. Malik and G. A. Souris, *NASA Contractor Report CR-1106: Project Gemini: A Technical Summary; Prepared by McDonnell Douglas* (Washington, DC: National Aeronautics and Space Administration, June 1968), 50–59. Of the 13 Gemini missions, batteries were used as the sole power source for 6 flights.

47. Jim Maloney, "Gemini Fuel Cell Water Won't Be for Drinking," *Houston Post*, September 16, 1965.

48. Malik and Souris, *NASA Contractor Report*, 50–59.

49. Smithsonian Institution Press Release, "Smithsonian to Get General Electric Fuel-Cell Power Source; Identical with Those Used in the Successful Gemini VII Mission," January 10, 1966, NAM/NASM 1673, Cat. Nos. 1966–0646 and 0647, pp. 0028–0029, National Air and Space Museum Archive.

50. Cohen interview by Voorzimmer, February 21, 1967, 1.

51. Simon interview by Mauer, 7.

52. Perry and Fuller, "Historical Perspective of Fuel Cell Technology," S64; Keith B. Prater, "The Renaissance of the Solid Polymer Fuel Cell," *Journal of Power Sources* 29 (1990): 239.

53. Bacon notes on meeting with Cohn, July 17, 1974, Bacon MS, Section G, Correspondence, Fuel Cell Correspondence, G.150; Manned Spacecraft Center Announcement, "Organizational Changes to the Propulsion and Power Division Engineering and Development Directorate," November 14, 1972, Record Number 137246, Report Number 72-176, Program: Center, Location: GR 1015, University of Houston–Clear Lake, Neumann Library; Bacon notes on meeting with Cohn, July 17, 1974, Bacon MS, Section G, Correspondence, Fuel Cell Correspondence, G.150.

54. Manned Spacecraft Center Announcement, "Organizational Changes," November 14, 1972.

55. Mark E. Byrnes, *Politics and Space: Image Making by NASA* (Westport, CT: Praeger, 1994), 3.

56. Remarks of March 16, 1962, *Congressional Record*, 87th Cong. 2d sess., 1962, 108, No. 57, April 12, 1962, A2857–A2858. Johnson had also mentioned that Americans consumed $7.5 billion worth of cigars and cigarettes annually, but this was usually omitted in media reports.

57. Gilbert Burke, "Hitching the Economy to the Infinite," in *The Space Industry: America's Newest Giant* (Englewood Cliffs, NJ: Prentice-Hall, 1962), 96–97.

58. John E. Condon, "Practical Values of Space Exploration," October 10, 1962, Record Number 18530 IX: Technology Utilization, Addresses, Speeches, 3–6, NASA Headquarters Archive.

59. NASA News Release, James T. Dennison, "Contributions of Aerospace Research to the Business Economy," September 26, 1963, Record Number 18530 IX: Technology Utilization, Addresses, Speeches, 12, NASA Headquarters Archive.

60. See, for example, Howard Simons, "New Fuel Cell Approved by NASA to Supply Moon Trip Power, Water," *Washington Post*, December 27, 1963; Phillip S. Brimble, "Fuel Cells Valuable Power Source: Used in Gemini 5," *Kansas City Star*, August 19, 1965; "Commercial Uses of Fuel Cell Seen" *New York Times*, September 2, 1965; Maloney, "Gemini Fuel Cell Water Won't Be for Drinking," *Houston Post*, September 16, 1965; John Barbour, "Gemini Success Points Out Future of Fuel Cell Power," *Orlando Evening Star*, October 6, 1965; "Find Myriad of Uses for Fuel Cells: May Some Day Power Automobiles," *Chicago Tribune*, March 30, 1966; Robert C. Toth, "Spaceflights Provide Down-to-Earth Benefits," *Washington Post*, November 10, 1966.

61. "Dr. Wernher von Braun Tells How We'll Get Space Power from Fuel Cells," *Popular Science* (August 1964): 80, 171.

62. Wernher von Braun to Members of Foreign Investors Council, October 15, 1967, Wernher von Braun Papers, Box 127, File 9, United States Space and Rocket Center, Huntsville, Alabama.

63. Bruce L. R. Smith, *American Science Policy since World War II* (Washington, DC: Brookings Institution, 1990), 50–56, 84.

64. James E. Webb, NASA's second administrator, noted approvingly that between October 1, 1960, and June 30, 1961, the proportion of the agency's funds transferred to private organizations had risen from 80 to 92 cents on the dollar; see Webb, *National Aeronautics and Space Administration, Fifth Semi-annual Report to Congress* (Washington, DC: United States Government Printing Office, 1962), iii–iv. By way of comparison, the Department of Defense has historically spent well over half of its budget on internal operations (known as "consumption" and consisting of personnel, operations, and maintenance) and the balance on goods and services produced largely or wholly by contractors (known as "investment" and consisting of procurement, research and development, military construction, and nuclear warheads); see http://www.whitehouse.gov/omb/budget/Historicals/.

65. Smith, *American Science Policy,* 50–56.

66. Lyndon B. Johnson, *The Vantage Point: Perspectives of the Presidency, 1963–1969* (New York: Holt, Rinehart and Winston, 1971), 279.

67. Richard Hirsch and Joseph John Trento, *The National Aeronautics and Space Administration* (New York: Praeger Publishers, 1973), 66; Bromberg, *NASA and the Space Industry,* 2. The former NACA research centers were Ames in lower San Francisco Bay, Langley Field in Virginia, and Lewis at the Cleveland Airport.

68. Howard J. Schwartz, "Chemical Systems," in *Space Power Review,* April 18–20, 1963, Summary, Record Number 13716: Space Power Review, 4, NASA Headquarters Archive. The review panel did, however, speculate on the roles a future fuel cell might play. Conventional fuel cells that required their own fuel supplies limited the choice of spacecraft to types that had mission profiles similar to NASA's first generation of piloted craft. Nonreversible fuel cells could not be used for deep-space missions of long duration owing to the heavy load of fuel required. Alternative concepts included a fuel cell that could consume the tank residuals of a hydrazine-powered rocket, but the joint panel's real hope was for a regenerative cell reconstituting its reaction products using photovoltaic electricity. Lewis's fuel cell manager believed such a system would be useful mainly for a geosynchronous earth satellite because its 24-hour orbit offered a much longer period of solar exposure and electricity generation than the 90-minute low-earth satellite orbit. The best potential application of such a system, planners believed, was as a power plant for a moon base.

69. Howard J. Schwartz, "Battery and Fuel Cell Program," in *Space Power Program Review for Office of Advanced Research and Technology, Volume I: Summary, July 13–14, 1964,* Record Number 13717, Space Power Review, 8, NASA Headquarters Archive. Of a total budget of almost $83 million in fiscal year 1964, Lewis Research Center allocated only six contracts worth a mere $2 million for fuel cells. Electrochemical technology had a far lower priority than either solar or nuclear technologies, which received $10.9 and $67 million respectively in fiscal year 1964. Lewis had no full-scale fuel cell projects of any type. Such work as did occur was limited to improving components of existing fuel cell concepts, particularly the PEM type, to obtain better durability and was not directly related to the two major projects then under way for NASA human spacecraft. Lewis devoted only 16 professional technicians to fuel cells, compared with 174 and 69 in the nuclear and solar fields respectively. For fiscal year 1965, Lewis even proposed to slash fuel cell funding by almost half to $1.1 million; see Bernard Lubarksy, "Overall View of the Lewis Power Program," in *Space Power Program Review for Office of Advanced Research*

and Technology: Volume II: Presentations, July 13–14, 1964, Record Number 13717, Space Power Review, 6, NASA Headquarters Archive.

70. Schwartz, "Battery and Fuel Cell Program," 8.

71. "Allis-Chalmers Gets AAP Fuel Cell Work," *Roundup: NASA Manned Spacecraft Center* 8, no. 4 (December 6, 1968): 1. Some agency managers believed the Allis-Chalmers design was superior to the PC-3A-2. It avoided some of the thermal expansion and corrosion problems attending the use of a strong potassium hydroxide solution by employing an electrolyte at a lower concentration than used in the Pratt & Whitney model and containing it within an asbestos mesh. Operating at around 87°C instead of 204°C, the Allis-Chalmers unit weighed 185 pounds, 55 pounds less than the PC-3A-2 and, at 2,800 watts, produced more than twice the power.

72. George E. Mueller to Robert R. Gilruth, August 17, 1966, Record Number 13761: Propulsion, Auxiliary Power, Fuel Cells, 1961–1999, NASA Headquarters Archive.

73. Ernst M. Cohn, "Space Applications," in *Proceedings: 17th Annual Power Sources Conference, May 21–23, 1963* (Red Bank, NJ: PSC Publications Committee, 1963), 86.

74. Cohn, "NASA's Fuel Cell Program," in *Fuel Cell Systems: Symposia Sponsored by the Division of Fuel Chemistry at the 145th and 146th Meetings of the American Chemical Society, New York, Sept. 12–13, 1963*, Philadelphia (Washington, DC: American Chemical Society, 1965), 8.

75. Cohn, "Interdepartmental Energy Study: Chemical and Biochemical Fuel Cells," August 19, 1963, Record Number 13761: Propulsion: Auxiliary Power: Fuel Cells, 1961–1999, 13–16, 24–33, NASA Headquarters Archive.

76. The expression "reverse spin-off" seems to have been first used in 1974 by a graduate student at George Washington University's Department of Science, Technology, and Public Policy named Jim Maloney. Maloney employed it in a paper written for the Smithsonian Institution's National Air and Space Museum (NASM) as background research for an exhibit devoted to the socioeconomic benefits of air and spaceflight intended for the inauguration of the NASM's new facility as part of the bicentennial celebrations of 1976; Jim Maloney, unpublished essay, "Fuel Cells: A Case Study of Benefits from Flight," May 15, 1974, Record Unit 348, Box 3: United Aircraft, 36, Smithsonian Institution Archive, Washington, DC.

77. Designers and marketers of consumer products employed futurist and aerospace themes well before Sputnik, of course, but this trend probably reached its apogee during the Space Race; see Sean Topham, *Where's My Space Age? The Rise and Fall of Futuristic Design* (Munich: Prestel, 2003); and Joseph J. Corn and Brian Horrigan, *Yesterday's Tomorrows: Past Visions of the American Future* (New York: Summit Books, 1984).

78. Raymond A. Bauer, *Second-Order Consequences: A Methodological Essay on the Impact of Technology* (Cambridge, MA: MIT Press, 1969), 170.

CHAPTER 4   DAWN OF THE COMMERCIAL FUEL CELL

1. Martin V. Melosi, *Coping with Abundance: Energy and Environment in Industrial America* (Philadelphia: Temple University Press, 1985), 324–326; Walter A. Rosenbaum, *Energy, Politics and Public Policy*, 2nd ed. (Washington, DC: Congressional Quarterly Press, 1987), 20.

2. News release, Pratt & Whitney Aircraft, "Pratt & Whitney Aircraft Fuel Cell History," undated, Papers and Correspondence of Francis Thomas Bacon, Reference Code

NCUACS 68.6.97, Section B, Research and Development, B.1353, "Fuel Cell Developments in the USA, 1963–1974," Churchill College Archives Centre, University of Cambridge, Cambridge, England (hereafter cited as Bacon typescript [TS] or manuscript [MS]).

3. During a visit to the Pratt & Whitney fuel cell facility at East Hartford in July 1961, Francis Bacon reported that the company's efforts were focused mainly on military and aerospace applications. He noted that no serious work on commercial fuel cells was then under way and that the possibility that the company might enter that field was contingent on its successful development of military and quasi-military aerospace fuel cells. Francis Thomas Bacon, "Report on Visit to USA, from July 14–31, 1961, to See the Present State of Development of Fuel Cells," Bacon TS, Section B, Research and Development, Energy Conversion Limited, Reports, B.672, 9; J. N. Haresnape, E. A. Shipley, G. H. Townend and K.E.V. Willis, "Report of the Technical Working Party," January 3, 1962, Bacon TS, Section B, Research and Development, Energy Conversion Limited, Reports, B.774, 8, 15.

4. A.D.S. Tantram, "Visit to Leesona-Moos and Pratt & Whitney," March 16–20, 1964, Bacon TS, Section B, Research and Development, Energy Conversion Limited, Reports, B.746, 2.3, 4.1–4.3, 4.6.

5. A.D.S. Tantram, "Visit Report 149: Pratt & Whitney: March 25–28, 1968 (Molten Carbonate Cell Discussions), Sylvania: March 29, 1968 (Iron Electrodes)," Bacon TS, Section B, Research and Development, Energy Conversion Limited, Reports, B.762, 8.

6. Space Division of North American Rockwell Corporation, "Space Program Benefits," April 17, 1968, Record Number 012978: Inventions from Space Program, 37–38, NASA Headquarters Archive, Washington, DC.

7. T. M. Fry, Bacon's former collaborator, claimed that he had learned that the system of U.S. government contract work had resulted in the transfer to United Aircraft of supplementary disbursements worth several million dollars for engineering support in addition to payment for the original contracts; T. M. Fry, "Visit of Mr. Podolny," August 21, 1970, Bacon TS, Section B, Research and Development, Background Material, "Miscellaneous," B.1350, 1; Gerald J. Mossinghoff, Deputy Assistant Administrator (Policy), Office of Legislative Affairs, to Carl Swartz, Minority Staff, Committee on Science and Astronautics, House of Representatives, August 18, 1971, Record Number 013761: Propulsion: Auxiliary Systems: Fuel Cells, 1, NASA Headquarters Archive.

8. Pratt & Whitney/United Aircraft/Technologies Corporation produced aerospace fuel cells for NASA from 1966 until the late 1990s, building a total of 90 Apollo units and a further 25 for the Space Shuttle; Joseph M. King and Michael J. O'Day, "Applying Fuel Cell Experience to Sustainable Power Products," *Journal of Power Sources* 86 (2000): 17.

9. A. J. Appleby, "Issues in Fuel Cell Commercialization," *Journal of Power Sources* 58, no. 2 (February 1996): 155–156; A. J. Appleby and F. R. Foulkes, *Fuel Cell Handbook* (New York: Van Nostrand Reinhold, 1989), 62–63.

10. "Fuel Cells: Production by '75?" *Energy Digest*, July 22, 1971.

11. Fry, "Visit of Mr. W. H. Podolny," August 21, 1970.

12. Terri Aaronson, "The Black Box," *Environment* 13, no. 10 (December 1971): 10–18.

13. "A New Generation of Electric Generating Systems," Pratt & Whitney Aircraft, S 3462, May 1972, Bacon TS, Fuel Cell Developments in the USA, B.1352.

14. "Power Problems: Consumption Up, Supplies Short," *Congressional Quarterly Weekly Report* 29, no. 38 (September 18, 1971): 1940.

15. James E. Meeks, "Concentration in the Electric Power Industry: The Impact of Antitrust Policy," *Columbia Law Review* 72, no. 1 (January 1972): 66–76, 114.

16. "Fuel Cells: Production by '75?" 103–104.

17. Clark Martin, "Apollo Spurred Commercial Fuel Cell," *Aviation Week & Space Technology*, January 1, 1973, 56.

18. Ibid.

19. Part of a research project sponsored by the Smithsonian's National Air and Space Museum (NASM), the interviews were conducted by Jim Maloney in support of an exhibit the museum was preparing for its new facility then under construction in Washington for the 1976 bicentennial celebrations. Titled "Benefits from Flight," it was designed to showcase the economic and social effects of the aerospace industry, in short, to legitimize and promote the concept of spin-off. Influenced heavily by interviews with Pratt & Whitney officials, the NASM analysis concluded that the Apollo program played absolutely no role in spurring development of commercial fuel cells. Jim Maloney, unpublished essay, "Fuel Cells: A Case Study of Benefits from Flight," May 15, 1974, Record Unit 348, Box 3: United Aircraft, 29–30, Smithsonian Institution Archive, Washington, DC.

20. In an address to the Twentieth Century Club of Hartford in November 1968, George E. Mueller, chief of NASA's manned spaceflight program, remarked that in "activating" the fuel cell from dormancy for use in space, the agency had contributed to TARGET and deserved recognition when Pratt & Whitney began to produce commercial units for terrestrial use; George E. Mueller, November 11, 1968, Record Number 012978: Inventions from Space Program, NASA Headquarters Archive. In spring 1970, Louis Frey (R-Florida and member of the House Committee on Science and Astronautics) presented a report in Congress identifying direct and indirect spin-off benefits that repeated Mueller's phraseology; Extension of remarks, April 30, 1970, *Congressional Record*, 91st Cong. 2d sess., 1970, 116, no. 69 (May 1, 1970), E3785, E3788. In a House committee inquiry into NASA's role in fuel cell development held in August 1971, the agency's Office of Legislative Affairs claimed that the private sector's fuel cell design and engineering database, workforce, and facilities had been developed almost entirely as a result of the space program; Mossinghoff to Swartz, August 18, 1971, Record Number 013761: Propulsion: Auxiliary Systems: Fuel Cells, 1, NASA Headquarters Archive.

21. Ernst M. Cohn to William E. Rice, "Meeting with Pratt & Whitney on October 21, 1971," October 26, 1971, Record Number 4764, Report Number RPP, Program: Shuttle, Box 007-14, University of Houston–Clear Lake, Neumann Library.

22. "Why Utilities Are Backing the Fuel Cell: It Stretches Fuel Supplies and Eliminates Pollution; The Economics Look Good," *Business Week*, January 12, 1974, 28C–28D; Gene Smith, "9 Utilities Plan Large Fuel Cell: Program Set Up to Develop an Extra Power Source," *New York Times*, December 20, 1973, 61, 63; see also Appleby, "Issues in Fuel Cell Commercialization," 156; Karl Kordesch and Günter Simader, *Fuel Cells and Their Applications* (New York: Weinheim, 1996), 185.

23. Kordesch and Simader, *Fuel Cells and Their Applications*, 185.

24. "Why Utilities Are Backing the Fuel Cell," *Business Week*, 28C; "Large Fuel Cells Sought: 9 Electric Utilities to Help Finance Project," *Washington Post*, December 20, 1973, D11.

25. Podolny to Bacon, October 5, 1973, Bacon TS, Section B, Research and Development, "Leesona-Moos and P&W," "Leesona-Moos and P&W II," B.1149.

26. C. G. James, "Visit to Pratt & Whitney, South Windsor Engineering Facility, East Hartford, Connecticut," November 8, 1973, Bacon TS, Section B, Research and Development, Background Material, "Miscellaneous," B.1352, 1.

27. Nicholas R. Iammartino, "Project Independence: What, If, and When?" *Chemical Engineering* (March 18, 1974): 46–50.

28. Pietro S. Nivola, *The Politics of Energy Conservation* (Washington, DC: Brookings Institution, 1986), 1–7.

29. Melosi, *Coping with Abundance*, 260–280.

30. The National Aeronautics and Space Administration's Bureau of Industry Affairs and Technology Utilization arranged a meeting in summer 1974 between representatives of the electric utilities and NASA director James Fletcher to determine the place of the fuel cell in power generation and the role of government; A. O. Tischler to Robert Bell, July 5, 1974; Tischler to Fletcher, July 8, 1974; Tischler to Fletcher, August 2, 1974, Record Number 013761: Propulsion, Auxiliary Power Fuel Cells, Memos, News Releases, 1974–, NASA Headquarters Archive.

31. NASA News Release No. 75-182, "ERDA, NASA Sign Cooperative Agreement on Energy R&D," June 23, 1975, Record Number 012173VII: NASA/Other Agencies' Agreement, 1, 2, NASA Headquarters Archive.

32. Lawrence H. Thaller to F. L. Sola and P. R. Miller, "Subject: Comments on the Proposed ERDA Fuel Cell Technology Contract with UTC," January 14, 1976; Sola to R. D. Ginter, January 26, 1976; Record Number 013761: Propulsion, Auxiliary Power Fuel Cells, Memos, News Releases, 1974–, NASA Headquarters Archive.

33. Sola to Ginter, January 26, 1976.

34. Sol H. Gorland, "Subject: Energy Conservation Conference, December 1–3, 1975, Ft. Lauderdale, Florida," December 29, 1975, Record Number 018520: Energy Office, Read File 5.1.74, 1976, 3, NASA Headquarters Archive.

35. Marvin Warshay and Paul Prokopius, "Coordinated Fuel Cell System Programs for Government and Commercial Applications: Are We in a New Era?" January 1, 1996, NASA-TM-106699; E-9601; NAS 1.15:106699, CASI 19960015871, Accession Number 96N21650, 2–3, NASA Aerospace and Space Database.

36. Appleby and Foulkes, *Fuel Cell Handbook*, 83–103.

37. Rosenbaum, *Energy, Politics and Public Policy*, 7–8.

38. Union of Concerned Scientists, "Clean Energy Backgrounder, PURPA," http://www .ucsusa.org/clean_energy/clean_energy_policies/public-utility-regulatory-policy-act-purpa.html.

39. Appleby and Foulkes, *Fuel Cell Handbook*, 68.

40. Kordesch and Simader, *Fuel Cells and Their Applications*, 185.

41. Appleby, "Issues in Fuel Cell Commercialization," 158–160.

42. Appleby, "Fuel Cells for Tactical Battlefield Power," *IEEE AES Systems Magazine* (December 1991): 54.

43. Appleby and Foulkes, *Fuel Cell Handbook*, 61, 95.

44. King and O'Day, "Applying Fuel Cell Experience," 17.

45. U.S. Department of Energy, Onsite 40-Kilowatt Fuel Cell Power Plant Manufacturing and Field Test Program, DOE/NASA/0255-1, NASA CR-174988 (Washington, DC 1985), 1-1, 1-2, 2-4.

46. Ibid., 1-1, 2-14, 2-15. NASA continued to try to exploit its involvement in this project as evidence of the economic benefits of the space program. Its official press release focused on the use of phosphoric acid fuel cells in hospitals and commercial and residential establishments, claiming that these were "larger terrestrial versions of the systems used to generate electricity in the nation's manned space program." NASA News Release No. 82-127, "NASA Awards Fuel Cell Technology Development Contract," August 23, 1982, Record Number 013761: Propulsion, Auxiliary Power Fuel Cells, Memos, News Releases, 2, NASA Headquarters Archive.

47. Marvin Warshay, "Status of Commercial Fuel Cell Powerplant System Development," prepared for the U.S. Department of Energy for the 22nd Intersociety Energy Conversion Engineering Conference, Philadelphia, Pennsylvania, August 10–14, 1987, DOE/NASA/17088-5, NASA TM-89896, AIAA-87-9081, 3, NASA Aerospace and Space Database.

48. Appleby, "Issues in Fuel Cell Commercialization," 160.

49. UTC Power, "PureCell 200 Commercial Fuel Cell Power System," www.utcfuelcells.com/utcpower/products/index.shtm.

50. Warshay, "Status of Commercial Fuel Cell," 4; *Congressional Record*, 100th Cong. 1st sess., 1987, 133, Attachment C, 12, E 2135.

51. From 1978 to 2000, the Department of Energy's Office of Fossil Energy spent more money researching and developing utility fuel cells ($1.167 billion) than it spent on any other single power source or energy conversion technology except for direct coal lique-faction ($2.302 billion) and integrated gas combined cycle ($2.35 billion). Industry's cost share was 20 percent, 48 percent, and 48 percent respectively. See Robert W. Fri et al., *Energy Research at DOE: Was It Worth It? Energy Efficiency and Fossil Energy Research 1978 to 2000* (Washington, DC: National Academy Press, 2001), 48.

CHAPTER 5   FUELING HYDROGEN FUTURISM

1. Jules Verne, *The Mysterious Island* (New York: C. Scribner's Sons, 1919), 278–280; Max Pemberton, *The Iron Pirate: A Plain Tale of Strange Happenings on the Sea* (London: Cassell and Company, 1893).

2. Peter Hoffmann, *The Forever Fuel: The Story of Hydrogen* (Boulder, CO: Westview Press, 1981), 39.

3. Among hydrogen researchers, Bockris seemed the most interested in tracking the intellectual provenance of this expression. In late 1973, Bockris wrote to Bacon, suggesting that the origins of the concept lay in Bacon's idea of 1952–53 to use wind turbines to electrolyze water, which would then be run through a fuel cell. Bacon declined credit, claiming that the "hydrogen economy" was Derek Gregory's idea. Bacon added he had first encountered the notion of a solar-hydrogen economy in Eduard Justi's *Leitungsmechanismus und Energieumwandlung in Festkörfern* (Göttingen, 1965). Bockris then claimed that, inspired by a meeting with Bacon and Watson in 1952 or 1953, it had been he, Bockris, who had most fully articulated the idea of a hydrogen economy. John O'M. Bockris to Francis Thomas Bacon, November 12, 1973; Bacon to Bockris, December 10, 1973, Section G, Correspondence, Fuel Cell Correspondence G.148; Bockris to Bacon, January 15, 1974; Bockris to Bacon, February 9, 1974, G.149, Churchill College Archives Centre, University of Cambridge, Cambridge, England (hereafter cited as Bacon typescript [TS] or manuscript [MS]).

4. John O'M. Bockris, "The Electrochemical Future," in *Electrochemistry of Cleaner Environments*, ed. John O'M. Bockris (New York: Plenum Press 1972), 1–23.

5. Ferdinand E. Banks, *The Political Economy of Natural Gas* (London: Croom Helm, 1987), 66–67.

6. Pietro S. Nivola, *The Politics of Energy Conservation* (Washington, DC: Brookings Institution, 1986), 86–87.

7. D. P. Gregory et al., *A Hydrogen Energy System* (American Gas Association, 1973), I, vii–29.

8. Bockris sometimes spoke as if this were the prime advantage of a nuclear electricity economy employing hydrogen as an energy carrier; see "Hydrogen: Then and Now," in *Hydrogen Energy Progress V: Proceedings of the Fifth World Hydrogen Energy Conference, Toronto, Canada, July 15–20, 1984*, ed. T. N. Veziroglu and J. B. Taylor (New York: Pergamon Press, 1984), 4–5.

9. D. P. Gregory, D.Y.C. Ng, and G. M. Long, "The Hydrogen Economy," in *Electrochemistry of Cleaner Environments*, 263.

10. John Bockris and A. J. Appleby, "The Hydrogen Economy—an Ultimate Economy? A Practical Answer to the Problem of Energy Supply and Pollution," in *Environment This Month: The International Journal of Environmental Science* I, no. 1 (July 1972): 29–35.

11. Cesare Marchetti, "From the Primeval Soup to World Government: An Essay on Comparative Evolution," *International Journal of Hydrogen Energy* 2, no. 1 (1977): 1–5; Hoffmann, *The Forever Fuel*, 232–239. In 2010, Smil cited a model developed by Marchetti in 1977 as an example of the hazards of projecting future patterns of energy use. Smil viewed the model—a series of uniform waves of global primary energy substitutions from 1850 to 2100—as simplistic and deterministic, noting that actual patterns of primary energy use were influenced by an array of socioeconomic and political factors and were far less regular and much more complex. Marchetti's predictions of a rapid decline in the use of oil and coal and rapid growth in consumption of natural gas, noted Smil, were not borne out by events in the 1980s and 1990s, casting doubt on the physicist's estimates of the rise to dominance of nuclear power in the first half of the twenty-first century and then fusion and solar energy in the second; see Smil, *Energy Transitions: History, Requirements, Prospects* (Santa Barbara, CA: Praeger, 2010), 66–69.

12. Lawrence W. Jones, "Hydrogen: A Fuel to Run Our Engines in Clean Air," *Saturday Evening Post*, Spring 1972, 34.

13. NASA Glenn Research Center: Centaur—America's Workhorse in Space, http://www.nasa.gov/centers/glenn/about/history/centaur.html.

14. Blackwell C. Dunnam, "Air Force Experience in the Use of Liquid Hydrogen as an Aircraft Fuel," in *Hydrogen Energy Part B: Proceedings of the Hydrogen Economy Miami Energy (THEME) Conference, March 18–20, 1974*, ed. T. Nejat Veziroglu (New York: Plenum Press, 1975), 992–1006.

15. George M. Low, Deputy Administrator, to Robert C. Seamans, Jr., President, National Academy of Engineering, October 2, 1973, Record Number 013547, Special Collections, NASA Administrators, Congressional Documents, George M. Low Papers, Energy—Part 2, NASA Headquarters Archive, Washington, DC.

16. James C. Fletcher to Roy L. Ash, October 11, 1973, Record Number 009819: Energy, NASA Documentation, NASA Headquarters Archive.

17. Ad Hoc Panel of National Academy of Engineering, "Informal Review of NASA Energy Research and Development Proposals," November 19, 1973, Record Number 013547,

Special Collections, NASA Administrators, Congressional Documents, George M. Low Papers, Energy—Part 2, p. 2, NASA Headquarters Archive.

18. Ernst M. Cohn to Bacon, February 8, 1974, Bacon TS, Section G, Correspondence, Fuel Cell Correspondence, G.149.

19. Jones, "Hydrogen: A Fuel to Run Our Engines in Clean Air," and Staff, "Fuel of the Future" *Time* 46, September 11, 1972; Staff, "When Hydrogen Becomes the World's Chief Fuel," *Business Week* 2247, September 12, 1972, 98; L. Lessing, "The Coming Hydrogen Economy," *Fortune* 138, November 1972; D. P. Gregory, "The Hydrogen Economy," *Scientific American* 228, no. 2 (January 1973): 13.

20. Edward Teller, "Banquet Address," in *Hydrogen Energy Part A: Proceedings of the Hydrogen Economy Miami Energy (THEME) Conference*, ed. T. Nejat Veziroglu (New York: Plenum Press, 1975), 1–5.

21. The chief commercial methods for producing hydrogen were catalytic steam reforming, designed in the 1930s to convert light hydrocarbons, and the partial oxidation process, developed in the 1950s for heavier hydrocarbons.

22. W. Kerr and D. P. Majumdar, "Aqueous Homogeneous Reactor for Hydrogen Production," in *Hydrogen Energy Part A,* 167–181.

23. The normal temperature at the center of a pellet of uranium fuel in a zirconium-clad fuel rod, on the other hand, is about 900°C; see the Union of Concerned Scientists, http://allthingsnuclear.org/tagged/Japan_nuclear.

24. G. D. Sauter, "Hydrogen Energy—Its Potential Promises and Problems," in *First World Hydrogen Energy Conference Proceedings Vol. III, March 1–3, 1976, Coral Gables Florida,* ed. T. Nejat Veziroglu (Coral Gables, FL: University of Miami, 1976), 1C-5–1C-11.

25. Hoffmann, *The Forever Fuel,* 51–92.

26. Gregory, Ng, and Long, "The Hydrogen Economy," 229.

27. *Hydrogen Energy Part B,* 1327–1353.

28. J. R. Whitehead and J. B. Taylor, "The International Energy Agency Cooperative Programs on Hydrogen: Past Achievements and Future Potential," in *Hydrogen Energy Progress V,* 63–64.

29. Ibid., 65.

30. "Fort St. Vrain Power Station History," http://www.fsvfolks.org/FSVHistory_2.html.

31. Hoffmann, *The Forever Fuel,* 67–68.

32. E. Eugene Ecklund, "Federal Hydrogen Energy Activities in the United States of America," in *Hydrogen Energy Progress IV: Proceedings of the Fourth World Hydrogen Energy Conference, Pasadena, California, U.S.A., June 13–17, 1982,* vol. 4, ed. T. N. Veziroglu, W. D. Van Vorst, and J. H. Kelley (Oxford, UK: Pergamon Press, 1982), 1431–1434.

33. "Hydrogen Bus Refueling Study," *International Journal of Hydrogen Energy* 2, no. 1 (1977): 75–76; "Hydrogen: Another Solution to the Energy Crunch," *Mother Earth Magazine,* March/April 1979, http://www.motherearthnews.com/Renewable-Energy/1979-03-01/ Hydrogen-Another-Solution-To-The-Energy-Crunch.aspx.

34. R. M. Zweig and F. E. Lynch, "Hydrogen Vehicle Progress in Riverside California," in *Hydrogen Energy Progress VII, Proceedings of the Seventh World Hydrogen Energy Conference, Moscow, USSR, 25–29 September 1988,* vol. 3, ed. T. N. Veziroglu and A. N. Protsenko (New York: Pergamon Press, 1988), 1923–1944.

35. Helmut Buchner, "The Hydrogen/Hydride Energy Concept," in *Hydrogen Energy System: Proceedings of the Second World Hydrogen Energy Conference, Zurich, Switzerland, August*

21–24, *1978*, vol. 4, ed. T. N. Veziroglu and Walter Seifritz (Oxford, UK: Pergamon Press, 1979), 1749–1773; Hoffmann, *The Forever Fuel*, 105–138.

36. G. Beghi et al., "The Economics of Pipeline Transport for Hydrogen and Oxygen," in *Hydrogen Energy*, ed. T. N. Veziroglu (New York: Plenum, 1975), 545–560; R. A. Reynolds and W. L. Slager, "Pipeline Transmission of Hydrogen," in *Hydrogen Energy*, ed. T. N. Veziroglu (New York: Plenum, 1975), 533–543.

37. Anthony W. Thompson and I. M. Bernstein, "Selection of Structural Materials for Hydrogen Pipelines and Storage Vessels," *International Journal of Hydrogen Energy* 2, no. 2 (1977): 169.

38. E. A. Laumann, "The NASA Hydrogen Energy Systems Technology Study: A Summary," in *First World Hydrogen Energy Conference Proceedings Vol. III, March 1–3, 1976, Coral Gables Florida*, 1C-45–1C-69.

39. Edward M. Dickson, John W. Ryan, and Marilyn H. Smulyan, *The Hydrogen Energy Economy: A Realistic Appraisal of Prospects and Impacts* (New York: Praeger, 1977), 11–14, 64–72.

40. Laurence O. Williams, *Hydrogen Power: An Introduction to Hydrogen Energy and Its Applications* (London: Pergamon Press, 1980), 11.

41. Wolf Häfele, "Banquet Address: Hydrogen and the Big Energy Options," in *Hydrogen Energy System: Proceedings of the Second World Hydrogen Energy Conference*, 2281–2288.

42. D. A. Freiwald and W. J. Barattino, "Alternative Transportation Vehicles for Military-Base Operations," *International Journal of Hydrogen Energy* 6, no. 6 (1981): 631–636.

43. William D. Van Vorst, T. N. Veziroglu, and James H. Kelley, "Foreword," in *Hydrogen Energy Progress IV*.

44. Whitehead and Taylor, "The International Energy Agency Cooperative Programs on Hydrogen," 71; Bockris, "Hydrogen: Then and Now," 5; R. D. Champagne et al., "Transitions to Hydrogen," in *Hydrogen Energy Progress V*, 17–19.

45. Whitehead and Taylor, "The International Energy Agency Cooperative Programs on Hydrogen," 71–72.

46. The organizers of the Fourth World Hydrogen Energy Conference in 1982 reported that conferees were impressed by the announcement of Helmut Buchner, chief of Daimler's Solid State Physics Research Department, that the company intended to build "an actual test fleet" of hydrogen-fueled automobiles; William D. Van Vorst, J. H. Kelley, and T. N. Veziroglu, "WHEC-IV," *International Journal of Hydrogen Energy* 8, no. 11/12 (1983): 858–859; Helmut Buchner, "Hydrogen Use—Transportation Fuel," in *Hydrogen Energy Progress IV*, 3–29; H. Buchner and R. Povel, "The Daimler-Benz Hydride Vehicle Project," *International Journal of Hydrogen Energy* 7, no. 3 (1982): 259–266; K. Feucht, G. Hölzel, and W. Hurich, "Perspectives of Mobile Hydrogen Application," in *Hydrogen Energy Progress VII*, 1963–1974.

47. W.J.D. Escher, "Cooperative International Liquid Hydrogen Automotive Progress Report," *International Journal of Hydrogen Energy* 7, no. 6 (1982): 519; Walter F. Stewart, "Hydrogen as a Vehicular Fuel," in *Recent Developments in Hydrogen Technology Volume II*, ed. K. D. Williamson, Jr., and Frederick J. Edeskuty (Boca Raton, FL: CRC Press, 1986), 69–145.

48. R. L. LeRoy, "Industrial Water Electrolysis: Present and Future," *International Journal of Hydrogen Energy* 8, no. 6 (1983): 401–417.

49. Linda L. Gaines and Alan M. Wolsky, "Economics of Hydrogen Production: The Next Twenty-Five Years," in *Hydrogen Energy Progress V*, 260–261.

50. Yokio Ohta and Isao Abe, "Hydrogen Energy Research and Development in Japan," in *Hydrogen Energy Progress V*, 47–48.

51. Van Vorst, Kelley, and Veziroglu, "WHEC-IV," 859–860, 864.

52. Gaines and Wolsky, "Economics of Hydrogen Production," 261, 267.

53. Peter Hoffmann's authoritative 239-page popular history of hydrogen, published in 1981, devoted only half a page to fuel cells but included a considerably lengthier section on electrolysis technology; see *The Forever Fuel*.

54. For example, 49 of 186 published papers from the 1980 conference dealt largely with electrochemistry, as did 61 of 153 papers from the 1982 conference; see, respectively, *Hydrogen Energy Progress: Proceedings of the Third World Hydrogen Energy Conference held in Tokyo, Japan, June 23–26, 1980*, ed. T. N. Veziroglu, K. Fueki, and T. Ohta (Oxford, UK: Pergamon Press, 1980), and *Hydrogen Energy Progress IV*.

55. Bacon was invited to deliver the keynote address at the conversion and utilization division of the Fifth World Hydrogen Energy Conference in 1984. The occasion presented Bacon, by then nearly eighty and long since retired from active research, with an opportunity to validate his life's work on hydrogen fuel cell technology, as well as cast some doubt on hydrocarbon fuel cells, although with his characteristic tact. He reciprocated by underscoring the conference theme of early deployment, remarking that further research was unnecessary when there was plenty of existing useful knowledge that could be applied. Bacon, "The Development and Practical Application of Fuel Cells: Keynote Address," *International Journal of Hydrogen Energy* 10, no. 7/8 (1985): 423–430.

56. Hans Blank, "Solar-Hydrogen Demonstration Plant in Nuenburg vorm Wald/ Germany," in *Hydrogen Energy Progress IX: Proceedings of the Ninth World Hydrogen Energy Conference, Paris France, June 22–25, 1992*, vol. 2, ed. T. N. Veziroglu, C. Derive, and J. Pottier (Paris: Manifestations et Communications Internationales, 1992), 677–680; D. Reister and W. Strobl, "Current Development and Outlook for the Hydrogen-Fuelled Car," in ibid., 1201–1206.

57. At the first international hydrogen conference in 1974, General Electric researchers conformed to the hydrogen futurist canon, noting the ineluctability of a post–fossil fuels hydrogen age but also making a pitch for solid polymer electrolyte technology as the simplest way of producing hydrogen as a substitute for carbonaceous fuel during the height of the energy crisis; L. J. Nuttall, A. P. Fickett, and W. A. Titterington, "Hydrogen Generation by Solid Polymer Electrolyte Water Electrolysis," in *Hydrogen Energy Part A*, 441–455.

58. L. J. Nuttall and J. F. McElroy, "Status of Solid Polymer Electrolyte Fuel Cell Technology and Potential for Transportation Applications," *International Journal of Hydrogen Energy* 8, no. 8 (1983): 610–611.

59. A. J. Appleby and F. R. Foulkes, *Fuel Cell Handbook* (New York: Van Nostrand Reinhold, 1989), 169, 189.

60. Tom Koppel, *Powering the Future: The Ballard Fuel Cell and the Race to Change the World* (Toronto: John Wiley & Sons Canada, 1999), 34–35.

61. Keith Prater, "The Renaissance of the Solid Polymer Fuel Cell," *Journal of Power Sources* 29 (1990): 240; Koppel, *Powering the Future*, 58–90.

62. Koppel, *Powering the Future*, 66, 94.

63. A. J. Appleby, "The Electrochemical Engine for Vehicles," *Scientific American* 281, no. 1 (July 1999): 74–80.

64. "Hydrogen & Fuel Cell Letter, Hits 10th Year," *NHA Quarterly Newsletter* 1, no. 1 (Spring 1996), http://www.hydrogenassociatio n.org/newsletter/.

65. Hawaii Natural Energy Institute, "History of HNEI," http://www.hnei.hawaii.edu/history.asp; and "Hydrogen," http://www.hnei.hawaii.edu/hydrogen.asp.

66. Marvin Warshay, NASA, "Status of Commercial Fuel Cell Powerplant System Development," prepared for the U.S. Department of Energy for the 22nd Intersociety Energy Conversion Engineering Conference, Philadelphia, Pennsylvania, August 10–14, 1987, DOE/NASA/17088-5, NASA TM-89896, AIAA-87-9081, 3, NASA Aerospace and Space Database; *Congressional Record*, 100th Cong. 1st sess., 1987, 133, Attachment C, 12, E 2135.

67. A team at Brigham Young University led by the physicist Steven E. Jones also conducted similar cold-fusion experiments at around the same time but detected only extremely low levels of neutrons. As a number of sources report, competition for priority of discovery between Fleischmann and Pons and the University of Utah, on the one side, and Jones and Brigham Young University on the other, driven by the potential for fame and profit associated with a miracle power source, contributed to the escalation of expectations and informed the timing and scope of claims for cold fusion. See, for example, Charles Seife, *Sun in a Bottle: The Strange History of Fusion and the Science of Wishful Thinking* (New York: Viking, 2008); Gary Taubes, *Bad Science: The Short Life and Weird Times of Cold Fusion* (New York: Random House, 1993); and John R. Huizenga, *Cold Fusion: The Scientific Fiasco of the Century* (Rochester, NY: University of Rochester Press, 1992).

68. Huizenga, *Cold Fusion*, 30–33. This apparent disciplinary and regional divide quickly evaporated. Bard soon became a cold-fusion skeptic and was recruited by Huizenga to the Energy Research Advisory Board. Caltech's Nathan S. Lewis, one of the leading debunkers of cold fusion, was a professor of electrochemistry, and the University of Utah's involvement with cold fusion became increasingly unpopular among the general faculty, forcing the resignation of its pro-cold-fusion president Chase N. Peterson in early 1991.

69. See Huizenga, *Cold Fusion*, 55–56, 60; William J. Broad, "Stanford Reports Success," *New York Times*, April 19, 1989, A8. For Huggins's legacy in battery research and development, see Seth Fletcher, *Bottled Lightning: Superbatteries, Electric Cars, and the New Lithium Economy* (New York: Hill and Wang, 2011), 24–27, 48.

70. Taubes, *Bad Science*, 275–280, 267.

71. Seife, *Sun in a Bottle*, 131–132; Taubes, *Bad Science*, xix–xx, 113–114.

72. Public Law 101-566, 101st Cong. 2d sess. (November 15, 1990), *Spark M. Matsunaga Hydrogen Research, Development, and Demonstration Program Act of 1990*, Sec. 104: "Research and Development"; "Hydrogen Fuel Research Program OK'd," *Congressional Quarterly Almanac* 46, 1990, 101st Cong. 2d sess. (Washington, DC: Congressional Quarterly, 1991), 318.

73. *Congressional Record*, 101st Cong. 2d sess., v. 136: pt. 27, "*Matsunaga Hydrogen Research and Development Act*," House of Representatives, May 8, 1990 (Washington, DC: U.S. Government Printing Office, 1990), H2055–H2056.

74. *Congressional Record*, 101st Cong. 2d sess., v. 136: pt. 27, "Spark M. Matsunaga Hydrogen Research, Development, and Demonstration Act of 1990," House of Representatives, October 23, 1990 (Washington, DC: U.S. Government Printing Office, 1990), H11877.

CHAPTER 6   GREEN AUTOMOBILE WARS

1. Matthew L. Wald, "Going beyond Batteries to Power Electric Cars," *New York Times*, March 3, 1993, D2.

2. Matthew Paterson; see his *Automobile Politics: Ecology and Cultural Political Economy* (Cambridge: Cambridge University Press, 2007), 25. Paterson notes this sense of the term has been current among social scientists since at least the mid-1990s.

3. For an exemplary treatment of these interlinked technosocial themes, see John Urry, "The 'System' of Automobility," *Theory, Culture & Society* 21, no. 4/5 (2004): 25–39.

4. Appleby calculated that there was sufficient platinum to produce only 800,000 20-kilowatt vehicle power plants per year, less than 3 percent of world automobile production in 1986. Polymer membrane then cost about $400 per square meter. A. J. Appleby and F. R. Foulkes, *Fuel Cell Handbook* (New York: Van Nostrand Reinhold, 1989), 14, 187–189, 201; Karl V. Kordesch and Günter Simader, *Fuel Cells and Their Applications* (New York: VCH Publishers, 1996), 6–7.

5. Samuel P. Hays, *Explorations in Environmental History: Essays by Samuel P. Hays* (Pittsburgh: University Pittsburgh Press, 1998), 248–256.

6. D. M. Hart, *Forged Consensus: Science, Technology, and Economic Policy in the United States, 1921–1953* (Princeton, NJ: Princeton University Press, 1998); Ann Johnson, "The End of Pure Science: Science Policy from Bayh-Dole to the NNI," in *Discovering the Nanoscale* (Amsterdam: IOS Press, 2004), 219–222.

7. William J. Clinton and Albert Gore, Jr., "Technology for America's Economic Growth: A New Direction to Build Economic Strength" (Washington, DC: Government Printing Office, 1993), ntl.bts.gov/lib/jpodocs/briefing/7423.pdf.

8. David C. Mowery, "Collaborative R&D: How Effective Is It?" *Issues in Science and Technology* (Fall 1998): 37.

9. The term "greenwash" was likely coined by Kenny Bruno and Jed Greer in their eponymous 1992 booklet; see *The Greenpeace Book of Greenwash* (Greenpeace International, 1992), 21–23.

10. Several researchers have explored the phenomenon of corporate environmental propaganda or "greenwashing"; see, for example, Sharon Beder's *Global Spin: The Corporate Assault on Environmentalism* (White River Junction, VT: Chelsea Green, 1998), Keith Bradsher's *High and Mighty: SUVs: The World's Most Dangerous Vehicles and How They Got That Way* (New York: Public Affairs, 2002), and William Rollins's "Reflections on a Spare Tire: SUVs and Postmodern Environmental Consciousness," *Environmental History* 11, no. 4 (2006): 684–723.

11. David A. Kirsch, *The Electric Vehicle and the Burden of History* (New Brunswick, NJ: Rutgers University Press, 2000); Gijs Mom, *The Electric Vehicle: Technology and Expectations in the Automobile Age* (Baltimore: Johns Hopkins University Press, 2004).

12. Kirsch, *The Electric Vehicle and the Burden of History*, 17.

13. Mom, *The Electric Vehicle*, 170.

14. Ibid., 250–257.

15. This was the view of Richard H. Schallenberg, author of one of the first comprehensive histories of the battery. Schallenberg wrote that the "inherent" short range of the early battery-powered electric car was the chief reason for its replacement by the gasoline-engine automobile; see his *Bottled Energy: Electrical Engineering and the Evolution of Chemical Energy Storage* (Philadelphia: American Philosophical Society, 1982), 252.

16. Ronald Kline and Trevor Pinch, "Users as Agents of Technological Change: The Social Construction of the Automobile in the Rural United States," *Technology and Culture* 37, no. 4 (October 1996): 767–777.

17. Mom, *The Electric Vehicle*, 295–296.

18. Mom, *The Electric Vehicle*, 273; Kirsch, *The Electric Vehicle and the Burden of History*, 226.

19. James R. Huff and John C. Orth, "The USAMECOM-MERDC Fuel Cell Electric Power Generation Program," in *Fuel Cell Systems II: 5th Biennial Fuel Cell Symposium Sponsored by the Division of Fuel Chemistry at the 154th Meeting of the American Chemical Society, Chicago, Illinois, September 12–14, 1967* (Washington, DC: American Chemical Society, 1969), 323–326.

20. M. R. Andrew, W. J. Gressler, J. K. Johnson, R. T. Short, and K. R. Williams, "A Fuel-Cell/Lead-Acid Battery Hybrid Car," in *Fuel Cell Technology for Vehicles*, ed. Richard Stobart (Warrendale, PA: Society of Automotive Engineers, 2001), 9, 11.

21. Kordesch and Simader, *Fuel Cells and Their Applications*, 257, 290. The Cleveland Auto Show even presented Kordesch with a special award in recognition of his car. He was surprised to discover that the honor had been bestowed mainly on account of the vehicle's unusual appearance—six bright red, roof-mounted hydrogen fuel tanks—rather than for its unique hand-built power system. Interview with the author, July 2005, Technical University, Graz, Austria.

22. Craig Marks, Edward A. Rishavy, and Floyd A. Wyczalek, "Electrovan—a Fuel Cell Powered Vehicle," *Automotive Engineering Congress and Exposition, Detroit, Michigan, January 9–13, 1967* (New York: Society of Automotive Engineers, 1967), 9, 13.

23. The report noted that as of 1967 there were 40,000 electric commercial delivery vehicles and 60,000 electric materials- and product-handling vehicles in the United Kingdom; Internal Memo, Ministry of Technology, "Fuel Cell Working Party: Possible Uses of Fuel Cells and User-Requirements (Ref. Minute No. 4 of the 2nd Meeting of the Working Party on Fuel Cells)," January 11, 1967, Papers and Correspondence of Francis Thomas Bacon, Reference Code NCUACS 68.6.97, Section B, Research and Development, Research Centres, Laboratories and Sponsors, "Leesona-Moos and P&W," B.1168, Churchill College Archives Centre, University of Cambridge, Cambridge, England (hereafter cited as Bacon TS); Ministry of Technology, "Minutes of the Fourth Meeting of the Working Party on Fuel Cells Held at Ministry on Monday 23rd January 1967," Bacon TS, Section B, Research and Development, Research Centres, Laboratories and Sponsors, 'Leesona-Moos and P&W,' B.1169.

24. P. R. Hall and A. E. James, "Report No. 66/321A/RI: A Survey of Vehicles to Assess Their Suitability as Applications for the Fuel Cell," October 1966, Norris Brothers (Research & Development, Ltd.) for Energy Conversion, Ltd., Bacon TS, Section B, Research and Development, Energy Conversion Limited, Reports, B.777.

25. This comprised 112 vehicles distributed to 12 electricity boards in England and Wales between 1966 and 1976; Michael H. Westbrook, *The Electric Car: Development and Future of Battery, Hybrid and Fuel-Cell Cars* (London: Institution of Electrical Engineers, 2001), 20–21.

26. Tom McCarthy, *Auto Mania: Cars, Consumers and the Environment* (New Haven, CT: Yale University Press, 2007), 187.

27. "Detroit's 'Total Revolution': The Fuel Crunch and Federal Demands Speed Shift to Smaller Models," *Time*, March 19, 1979, http://www.time.com/time/magazine/article/0,9171,947023-1,00.html.

28. "Exxon: Searching for Another Game That Equals Oil in Size," "Interview with Clifton Garvin: 'The Quicker We Get at Synthetic Fuels, the Better We're Going to Be,'" *Business Week*, July 16, 1979, 80–84.

29. U.S. Public Law 94-413, 94th Cong. 2d sess. (September 17, 1976), *Electric and Hybrid Vehicle Research, Development and Demonstration Act of 1976*.

30. Maxine Savitz, "The Federal Role in Conservation Research and Development," in *The Politics of Energy Research and Development: Policy Studies* 3, ed. John Byrne and Daniel Rich (New Brunswick, NJ: Transaction Books, 1986), 111.

31. Kirsch, *The Electric Vehicle and the Burden of History*, 204–206.

32. Public Law 96-294, June 30, 1980, "Energy Security Act"; Walter A. Rosenbaum, *Energy, Politics, and Public Policy*, 2nd ed. (Washington, DC: CQ Press, 1987), 94.

33. Sudhir Chella Rajan, *The Enigma of Automobility: Democratic Politics and Pollution Control* (Pittsburgh, PA: University of Pittsburgh Press, 1996), 25–28.

34. For passenger cars and light-duty trucks at 50,000 miles, the law required that transitional low-emission vehicles (TLEV, hydrocarbon standard of 0.125 gram per mile [g/mi], carbon monoxide 3.4 g/mi, nitrogen oxide 0.4 g/mi) had to constitute 10, 15, and 20 percent of new car production in model years 1994, 1995, and 1996 respectively, low-emission vehicles (LEV, hydrocarbon 0.075 g/mi, carbon monoxide 3.4 g/mi, nitrogen oxide 0.2 g/mi) had to make up 25, 48, 73, 96, 90, 85, and 75 percent of new car production in 1997, 1998, 1999, 2000, 2001, 2002, and 2003 respectively, and ultralow-emission vehicles (ULEV, hydrocarbon 0.040 g/mi, carbon monoxide 1.7 g/mi, nitrogen oxide 0.2 g/mi) had to constitute 2 percent of new car production each year between 1997 and 2000, 5 percent in 2001, 10 percent in 2002, and 15 percent in 2003; see CARB, "Proposed Regulations for Low-Emissions Vehicles and Clean Fuels: Technical Support Document," August 13 1990, I-4–I-16.

35. See S 1630, "To amend the Clean Air Act to provide for attainment and maintenance of health protective national ambient air quality standards, and for other purposes," 101st Cong. 2d sess.; see also Hays, *Explorations*, 262.

36. See, for example, Michael Shnayerson, *The Car That Could: The Inside Story of GM's Revolutionary Electric Vehicle* (New York: Random House, 1996), 48, 55; Jack Doyle, *Taken for a Ride: Detroit's Big Three and the Politics of Pollution* (New York: Four Walls, Eight Windows, 2000), 276; John M. DeCicco, "The 'Chicken or Egg' Problem Writ Large: Why a Hydrogen Fuel Cell Focus Is Premature," in *The Hydrogen Energy Transition: Moving toward the Post-petroleum Age in Transportation*, ed. Daniel Sperling and James S. Cannon (Burlington, MA: Elsevier Academic Press, 2004), 215, 218; Gustavo Oscar Collantes, "The California Zero-Emission Vehicle Mandate: A Study of the Policy Process, 1990–2004" (PhD diss., University of California, Davis, 2005), 33–34.

37. There has been no scholarly consensus on Smith's motives. In his 1995 book *Future Drive: Electric Vehicles and Sustainable Transportation* (Washington, DC: Island Press, 1995), Daniel Sperling cites the May 8, 1994, report of David Sedgwick and Bryan Gruley of the *Detroit News and Free Press* that following Smith's January 3, 1990, press conference, the CEO remarked that he hoped "you guys aren't going to make us build that car, are you?" But elsewhere in the book Sperling refers to Smith as a "strong electric vehicle advocate"; see *Future Drive*, 38–39. Doyle interprets Smith's public statements as a clear commitment to produce electric automobiles; Doyle, *Taken for a Ride*, 276. Shnayerson is less explicit about Smith's motives but suggests public relations was an important consideration; Shnayerson, *The Car That Could*, 1–27. In his 2005 doctoral dissertation, Collantes interprets Smith's statements as pure rhetoric that

was misinterpreted by those who did not understand GM's "public relations style"; see "The California Zero-Emission Vehicle Mandate," 32–33.

38. Neal Templin, "Auto Makers Strive to Get Up to Speed on Clean Cars for the California Market," *Wall Street Journal*, March 26, 1991, B1.

39. Shnayerson, *The Car That Could*, 259.

40. Although Bruno and Greer referred to General Motors in their study, they did not mention any specific greenwashing strategies employed by the company; see *The Greenpeace Book of Greenwash*, 21–23.

41. This contrasted notably with Honda, which shrouded its automobile experiments in secrecy; George Harrar, "The 'Concept Car' Pushes Change," *New York Times*, July 1, 1990, F5.

42. GM Display Ad 203, "Earth Day 1990: General Motors Marks 20 Years of Environmental Progress," *New York Times*, April 22, 1990, SM13.

43. In his January 3 press conference, Smith had merely stated that GM had developed a "producible" electric car and, on April 18, that the company had set itself the goal of becoming the first firm since the early days of the auto industry to "mass-produce" an electric automobile; see Richard W. Stevenson, "GM Displays the Impact, an Advanced Electric Car," *New York Times*, January 4, 1990, D1, D17; Doron P. Levin, "GM to Begin Production of Battery-Powered Car," *New York Times*, April 19, 1990, D5. Of the two major New York–based national newspapers, the *Wall Street Journal* was more skeptical of Smith's claims; see Rick Wartzman, "GM Unveils Electric Car with Lots of Zip but Also a Battery of Unsolved Problems," *Wall Street Journal*, January 4, 1990, A1; Joseph P. White, "GM Says It Plans an Electric Car, but Details Are Spotty," *Wall Street Journal*, April 19, 1990, B1.

44. Matthew L. Wald, "A Tough Sell for Electric Cars: Technology Lagging as Markets Emerge," *New York Times*, November 26, 1991, D1.

45. Collantes, "The California Zero-Emission Vehicle Mandate," 37.

46. Shnayerson, *The Car That Could*, 82–83; "3 Auto Makers in Battery Plan," *New York Times*, February 1, 1991, D3.

47. Wald, "A Tough Sell for Electric Cars."

48. In the ongoing green automobile intrigue, Ford displayed its nonfunctioning Connecta concept vehicle in January 1992 in an effort to mislead GM on its intentions in electric vehicle development; Shnayerson, *The Car That Could*, 100–101.

49. "Ford Planning an Electric Car," *New York Times*, April 11, 1991, D4.

50. Michael M. Thackeray, "20 Golden Years of Battery R&D at CSIR, 1974–1994: A Contribution to the History of Science in South Africa," unpublished manuscript.

51. Shnayerson, *The Car That Could*, 99, 121.

52. Matthew L. Wald, "Government Dream Car: Washington and Detroit Pool Resources to Devise a New Approach to Technology," *New York Times*, September 30, 1993, A1.

53. Brent D. Yacobucci, "The Partnership for a New Generation of Vehicles: Status and Issues," *Congressional Research Service Report RS20852*, January 22, 2003, 1–2, http://wikileaks.org/wiki/CRS-RS20852; Trevor O. Jones et al., *Review of the Research Program of the Partnership for a New Generation of Vehicles, Sixth Report* (Washington, DC: National Academy Press, 2000), 14–15.

54. Robert M. Chapman, *The Machine That Could: PNGV, a Government-Industry Partnership* (Santa Monica: RAND, 1998), 9; Daniel Sperling, "Public-Private Technology R&D

Partnerships: Lessons for U.S. Partnership for a New Generation of Vehicles," *Transport Policy* 8 (2001): 248.

55. Doyle, *Taken For a Ride*, 370–394; Shnayerson, *The Car That Could*, 164.

56. Sperling, *Future Drive*, 143.

57. Sperling, "Public-Private Technology R&D Partnerships," 251.

58. J. Byron McCormick and James R. Huff, "The Case for Fuel-Cell–Powered Vehicles," *Technology Review* (August/September 1980): 54–65.

59. N. P. Rossmeissl and L. A. Waltemath, "The United States Department of Energy Hydrogen Program," in *Hydrogen Energy Progress XI: Proceedings of the Eleventh World Hydrogen Energy Conference, Stuttgart, Germany, June 23–28, 1996*, vol. 1, ed. T. N. Veziroglu, C. J. Winter, J. P. Baselt, and G. Kreysa (Frankfurt am Main: Schön & Wetzel, 1996), 31–32.

60. The DOE's fiscal year 1993 budget allocated $12 million for such work; see Pandit Patil, "Fuel Cell Vehicle Technology for the Twenty-First Century," in *Proceedings: First Annual World Car 2001 Conference, University of California at Riverside, Riverside, CA*, 113.

61. E-mail communication with Dr. John M. DeCicco, January 3, 2011.

62. Henry Kelly and Robert H. Williams, "Fuel Cells and the Future of the U.S. Automobile," draft (Princeton University: Center for Energy and Environmental Studies, December 1992). I thank Dr. Kelly for providing this document. A nearly identical paper was published by Robert H. Williams as "Fuel Cells, Their Fuels, and the U.S. Automobile" in *Proceedings: First Annual World Car 2001 Conference*, 73–75.

63. Robert W. Fri et al., *Energy Research at DOE: Was It Worth It? Energy Efficiency and Fossil Energy Research 1978 to 2000* (Washington, DC: National Academy Press, 2001), 154.

64. E-mail communication with the author, January 31, 2011. Interestingly, Sperling believed, at least in 1995, that fuel cells had been the "impetus" for the creation of the PNGV; see *Future Drive*, 84.

65. Tom Koppel, *Powering the Future: The Ballard Fuel Cell and the Race to Change the World* (Toronto: John Wiley & Sons Canada, 1999), 115–116.

66. Ibid., 126–129, 165. Between 1994 and 1996, Ballard reported revenue of CAD$63.3 million. In this period, the company had contracts for automotive fuel cells with the Chicago and Vancouver transit authorities ($8 million and $8.6 million respectively), Daimler-Benz ($2.3 million), Honda ($2 million), Volkswagen and Volvo ($1.2 million), General Motors and the U.S. Department of Energy ($6 million), and the U.S. Department of Transportation's Federal Transit Administration ($8.1 million). The Canadian Department of National Defense and the German shipbuilder Howaldtswerke-Deutsche Werft (HDW) had contracts with Ballard for submarine fuel cells worth $3.7 million and $9.3 million respectively. Finally, Ballard had revenue from its battery division totaling $8.64 million ($2.64 million from the sale of batteries and $6 million from the sale of the battery division itself in 1995). Ballard Power Systems Inc., *Annual Report 1996* (1997), 21–25. Ballard's major source of revenue into the mid-2000s, aside from equity sales, remained precommercial automotive fuel cell power plants for demonstration programs. The company did not develop a large stationary fuel cell generator for field trials until mid-1997 and did not begin sales of precommercial demonstrators for trials until 1999. Ballard Power Systems Inc., *Annual Report 1999* (2000), 26.

67. A. J. Appleby, "Issues in Fuel Cell Commercialization," *Journal of Power Sources* 69 (1996): 172.

68. Peter Verburg, "Is There Something We Don't Know?" *Canadian Business* 71, no. 3 (February 27, 1998): 29.

69. Charles Stone and Anne E. Morrison, "From Curiosity to 'Power to Change the World®,'" *Solid State Ionics* 152–153 (2002): 5–7. Geoffrey Ballard claimed that the company's first fuel cell electric bus demonstrator, unveiled in 1993, had been conceived in a deal between him and British Columbia energy minister Jack Davis in which Davis agreed to provide funds in exchange for a "green photo-op" to help boost the sagging political fortunes of then-premier Bill Vander Zalm; see Koppel, *Powering the Future*, 147–148.

70. For example, see David L. Levy and Sandra Rothenberg, "Heterogeneity and Change in Environmental Strategy: Technological and Political Responses to Climate Change in the Global Automobile Industry," in *Organizations, Policy and the Natural Environment: Institutional and Strategic Perspectives*, ed. Andrew J. Hoffman and Marc J. Ventresca (Stanford, CA: Stanford University Press, 2002), 178–180; and John J. Mikler, "Varieties of Capitalism and the Auto Industry's Environmental Initiatives: National Institutional Explanations for Firms' Motivations," *Business and Politics* 9, no. 1 (2007): 1–38.

71. National Highway Traffic Safety Administration, "Summary of CAFE Fines Collected, December 13, 2007."

72. James P. Womack and Daniel T. Jones held that the German manufacturing colossus belonged to a national corporate-industrial culture that stressed "deep technical knowledge" within an organizational structure based on rigidly defined and hierarchized functions. Professional advancement was, hence, attained by ascending the "functional ladder." The result was that German automakers generally had numerous hermetic engineering communities that rarely communicated with each other. This bred technological esprit but also isolated functional units, creating redundancies and high labor costs. James P. Womack and Daniel T. Jones, "From Lean Production to the Lean Enterprise," *Harvard Business Review* (March–April 1994): 97–98.

73. "Daimler's New Driver Won't Be Making Sharp Turns," *Business Week*, July 4, 1994, http://www.businessweek.com/archives/1994/b337973.arc.htm. Appointed chair of the Daimler-Benz board in 1987, Reuter continued the corporate policy of acquiring aerospace assets. Under the previous chair Werner Breitschwerdt, Daimler-Benz had purchased Dornier and bought out engineering firm MAN's 50 percent stake in MTU, their aero-engine joint venture, in 1985. In 1989, Daimler-Benz acquired the storied aircraft house MBB and merged all its aviation assets along with electrical equipment maker AEG into Deutsche Aerospace AG (DASA).

74. Thackeray, "20 Golden Years."

75. Wald, "A Tough Sell for Electric Cars."

76. J. Zieger, "HYPASSE: Hydrogen Powered Automobiles Using Seasonal and Weekly Surplus of Electricity," in *Hydrogen Progress X: Proceedings of the Tenth World Hydrogen Energy Conference, Cocoa Beach, Florida, U.S.A., June 20–24, 1994*, vol. 3, ed. D. L. Block and T. N. Veziroglu (International Association for Hydrogen Energy, 1994), 1367–1376.

77. One Daimler researcher wrote that the company's automotive fuel cell program was a major challenge, an opportunity to "potentially influence" the global automobile market with new technology and, consequently, was "highly attractive for research and development people"; see W. Dönitz, "Fuel Cells for Mobile Applications: Status, Requirements, and Future Application Potential," in *Hydrogen Energy Progress XI*, vol. 3, 1624.

78. John M. DeCicco, *Fuel Cell Vehicles: Technology, Market and Policy Issues* (Warrendale, PA: Society of Automotive Engineers, 2001), 72.

79. Oscar Suris, "Daimler-Benz Unveils Electric Vehicle, Claiming a Breakthrough on Fuel Cells," *Wall Street Journal,* April 14, 1994, B2.

80. Steven G. Chalk, Pandit G. Patil, and S. R. Venkateswaran, "The New Generation of Vehicles: Market Opportunities for Fuel Cells," *Journal of Power Sources* 61 (1996): 10; "Ford, Chrysler Win Auto Fuel-Cell Work," *Wall Street Journal,* July 13, 1994, B2.

81. Shnayerson, *The Car That Could,* 187–191. Interestingly, one of the justifications research and development chief Kenneth Baker advanced to top GM leaders in his pitch to restart the Impact program, reports Shnayerson, was that it could help develop other advanced automotive drive technologies including hybrid and, even more intriguingly, fuel cell drive. In terms reminiscent of those used by Williams and Kelly in their 1992 paper, Baker remarked that this electrochemical system would operate as long as it was supplied with fuel and could be packaged like a gasoline system, thus giving the fuel cell electric car the range and convenience of existing automobiles.

82. David A. Kirsch similarly noted this "odd twist" in the negotiating stance of the automakers; see *The Electric Vehicle and the Burden of History,* 207.

83. Brad Heavner, *Pollution Politics 2000: California Political Expenditures of the Automobile and Oil Industries, 1997–2000* (Santa Barbara, CA: California Public Interest Research Group Charitable Trust, 2000), 7–8; Deborah Salon, Daniel Sperling, and David Friedman, "California's Partial ZEV and LEV II Program," Institute of Transportation Studies, University of California, Davis, CA, 1999, 2.

84. Lawrence M. Fisher, "GM, in a First, Will Sell a Car Designed for Electric Power This Fall," *New York Times,* January 5, 1996, A10.

85. Lawrence M. Fisher, "California Is Backing Off Mandate for Electric Car: Board Finds Shortcomings in Technology," *New York Times,* December 26, 1995, A14.

86. Trevor O. Jones et al., *Review of the Research Program of the Partnership for a New Generation of Vehicles: Second Report* (Washington, DC: National Academy Press, 1996), 53–54.

87. Stone and Morrison, "From Curiosity to 'Power to Change the World®,'" 8.

88. Nick Nuttall, "Breathtaking . . . the Vehicle Powered by Air," *Times* (London), May 15, 1996, Home News 7.

89. Brandon Mitchener and Tamsin Carlisle, "Daimler, Ballard Team to Develop Fuel-Cell Engine," *Wall Street Journal,* April 15, 1997, B8.

90. Matthew L. Wald, "Ford Plans Zero-Emission Fuel Cell Car: Electric Vehicle Prototype May Be Ready for Evaluation by 2000," *New York Times,* April 22, 1997, D2.

91. Nick Nuttall, technology correspondent for the *Times,* suggested that fuel cells operated equally well on any hydrogen-rich fuel including liquid hydrogen, methanol, ethanol, and gasoline. His report on the rollout of NECAR II was also interesting in its observation of the vehicle's effect on British researchers, who responded with a variation of the national techno-declensionist narrative. Invoking William Grove, some expressed disappointment that a "British invention" was being exploited abroad, despite the fact that Grove's gaseous voltaic battery bore little practical resemblance to the modern PEM fuel cell, which was first developed in the United States. See *Times* (London), May 15, 1996, Home News 7.

92. Christopher E. Borroni-Bird, "Fuel Cell Commercialization Issues for Light-Duty Vehicle Applications," *Journal of Power Sources* 61 (1996): 42.

93. Matthew L. Wald, "Three Guesses: The Fuel of the Future Will Be Gas, Gas, or Gas," *New York Times*, October 16, 1997, G16.

94. Trevor O. Jones et al., *Review of the Research Program of the Partnership for a New Generation of Vehicles, Third Report* (Washington, DC: National Academy Press, 1997), 65; Steven G. Chalk, James F. Miller, and Fred W. Wagner, "Challenges for Fuel Cells in Transport Applications," *Journal of Power Sources* 86 (2000): 44.

95. Jones et al., *Review of the Research Program of the Partnership for a New Generation of Vehicles, Sixth Report*, 76.

96. Matthew L. Wald, "In a Step toward a Better Electric Car, Company Uses Fuel Cell to Get Energy from Gasoline," *New York Times*, October 21, 1997, A14.

97. Jason Mark, "Clean Car's Wrong Turn," *New York Times*, October 26, 1997, WK14. Interestingly, Mark had served as a referee for Gregory P. Nowell's 1998 report for the American Methanol Institute (*Looking beyond the Internal Combustion Engine: The Promise of Methanol Fuel Cell Vehicles*, www.methanol.org/pdf/amipromise.pdf) and was prominently quoted twice, lauding fuel cell technology in general (p. 16) and condemning gasoline reforming (p. 30). At no point did Mark explicitly endorse methanol reforming, although Nowell's placement of the quotes suggested otherwise.

98. Doyle, *Taken for a Ride*, 259.

99. In November 1994, Ford told dealers its converted Ranger pickup truck would cost around $20,000 but informed the press that it would cost $30,000; Doyle, *Taken for a Ride*, 316–320.

100. Earle Eldridge and Traci Watson, "Carmakers Lobby against Global Warming Treaty," *USA Today*, October 3, 1997, B1.

101. Valerie Reitman, "Ford Is Investing in Daimler-Ballard Fuel-Cell Venture," *Wall Street Journal*, December 16, 1997, 1; Anthony DePalma, "Ford Joins in a Global Alliance to Develop Fuel-Cell Auto Engines," *New York Times*, December 16, 1997, D1.

102. Stuart F. Brown, "The Automakers' Big-Time Bet on Fuel Cells," *Fortune*, March 30, 1998, 122D.

103. Trevor O. Jones et al., *Review of the Research Program of the Partnership for a New Generation of Vehicles, Fourth Report* (Washington, DC: National Academy Press, 1998), 4–5, 38.

104. DeCicco, *Fuel Cell Vehicles*, 57, Brown, "The Automakers' Big-Time Bet on Fuel Cells," 122D.

105. During a visit to Daimler's test facility on the outskirts of Nabern in March 1998, where he was give a test-drive in the NECAR II, the journalist Stuart F. Brown did implicitly acknowledge the potential for confusion to arise from the company's involvement in a number of fuel systems; see *Fortune*, March 30, 1998, 122D.

106. David P. Wilkinson and Alfred E. Steck, "General Progress in the Research of Solid Polymer Fuel Cell Technology at Ballard," in *Proceedings of the Second International Symposium on New Materials for Fuel Cell and Modern Battery Systems: New Materials for Fuel Cells and Modern Battery Systems II, Montréal, Canada, July 6–10, 1997*, ed. O. Savadogo and P. R. Roberge (Montréal: École Polytechnique de Montréal, 1997), 32.

107. "Intel on Wheels: Can a Small Canadian Company Overthrow the Internal-Combustion Engine?" *Economist*, October 31, 1998, 69–70; for Rasul's comparison between Ballard and Intel, see Verburg, "Is There Something We Don't Know?" 29–30.

108. Salon, Sperling, and Friedman, "California's Partial ZEV and LEV II Program," 4–8.

109. Jeffrey Ball, "DaimlerChrysler Unveils Prototype Car Using Fuel Cell, Seeks Sales in 5 Years," *Wall Street Journal*, March 18, 1999, 1. The title of first fuel cell electric automobile to be driven on public roads in the United States actually belongs to Karl Kordesch and his converted Austin A40. In April 1998, another fuel cell conversion, a Danish Kewet two-seater, took to U.S. public roads. A project of Humboldt State University's Schatz Energy Research Center (SERC), it was funded by the Department of Energy, California's South Coast Air Quality Management District, and the southern California city of Palm Desert; Environmental News Network staff, "First U.S. Fuel Cell Car Hits the Road," *CNN interactive*, April 29, 1998, http://edition.cnn.com/EARTH/9804/29/fuel.cell.car/index.html.

110. DeCicco, *Fuel Cell Vehicles*, 73.

111. Doyle, *Taken for a Ride*, 426.

112. Andrew Pollack, "Where to Put the Golf Clubs? Right Next to the Hydrogen!" *New York Times*, May 19, 1999, G20.

113. Jeffrey Ball, "Road Test: Auto Makers Are Racing to Market 'Green' Cars Powered by Fuel Cells—DaimlerChrysler Sets Deadline to Sell a Version by 2004; Happy Times for Ballard, but 'Who Buys One?'" *Wall Street Journal*, March 15, 1999, A1. Like many other commentators, Ball made a false distinction between the "electric" car and the "fuel cell" car.

114. Shnayerson, *The Car That Could*, 180, 202–204. Batteries faced problems similar to those of fuel cell stacks when wired in series. If one cell discharged faster than the others, it could reverse the polarity of its neighbors, neutralizing them and damaging the entire power source.

115. Doyle, *Taken for a Ride*, 427.

116. An additional problem was posed by this fuel's propensity to burn with an invisible flame. Because of this, a luminosity agent would likely have to added before the fuel could become widely commercialized, placing additional stress on the electrocatalytic reformer and shortening the lifetime of the system. Richard K. Stobart, "Fuel Cell Power for Passenger Cars: What Barriers Remain?" in *Fuel Cell Technology for Vehicles*, ed. Richard Stobart (Warrendale, PA: Society of Automotive Engineers, 2001), 14.

117. Paul J. Berlowitz and Charles P. Darnell, "Fuel Choices for Fuel Cell Powered Vehicles," in *Fuel Cell Technology for Vehicles*, 50; see also Borroni-Bird, "Fuel Cell Commercialization Issues," 42.

118. DeCicco, *Fuel Cell Vehicles*, 61.

119. Dave Nahmias, *Fuel Choice for Fuel Cell Vehicles: Report to the Department of Energy from Its Hydrogen Technical Advisory Board* (National Hydrogen Association, 1999).

120. Sean Casten, Peter Teagan, and Richard Stobart, "Fuels for Fuel Cell-Powered Vehicles," in *Fuel Cell Technology for Vehicles*, 61–62.

121. Jones et al., *Review of the Research Program of the Partnership for a New Generation of Vehicles, Sixth Report*, 85–87.

CHAPTER 7   ELECTROCHEMICAL MILLENNIUM

1. Sharon Beder, *Global Spin: The Corporate Assault on Environmentalism* (White River Junction, VT: Chelsea Green, 2002), 238.

2. Lovins seems to have first expounded on fuel cells at length (with coauthors Paul Hawken and L. Hunter Lovins) in *Natural Capitalism: Creating the Next Industrial Revolution* (Boston: Back Bay Books, 2000).

3. Interview by Jann S. Wenner with Will Dana, *Rolling Stone* 853, November 9, 2000, http://www.jannswenner.com/Archives/Al_Gore.aspx; Gore also claimed that the PNGV had helped stimulate the current "massive cutthroat competition" in fuel cell automobility among the major manufacturers.

4. N. P. Rossmeissl and L. A. Waltemath, "The United States Department of Energy Hydrogen Program," in *Hydrogen Energy Progress XI: Proceedings of the Eleventh World Hydrogen Energy Conference, Stuttgart, Germany, June 23–28, 1996*, vol. 1., ed. T. N. Veziroglu, C. J. Winter, J. P. Baselt, and G. Kreysa (Frankfurt am Main: Schön & Wetzel, 1996), 27–35.

5. Warren E. Leary, "Use of Hydrogen as Fuel Is Moving Closer to Reality," *New York Times*, April 16, 1995, 15.

6. Rossmeissl and Waltemath, "The United States Department of Energy Hydrogen Program," 27–35.

7. Leary, "Use of Hydrogen as Fuel Is Moving Closer to Reality," 10; *Congressional Quarterly Almanac*, vol. 51, 1995, 104th Cong. 1st sess. (Washington, DC: Congressional Quarterly, 1996), 5–29.

8. U.S. Public Law 104-271, 104th Cong. 2d sess. (October 9, 1996), *Hydrogen Future Act of 1996*, Sec. 103, "Hydrogen Research and Development," Sec. 201, "Integration of Fuel Cells with Hydrogen Production Systems," 110 STAT.3306–3307.

9. House Committee on Science, *Fuel Cells: The Key to Energy Independence?* 107th Cong. 2d sess., June 24, 2002, serial no. 107-83, 13–14.

10. As of 1997, the NHA consisted of Air Products and Chemicals (founder, 1989), Praxair (founder), Air Liquide Advanced Technology (founder), GTI (founder), University of Hawaii–Natural Energy Institute (founding member), NASA (founding honorary member), U.S. DOE (founding honorary member), Iwatani International (gas trader, 1990), Ballard (1991), Sacramento Municipal Utility District (1992), Proton Energy Systems (PEM electrolyzers, 1996), Sandia National Laboratories (1996), Savannah River National Laboratories (tritium, 1997), Humboldt State U-SERC (hydrogen fuel cells, 1996), Hydrogen 2000 (founded 1995 by media, scientists, 1997).

11. James S. Cannon, "End of an Era for Hydrogen at the PNGV," *NHA Quarterly Newsletter* 2, no. 1 (Winter 1997), http://www.hydrogenassociation.org/newsletter/.

12. Carol Bailey et al., "The United States Department of Energy's Hydrogen Program—Perspectives from the Hydrogen Technical Advisory Panel (HTAP)," in *Hydrogen Energy Progress XII: Proceedings of the Twelfth World Hydrogen Energy Conference, Buenos Aires, Argentina, June 21–26, 1998*, vol. 1, ed. J. C. Bolcich and T. N. Veziroglu (Buenos Aires: Editorial Asociación Argentina del Hidrógeno, 1998), 396–404.

13. *Congressional Quarterly Almanac*, vol. 51, 1995, 5–29.

14. Robert S. Walker, "Hydrogen Is Regaining Its Status as an Alternative Fuel," adapted by the *Minneapolis Star-Tribune*, October 4, 2000, from Walker's speech in Dublin, Ireland, at the International Symposium on Automotive Technology and Automation, http://www.wexlerwalker.com/hydrogen.htm. Walker's speech, a virtual manifesto of the political uses of hydrogen, corresponded well with the activities of his lobbying firm. One of its specialties was guiding industry through the maze of environmental and product regulation legislation, and it boasted of stopping "bad legislation in its tracks." The firm's client list in its "Environment and Energy-Related" category included Caterpillar, British Petroleum, General Motors, and the Alliance of Automobile Manufacturers; http://www.wexlerwalker.com/environ.htm.

15. Gregory L. White, "GM Stops Making Electric Car, Holds Talks with Toyota," *Wall Street Journal*, January 12, 2000, 1.

16. Nelson D. Schwartz, "Meet the New Market Makers: They're Young, They're Rich, and They Couldn't Care Less about Graham & Dodd. But They're the Ones Driving Those Insane Tech Stocks, and They're Not Going Away," *Fortune* 141, no. 4 (February 21, 2000): 90–97; Sana Siwolop, "A Back Door Is Open to the Fuel Cell Party," *New York Times*, March 5, 2000, BU9.

17. Barnaby J. Feder, "Almost an Energy Alternative: Fuel Cells Hold Promise, but Problems Remain," *New York Times*, May 27, 2000, C1.

18. California Environmental Protection Agency Air Resources Board (cited hereafter as CARB), "Fact Sheet: 2003 Zero Emission Program Change," www.arb.ca.gov/msprog/zevprog/factsheets/2003zevchanges.pdf.

19. Richard Littlemore, "Hydrogen Bombs Out: Many Investors Who Thought Ballard Power's Fuel-Cell Technology Was a Miracle Last Year Have Lately Been Bailing Out," *National Post Business*, April 2001, 54.

20. David Ludlum, "Fuel Cell Companies Offer Choice and Risk," *New York Times*, December 23, 2001, BU9.

21. Matthew L. Wald, "Another GM Investment in Fuel Cell Development: A Hydrogen Approach to Electric Cars," *New York Times*, June 14, 2001, C10.

22. Robert W. Fri et al., *Energy Research at DOE: Was It Worth It? Energy Efficiency and Fossil Energy Research 1978 to 2000* (Washington, DC: National Academy Press, 2001), 23, 33–36, 48, 53–54.

23. Ibid., 155.

24. "Solid State Energy Conversion Alliance," U.S. Department of Energy, "Industry Teams Begin Quest for Low-Cost, Breakthrough Fuel Cell," press release, August 8, 2001.

25. David Stipp, "The Coming Hydrogen Economy: Fuel Cells Powered by Hydrogen Are about to Hit the Market," *Fortune* 144, no. 9 (November 12, 2001); U.S. Department of Energy, *Budget Highlights: Fiscal Year 2002 Budget Request*, April 2001, 16.

26. Jeffrey Ball, "The White House Energy Plan: Auto Makers May Get Some Tax Help but Could Face Tougher Fuel Standards," *Wall Street Journal*, May 18, 2001, A6.

27. Jeffrey Ball, "Bush Shifts Gears on Car Research Priority," *Wall Street Journal*, January 9, 2002, C14.

28. Jeffrey Ball, "Fuel Cell Makers Get a Big Boost from Bush's Auto Subsidy Plan," *Wall Street Journal*, January 10, 2002, B2.

29. Neela Banerjee and Danny Hakim, "U.S. Ends Car Plan on Gas Efficiency, Looks to Fuel Cells," *New York Times*, January 9, 2002, A1.

30. "GM Proposes Effort by the White House on Auto Fuel Cells," *Wall Street Journal*, February 21, 2002, B14.

31. See CARB, "Fact Sheet"; National Resources Defense Council, "Press Release: California's Clean Car Program under Attack," February 13, 2003, http://www.nrdc.org/media/pressreleases/030213.asp; Collantes, "The California Zero-Emission Vehicle Mandate," 112–113.

32. Jeffrey Ball, "Evasive Maneuvers: Detroit Again Tries to Dodge Pressures for a 'Greener' Fleet; Oil Fears since September 11 Add Urgency to Latest Round of Gas-Mileage Politics; 'Supercars' and Fuel Cells," *Wall Street Journal*, January 28, 2002, A1.

33. "Spencer Abraham's Dream Car," *New York Times*, January 14, 2002, G3.

34. Danny Hakim, "Dream Car: A Dressed-Up Skateboard," *New York Times*, January 10, 2002, G3; Stuart F. Brown, "A Wild Vision for Fuel-Cell Vehicles," *Fortune* 145, no. 7 (April 1, 2002): 72.

35. Ball, "Fuel Cell Makers Get a Big Boost."

36. Matthew L. Wald, "A Fuel Cell Initiative Too Costly for Use in Cars," *New York Times*, January 15, 2002, C4.

37. General Motors press release, "General Motors to Display AUTOnomy Concept at World Hydrogen Energy Conference," June 6, 2002.

38. Steve Lohr, "New Economy: Fuel-Cell Work Is Helping Build Research Prowess for Detroit," *New York Times*, July 15, 2002, C3.

39. Danny Hakim, "Smokestack Visionary: General Motors Isn't Known for Being Green, but Larry Burns, Its Fuel-Cell Guru, Is Working on the Most Ambitious Project Yet to Create the Future's Environment-Friendly Car," *New York Times*, September 29, 2002, F100.

40. Honda Motor Company press release, "Honda Fuel Cell Vehicle First to Receive Certification; Honda FCX Slated for Commercial Use," July 24, 2002, http://world .honda.com/news/2002/printerfriendly/4020724.html; Joseph B. White, "Honda Will Introduce Fuel-Cell Cars," *Wall Street Journal*, July 25, 2002, D4.

41. George W. Bush, "President Delivers 'State of the Union,'" January 28, 2003, http:// www.whitehouse.gov/news/releases/2003/01/20030128-19.html; George W. Bush, "Hydrogen Fuel Initiative Can Make 'Fundamental Difference': Remarks by the President on Energy Independence," February 6, 2003, http://www.whitehouse.gov/ news/releases/2003/02/20030206-12.html.

42. Ryan Lizza, "A Green Car That the Energy Industry Loves," *New York Times*, February 2, 2003, WK3.

43. "Hydrogen Car Hype," *Wall Street Journal*, January 30, 2003, A14.

44. Jeffrey Ball, "State of the Union: The Day After; Critics Call Fuel Cells Long Shot," *Wall Street Journal*, January 30, 2003, A10.

45. House Committee on Science, *The Path to a Hydrogen Economy: Hearing before the Committee on Science*, 108th Cong. 1st sess., March 5, 2003, serial no. 108-4, 21–22, 95–96, 124–125.

46. Andrew Pollack, "Where to Put the Golf Clubs? Right Next to the Hydrogen!" *New York Times*, May 19, 1999, G20.

47. House Committee on Science, *The Path to a Hydrogen Economy*, March 5, 2003, 54–56, 68–69, 78, 146–147, 179–180.

48. Gustavo Oscar Collantes, "The California Zero-Emission Vehicle Mandate: A Study of the Policy Process, 1990–2004" (PhD diss., University of California, Davis, 2005), 113–117; CARB, "Fact Sheet," 1–2.

49. EV Canada, "Out of Production," http://www.evcanada.org/outprod.aspx.

50. Collantes, "The California Zero-Emission Vehicle Mandate," 113–117.

51. A hydrogen committee convened by the National Academy of Sciences' National Research Council in the fall of 2002 suggested the 4,000 to 5,000–hour standard had not been demonstrated under realistic operating conditions as of 2004; Michael P. Ramage et al., *The Hydrogen Economy: Opportunities, Costs, Barriers, and R&D Needs* (Washington, DC: National Academies Press, 2004), 26.

52. Rebecca L. Busby, *Hydrogen and Fuel Cells: A Comprehensive Guide* (Tulsa, OK: PennWell, 2005), 256–260. In comparison, the makers of the most advanced battery-powered electric vehicles of the early 2000s claimed ranges of as high as 190 kilometers (Honda EV

Plus, Nissan Altra EV) to 200 kilometers (Toyota RAV 4); Michael H. Westbrook, *The Electric Car: Development and Future of Battery, Hybrid and Fuel-Cell Cars* (London: Institution of Electrical Engineers, 2001), 133.

53. Charles Stone, presentation, First Annual Society of Chemical Industry–Chemical Heritage Foundation Innovation Day, Philadelphia, PA, September 14, 2004.

54. Vaclav Smil, *Energy at the Crossroads: Global Perspectives and Uncertainties* (Cambridge, MA: MIT Press, 2003), 304–309.

55. Matthew L. Wald, "Report Questions Bush Plan for Hydrogen-Fueled Cars," *New York Times*, February 6, 2004, A20; Joseph J. Romm, *The Hype about Hydrogen: Fact and Fiction in the Race to Save the Climate* (Washington, DC: Island Press, 2004).

56. Matthew L. Wald, "Will Hydrogen Clear the Air? Maybe Not, Say Some," *New York Times*, November 12, 2003, C1.

57. Loyola de Palacio, "New and Original Ways for Hydrogen to Reach the Market," speech at the first meeting of the High Level Group for Hydrogen and Fuel Cells, Brussels, October 10, 2002, http://europa.eu/rapid/pressReleasesAction.do?reference=SPEECH/02/470&format=HTML&aged=0&language=EN;&guiLanguage=en; Mark Landler, "Europe and America, Partners (Sort of)," *New York Times*, July 27, 2003, WK6.

58. Paul Meller, "Europe and U.S. Will Share Research on Hydrogen Fuel," *New York Times*, June 17, 2003, W1.

59. Ramage et al., *The Hydrogen Economy*, 1–10, 116–120.

60. Wald, "Report Questions Bush Plan for Hydrogen-Fueled Cars."

61. Ramage et al., *The Hydrogen Economy*, 2.

62. See Peter Hoffmann, *Tomorrow's Energy: Hydrogen, Fuel Cells, and the Prospects for a Cleaner Planet* (Cambridge, MA: MIT Press, 2001), ix. Hoffmann had hardly mentioned fuel cells in his 1981 work; see *The Forever Fuel: The Story of Hydrogen* (Boulder, CO: Westview Press, 1981), 136–138.

63. Review papers by company officers assessing Ballard fuel cells typically focused on current density, rarely mentioning durability; see, for example, Keith B. Prater's "Polymer Electrolyte Fuel Cells: A Review of Recent Developments," *Journal of Power Sources* 51 (1994): 129–144; Charles Stone and Anne E. Morrison, "From Curiosity to 'Power to Change the World®,'" *Solid State Ionics* 152–153 (2002): 1–13.

64. Ballard made the multifuel claim at least as far back as 1997 and maintained it until 2002; Ballard Power Systems, Inc., *Annual Reports*, 1997–2002.

65. Charles Stone, presentation, First Annual Society of Chemical Industry–Chemical Heritage Foundation Innovation Day, Philadelphia, PA, September 14, 2004.

66. Lynne Olver, "The Forklift Frontier: Fuel Cells Have Finally Found a Market Where They're Viable Right Now: Forklifts," *National Post*, November 3, 2005, FP9.

67. Stone, presentation, First Annual Society of Chemical Industry–Chemical Heritage Foundation Innovation Day.

68. Tom Koppel, *Powering the Future: The Ballard Fuel Cell and the Race to Change the World* (Toronto: John Wiley & Sons Canada, 1999), 141; Jeffrey Ball, "Road Test: Auto Makers Are Racing to Market 'Green' Cars Powered by Fuel Cells—DaimlerChrysler Sets Deadline to Sell a Version by 2004; Happy Times for Ballard, but 'Who Buys One?'" *Wall Street Journal*, March 15, 1999, A1.

69. "Ballard Sells Automotive Fuel Cell Business as Hydrogenics Streamlines, Drops Test Division," *Fuel Cells Bulletin* 12, December 2007.

70. Although he exhaustively examines the question of individual identity and auto-mobility, Matthew Paterson offers no explicit theorization of this phenomenon as it pertains to the green automobile project; see his *Automobile Politics: Ecology and Cultural Political Economy* (Cambridge: Cambridge University Press, 2007), 223.

71. This is a crucial distinction that Gartman glosses. American automakers were extremely flexible in the modes of commercially producing the internal-combustion-powered automobile, replacing workers with robots, adopting Japanese lean-manufacturing models, and decentralizing and relocating production offshore. But as we have seen, Detroit was completely committed to internal combustion power and hostile to the alternatives; see David Gartman, "Three Ages of the Automobile: The Cultural Logics of the Car," *Theory, Culture and Society* 21 (2004): 193, 186.

72. Paterson, *Automobile Politics*, 221–222.

73. "GM, Daimler Plan Fuel-Cell Vehicles," *Wall Street Journal*, March 31, 2005, 1.

74. Tom McCarthy, *Auto Mania: Cars, Consumers and the Environment* (New Haven, CT: Yale University Press, 2007), 234–235.

75. William Rollins, "Reflections on a Spare Tire: SUVs and Postmodern Environmental Consciousness," *Environmental History* 11, no. 4 (October 2006): 686, 711.

76. Bill Vlasic, "G.M. Shifts Focus to Small Cars in Sign of Sport Utility Demise," *New York Times*, June 4, 2008, http://www.nytimes.com/2008/06/04/business/04motors.html?scp=4&sq=volt,%20rick%20wagoner&st=cse; "G.M. and Chrysler Explore Merger," *New York Times*, October 10, 2008, http://dealbook.blogs.nytimes.com/2008/10/10/gm-and-chrysler-explore-merger/?scp=5&sq=GM%20and%20Ford&st=cse.

77. Martin Fackler, "Latest Honda Runs on Hydrogen, Not Petroleum," *New York Times*, June 17, 2008, http://www.nytimes.com/2008/06/17/business/worldbusiness/17fuelcell.html.

78. State of California, http://www.hydrogenhighway.ca.gov/; California Fuel Cell Partnership, "Progress," http://www.fuelcellpartnership.org/progress.

79. Daimler, Technology & Innovation News, "OMV Opens Baden-Württemberg's First Public Hydrogen Filling Station for Emission-Free Mobility of the Future," June 17, 2009, http://www.daimler.com/dccom/0-5-7145-1-1215398-1-0-0-0-0-0-11979-0-0-0-0-0-0-0-0.html.

80. Alan Ohnsman, "Toyota Says It's Now Turning a Profit on the Hybrid Prius," *Los Angeles Times*, December 19, 2001, http://articles.latimes.com/2001/dec/19/autos/hy-prius19.

81. "Worldwide Prius Cumulative Sales Top 2M Mark; Toyota Reportedly Plans Two New Prius Variants for the U.S. by the End of 2012," Green Car Congress, October 7, 2010, http://www.greencarcongress.com/2010/10/worldwide-prius-cumulative-sales-top-2m-mark-toyota-reportedly-plans-two-new-prius-variants-for-the-.html#more.

82. Keith Naughton, "Bob Lutz: The Man Who Revived the Electric Car," *Newsweek*, December 22, 2007, http://www.newsweek.com/2007/12/22/bob-lutz-the-man-who-revived-the-electric-car.html.

83. Haruo Ikehara, "Toyota's Plug-in Hybrid: Debut of Prototype Is Near," *Nikkei Business Online*, January 29, 2007, http://business.nikkeibp.co.jp/article/eng/20070129/117846/.

84. Toyota Press Release, "Toyota Advances Plug-in Hybrid Development with Partnership Program," July 25, 2007, http://pressroom.toyota.com/pr/tms/toyota/TYT2007072552930.aspx.

85. CARB, "Fact Sheet: The Zero Emission Vehicle Program—2008," May 6, 2008; "California Cuts ZEV Mandate in Favor of Plug-In Hybrids," *Wired*, March 27, 2008, http://www.wired.com/autopia/2008/03/the-california/.

86. U.S. Environmental Protection Agency, "Summary of Current and Historical Light-Duty Vehicle Emissions Standards," April 2007; HR 6, Energy Independence and Security Act of 2007.

87. U.S. Department of Energy-Loan Programs Office, http://loanprograms.energy.gov/.

88. John M. Broder, "Obama to Toughen Rules on Emissions and Mileage," May 18, 2009, *New York Times*, http://www.nytimes.com/2009/05/19/business/19emissions.html.

89. Seth Fletcher, *Bottled Lightning: Superbatteries, Electric Cars, and the New Lithium Economy* (New York: Hill and Wang, 2011).

CONCLUSION

1. David E. Nye, *America as Second Creation: Technology and Narratives of New Beginnings* (Cambridge, MA: MIT Press, 2003), 1–37.

2. David E. Nye, *Consuming Power: A Social History of American Energies* (Cambridge, MA: MIT Press, 1998), 222.

3. Eric S. Hintz, "Portable Power: Inventor Samuel Ruben and the Birth of Duracell," *Technology and Culture* 50, no. 1 (January 2009): 26–27. The history of electrochemical energy conversion is sprinkled with "lone wolf" inventors such as Planté, Edison, Bacon, Kordesch, Ballard, Ovshinsky, Thackeray, and Chiang. Of course, developing and commercializing these technologies required dense networks of researchers and sponsors.

4. Harvey Brooks, "The Evolution of U.S. Science Policy," in *Technology, R&D, and the Economy*, ed. Bruce L. R. Smith and Claude E. Barfield (Washington, DC: Brookings Institution and American Enterprise Institute, 1996), 21–23.

5. Breakthrough Technologies Institute, "Fuel cells at the Crossroads: Attitudes Regarding the Investment Climate for the U.S. Fuel Cell Industry and a Protection of Industry Job Creation Potential," Report No. ANL/OF-00405/300, 2003, 1–11, 26–31.

6. Scott L. Montgomery, *The Powers That Be: Global Energy for the Twenty-First Century and Beyond* (Chicago: Chicago University Press 2010), 187.

7. Jeremy Rifkin, *The Hydrogen Economy: The Creation of the World-Wide Energy Web and the Redistribution of Power on Earth* (New York: Jeremy P. Tarcher/Penguin, 2002), 37–57, 185–215.

8. Vaclav Smil, *Energy Myths and Realities: Bringing Science to the Energy Policy Debate* (Washington, DC: AEI Press, 2010), 7.

9. For example, Ballard's improvements in current density during the 1990s forms an S-curve. Between 1989 and 1992, current density was boosted from 85 to 140 watts per liter. From 1992 to 1994, current density doubled from 140 to 290 watts per liter. It nearly doubled again between 1994 and 1995 to 570 watts per liter and then again between 1995 and the middle of 1996 to 1,000 watts per liter. Thereafter, advances came at a much slower rate; current density was boosted by only 20 percent, to 1,200 watts per liter, at the end of 1997 and by a little over 10 percent between the end of 1997 and 2000 to 1,310 watts per liter. Charles Stone and Anne E. Morrison, "From Curiosity to 'Power to Change the World®,'" *Solid State Ionics* 152–153 (2002): 7–8.

10. David Edgerton has written that futurism had become passé by the turn of the millennium, pointing to the complex interplay between old and new technological systems in the present; see *The Shock of the Old: Technology and Global History since 1900* (New York: Oxford University Press, 2007), x. The technopolitics of fuel cells, hydrogen, and

nanotechnology demonstrate otherwise. It was precisely this interplay—efforts to use advanced technologies to reconcile the contradictions of technologies-in-use—that gave rise to futurism in this period, at least in the United States.

11. See, for example, Cyrus C. M. Mody, "Introduction," in *Perspectives on Science* 17, no. 2 (Summer 2009): 111–122; Jason Gallo, "The Discursive and Operational Foundations of the National Nanotechnology Initiative in the History of the National Science Foundation," in *Perspectives on Science* 17, no 2 (Summer 2009): 174–211; Mihail C. Roco, "National Nanotechnology Initiative—Past, Present, and Future," in *Handbook on Nanoscience, Engineering, and Technology* (Taylor and Francis, 2007).

12. W. Patrick McCray, "Will Small Be Beautiful? Making Policies for Our Nanotech Future," *History and Technology* 21, no. 2 (June 2005): 192.

13. Here I adopt Jennifer R. Fishman's interpretation of Pierre Bourdieu's understanding of exchange networks as applied to the process of commercializing new drugs: see her "Manufacturing Desire: The Commodification of Female Sexual Dysfunction," *Social Studies of Science* 34, no. 2 (April 2004): 191, 203.

14. McCray, "Will Small Be Beautiful?" 192–193.

15. See, for example, A. Paul Alivisatos et al., *Nanoscience Research for Energy Needs: Report of the National Nanotechnology Initiative Grand Challenge Workshop, March 16–18, 2004* (Washington, DC: National Nanotechnology Coordinating Office, U.S. Department of Energy, 2004), 4–16; Richard E. Smalley, one of the codiscoverers of buckminsterfullerenes and a pioneer in the field of carbon nanotubes, frequently linked nanotechnology and hydrogen and fuel cell technology in the latter part of his career; see, for example, Tom Bearden's PBS interview with Smalley on October 20, 1993, "The Future of Fuel: Advances in Hydrogen Fuel Cell Technology," *Online News Hour*, http://www.pbs.org/newshour/science/hydrogen/smalley.html.

16. This idea as applied to nanotechnology was explored in depth by K. Eric Drexler in his classic *Engines of Creation: The Coming Era of Nanotechnology* (New York: Anchor Books, 1986). For an excellent analysis of nanoteleology, see Cyrus C. M. Mody, "Small, but Determined: Technological Determinism in Nanoscience," *International Journal for Philosophy of Chemistry* 10, no. 2 (2004): 101–130.

17. William J. Mitchell, Christopher E. Borroni-Bird, and Lawrence D. Burns, *Reinventing the Automobile: Personal Urban Mobility for the 21st Century* (Cambridge, MA: MIT Press, 2010).

18. Ethan Stewart, "Energy Switcheroo: El Estero Wastewater Treatment Plant to Get Internal Combustion Engine," *Santa Barbara Independent*, September 22, 2010, http://www.independent.com/news/2010/sep/22/energy-switcheroo/.

19. Gijs Mom, *The Electric Vehicle: Technology and Expectations in the Automobile Age* (Baltimore: Johns Hopkins University Press, 2004), 301.

INDEX

ABOUT THE AUTHOR

MATTHEW N. EISLER obtained his doctorate in history at the University of Alberta in April 2008. He was the 2008–2009 Harris Steel Postdoctoral Fellow at the Department of History at the University of Western Ontario before taking a postdoctoral post at the Center for Nanotechnology in Society at the University of California at Santa Barbara in fall 2009. He is currently a Research Fellow at the Chemical Heritage Foundation. Dr. Eisler's interests lie at the intersection of the history of science, technology, and environment, especially after 1945. He focuses on the political economy, culture, and discourse of science-based innovation and research and development policies and their relationship with and impact on industrialism, consumerism, and the environment, particularly in the energy and transportation sectors.